TOPICS IN COSMIC-RAY ASTROPHYSICS

Horizons in World Physics

Horizons in World Physics
Vol. 230

TOPICS IN COSMIC-RAY ASTROPHYSICS

MICHAEL A. DUVERNOIS (EDITOR)

Nova Science Publishers, Inc.
Commack, New York

Editorial Production: Susan Boriotti
Office Manager: Annette Hellinger
Graphics: Frank Grucci and Jennifer Lucas
Information Editor: Tatiana Shohov
Book Production: Donna Dennis, Patrick Davin, Christine Mathosian,
Tammy Sauter and Lynette Van Helden
Circulation: Maryanne Schmidt
Marketing/Sales: Cathy DeGregory

Library of Congress Cataloging-in-Publication Data

DuVernois, Michael
    Topics in Astrophysics / Michael DuVernois.
            p.   cm. -- Horizons in World Physics; v. 230
    Includes bibliographical references and index.
    ISBN 1-56072-658-X
    1. Astrophysics. 2. Cosmic rays... I. Titles. II. Series.
QC1.A4114 vol. 230   99-14709
[QB461]
630 s--dc21   CIP
[523.01'97223

Copyright © 1999 by Nova Science Publishers, Inc.
    6080 Jericho Turnpike, Suite 207
    Commack, New York 11725
    Tele. 516-499-3103   Fax 516-499-3146
    e-mail: Novascience@earthlink.net
    e-mail: Novascil@aol.com
    Web Site: http://www.nexusworld.com/nova

Printed in the United States of America

Contents

The discipline of cosmic-ray astrophysics has been experimentally dominated for much of its history. Instrumental advances have been responsible for the majority of advances in the field. New and proposed instruments and techniques are presented here.

III. **Galactic cosmic rays: acceleration and propagation**

Low-energy, Galactic cosmic rays have been studied quite closely, and detailed models of their source, acceleration, and transport in the Galaxy are actively studied. The crucial propagation and acceleration parameters of these models can be addressed from the data.

IV. Galactic cosmic rays: sources 169

The composition of the Galactic cosmic-ray source provides a linkage between the cosmic rays and Galactic astronomy. The chemical evolution of the Galaxy is constrained by cosmic-ray data and the cosmic rays may contain telltale signs of their site of nucleosynthesis.

About the Editor

Michael A. DuVernois was born and raised in central Connecticut. His interests have always been varied, but his science career started with amateur astronomy. A student internship with the local planetarium and observatory sealed this path and lead him to study physics. He attended the Georgia Institute of Technology as an undergraduate and the University of Chicago as a graduate student. Although he originally wanted to be a general relativity theorist, better sense prevailed, and experimental astrophysics it was. His Ph.D. thesis was on the radioactive isotopes of manganese in the Galactic cosmic rays under the supervision of John A. Simpson. He is the author of a number of papers on cosmic-ray astrophysics.

Currently at the Pennsylvania State University, he is currently working on the HEAT balloon-borne magnet spectrometer, the Pierre Auger giant air-shower observatory, the CREAM balloon thin ionization calorimeter, and the study phase for the NASA space-based ACCESS experiment. Other interests include music (both performing and collecting), travel, photography, books, wine, his wife, and cooking (in no particular order).

I. Introduction

The cosmic radiation has been an active field of study at least since the heroic balloon flights of Viktor F. Hess in the first decade of this century. In the earliest days, cosmic ray physics meant a study of the basic properties of electricity and magnetism. Later, cosmic ray physics was particle physics until the large accelerators were built. Still later, it became astrophysics---studying the Galactic sources of the lower energy cosmic rays, the magnetic fields in the heliosphere and the Galaxy, and the acceleration mechanisms in supernova shocks. Today, cosmic-ray astrophysics touches on the nuclear astrophysics of stars and supernova, particle physics at energies above those achievable by terrestrial accelerators, the cosmology of the microwave and IR backgrounds, the Galactic physics of chemical evolution and interstellar medium processes, and unexplored physics at extremely high energies. The experimental techniques are wide-ranging as well, with experiments that could be held in one hand and huge air-shower arrays covering thousands of square kilometers. Cosmic-ray experiments have flown on the Space Shuttle, been onboard almost every interplanetary spacecraft, exist at the South Pole, and are flown every year on balloons launched in the desert southwest of the United States, Antarctica, Australia, and northern Canada. Other experiments lie on high mountain plateaus in Tibet, in urban areas of Japan, and under mountains.

It is a field which has in recent years been undergoing a renaissance in many ways. Balloon payloads which had previously been limited to 1-2 day flights are now routinely getting 10-20 day Antarctic orbital flights. Similar flights in the Arctic have been demonstrated, but have yet to become a standard capability. In less than two years, the TIGER experiment from Washington University will be launched on a balloon with a goal of 100 days at float altitude. There are a number of satellite-based cosmic ray instruments, and, although the lead-time is long, more are planned for the future. Current satellite experiments have very high elemental and isotopic resolving power, and their measurements have, for the moment, outstripped our theoretical understanding of the cosmic-ray source and propagation processes. Future plans include very ambitious space station-based instruments to push direct measurements up in energy by several orders of magnitude. The largest ground-based air-shower array yet, the Pierre Auger Observatory, is about to begin construction in Argentina, with a second site planned in Utah. At the moment in Utah, the HiRes Fly's Eye experiment continues the air fluorescence measurements of the highest energy cosmic rays.

In this volume we limit our discussion to the charged cosmic rays, primarily nuclei, from Galactic and extra-Galactic sources. The one exception is the paper of Dwyer, which addresses particle acceleration in the Earth's bow shock. This paper is important as it directly ties our observations of shock acceleration *in situ* with the theoretical expectations of shock acceleration of the cosmic rays elsewhere. There is no coverage of neutrino astrophysics, gamma-ray astronomy, or solar/heliospheric cosmic rays. All of these are important and interesting subjects with significant scientific, technical, and personnel overlaps with the fields covered here, but are outside the scope of this volume.

With the great diversity of the field of even charged-particle cosmic-ray astrophysics, no one volume can pretend to cover the entire discipline. This volume is a survey of important topics in cosmic-ray astrophysics, covered from a variety of different viewpoints. As such, it represents the opinions and prejudices of the editor and contributors.

II.   Cosmic-ray instrumentation

The SilEye apparatus for the study of Cosmic ray on the MIR Space Station

A. Morselli, P.Picozza on the behalf of SilEye Collaboration
Dept. of Physics, II Univ. of Rome "Tor Vergata" and INFN, Italy

Since the Apollo missions it is known that the crews, after some minutes of dark adaptation, observed brief flashes of white light or pencil-thin streaks of light . The first report of *"Light Flash"* (LF) dates back to 1969 by astronaut Edwin Aldrin, on board Apollo-11 [1]. Subsequently other crews reported similar experiences. (LF) consist of unexpected visual sensations and appear to the astronaut as faint spots or streaks of light in closed eyes after a period of dark adaptation. Dedicated observation programs were performed on Apollo, Skylab, Apollo-Soyuz missions [2] and, more recently, on MIR Space Station. Since 1995 the experiment SilEye has been working inside the Space Station MIR. It involves Russian and Italian Institutions as well as other European scientific partners, and the Russian Space Corporation that has granted the use of the Space Station and the collaboration of its crews. The objective of the experiment is the study of biophysical tasks related to the radiation environment inside MIR with particular attention to the phenomenon of "Light Flashes". The frequency of LF depends on orbit parameters, especially on the latitude and grows in polar areas and in the area of the South Atlantic Anomaly. LF are practically absent on the equator, where the charged particles' flux is at minimum. There are different hypothesis on the generation mechanisms of visual effects, like direct interaction of charged particles with the retina of the eye by ionization [3] or Cherenkov effect in the ocular bulb or indirect effect from proton knocked out by protons [4]. It was also suggested that scintillation in the eye lens could cause the observed LF. For a review see [5]. Although the cause of LF is still not known, the most probable mechanism involves the passage of a cosmic charged particle through the cosmonaut visual system. Therefore, it is extremely important to determine simultaneously time, nature, energy and trajectory of the particle passing through the cosmonaut eyes, as well as the astronaut's LF perception time. This kind of measurements were not fulfilled by all the previous experiments in space. The first prototype of the detector SilEye-1 [6], [7] was placed on the Station MIR in October 1995. During the last two years 25 measurement sessions involving 6 astronauts have been performed and more than fifty LF's have been recorded. Following this mission, a new detector, SilEye-2, has been developed and placed on MIR in 1997 in order to have a more accurate nuclear identification in selected energy range [8].

1 Apparatus description and capabilities

The detector array of SilEye-2 is derived from the technology developed for the construction of NINA Cosmic ray space detector [9] and consists of six silicon views, each composed of a $6 \times 6 \times 0.038$ cm^3 silicon wafer, divided in 16 strips 3.6 mm wide. Two views, orthogonally attached, constitute a plane, for a total of three planes, 96 strips and an active thickness of 2.28 mm. The distance between the silicon planes is 49mm in SilEye-1 and 14 mm in SilEye-2. The hit strips in different layers determine the particle position and direction. In SilEye-2 detector two passive absorbers of 1 mm Fe each are inserted between the planes. SilEye-1 provides track positional information (with angular accuracy of 3 degrees), whereas SilEye-2 may also measure particle energy losses from 0.25 to more than 250 MeV: with these information nuclear species in the energy range 40-200MeV/n can be identified.

The detector is placed in the front and on one side of the astronaut in SilEye-1, and only on one side of the astronaut for SilEye-2; size and mass are reduced (SilEye-2 max dimension 26.4 cm and mass of 5.5 Kg). Data acquisition is performed by a portable PC with PCMCIA acquisition card which is also linked to a joystick with a button (the PC keyboard for SilEye-1) to be pressed by the astronaut when he observes a LF. Data come therefore from two independent sources: the particle track recorded by the silicon detector and the observation of the LF by the astronaut. The helmet has a mask that shields the astronaut's eyes from light; in SilEye-2, also, three internal LEDs allow to check the correct position of the detector, the dark adaptation of the observer and his reaction time.

2   Beam test performances

The calibration of the device SilEye2 was carried out on a proton beam from the CELSIUS storage ring at TSL, Uppsala. The measurements were done at two different proton energies: 48 MeV and 70 MeV.

We have simulated, with standard Geant 3.21 Monte Carlo program, all the beam conditions with different energies and absorbers. In Fig. 1 are plotted the energy response (on the left) and the energy resolution (R.M.S./E) (on the right) of SilEye2 as a function of reconstructed proton energy for events with straight tracks in both views which have declination of middle point in both views no more than one strip from the fitted line. Events with two or more hitted strips in one layer were excluded. In Fig.2 is shown a Monte Carlo study on the particle separation capabilities for several nuclei of interest, for impinging particles with low and high energy respectively. The incident energy range is chosen randomly between 50 MeV/n - 1 GeV/n. It should be noted that even in the region with (E3-E1)$\leq$ 0.2 MeV (corresponding to particles with high kinetic energy) it is possible to discriminate the nuclei, using Etot as the only parameter. In Fig. 3 (on the left) the distribution of the total detected energy versus the difference between the energy lost in the last plane and in the first view is plotted. It can be seen that the protons' energies are very well separated.

3   SilEye-1 data on LF

Data acquired with SilEye-1 apparatus come from 25 sessions each of 90 minutes average duration (one orbit). A dark adaptation sequence of 15 minutes precedes

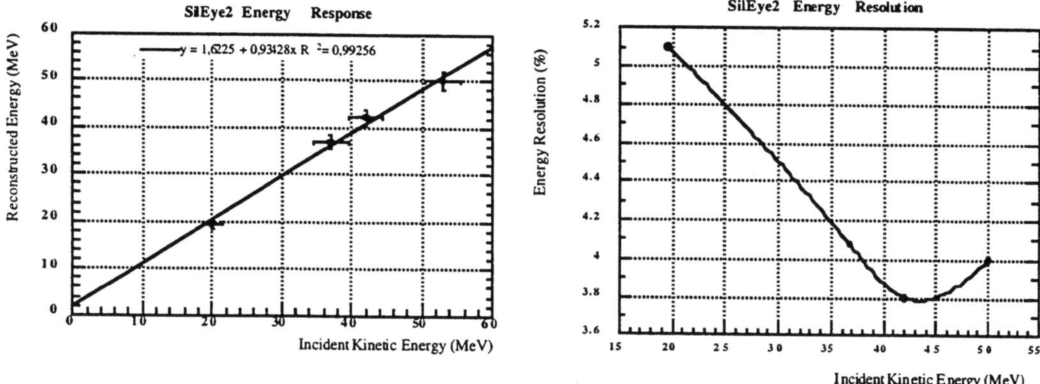

Figure 1: *Energy response of SilEye2 as a function of reconstructed proton energy (on the left) and as a function of reconstructed proton energy (on the right).*

each session. Most of the sessions took place in the crew's cabins in the main module of the Station, other session in the *Kwant* or in the "D" module. In Figure 3(on the right) are shown two typical result typologies. Particle flux has been averaged over 60 seconds intervals: it is possible to observe the increase in the polar regions and in the South Atlantic Anomaly (SAA). In SAA the real flux is order of magnitudes above the maximum acquisition rate (SilEye-1 maximum acquisition rate was about $25\ Hz \simeq 1500$ Ev/min); the black point on the time axis represent the LF observed by the astronaut during the sessions. To perform a quantitative data analysis on the relationship between LF and particle flux, each orbit has been divided in portions of $10°$ of latitude. Particle and LF frequencies have been obtained averaging the results obtained at equal latitude in the SAA over all 11 orbits, for the three astronauts involved. SAA data have been treated separately. The results are shown in figure 4: we can observe a proportionality relation between particle flux and LF observation (the probability of a chance correlation is p< 0.02) except in the SAA region. From this plot it is possible to see that a rather small growth of registration of LF rate in the SAA (about 2 σ) is observed while it is known that the proton flux in the SAA increases several order of magnitudes in comparison with the equator. The likely conclusion is then that protons are not the main LF source in orbit, and it seems more probable that heavy ions are the initiators.

4   MIR nuclear abundance with SilEye-2

In the first three sessions with SilEye-2 instrument on MIR station, in three hours of work, about 19000 particle tracks have been collected. Figure 5 shows relative measured abundance ratio between nuclei and protons, as a function of the atomic number Z of the nuclei, for all data collected. The measured spectrum is very different from the usual cosmic ray nuclei distribution [10], due to the abundance of recoil nuclei from the body of the Space Station. More accurate measurements

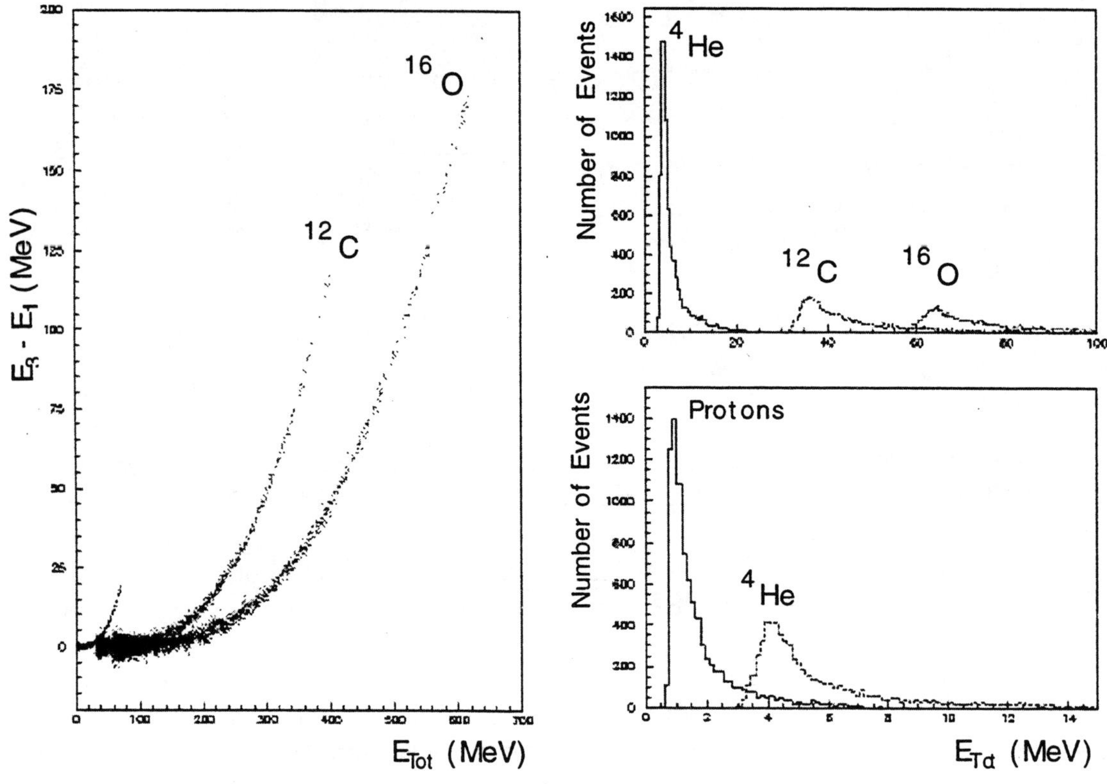

Figure 2: *Discrimination of nuclei with low incident energy (on the left) and with high incident energy (E3-E1<0.2 MeV) on the right ; Etot is the total energy released by the particles in the whole detector; E3-E1 is the released energy difference between the third and the first plane.*

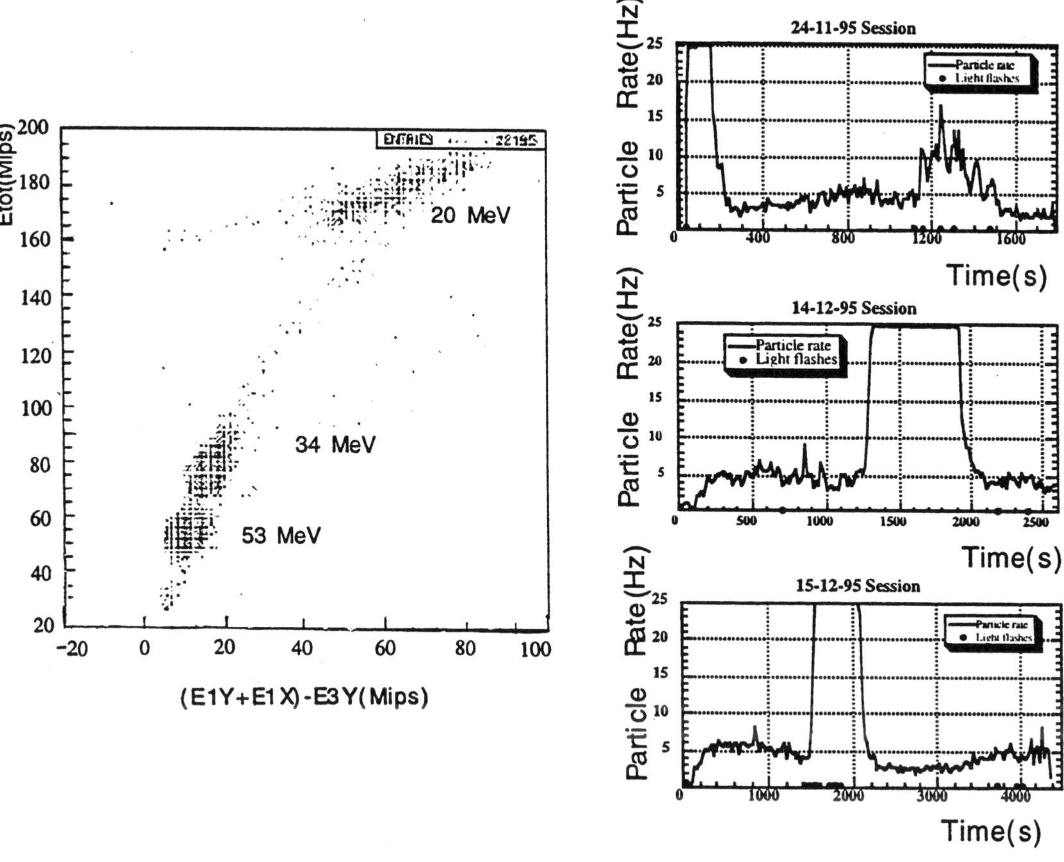

Figure 3: *Distribution of the total detected energy versus the difference between the energy loss in the last plane and in the first view (on the left). Trigger rate vs. the different position in the orbit.*

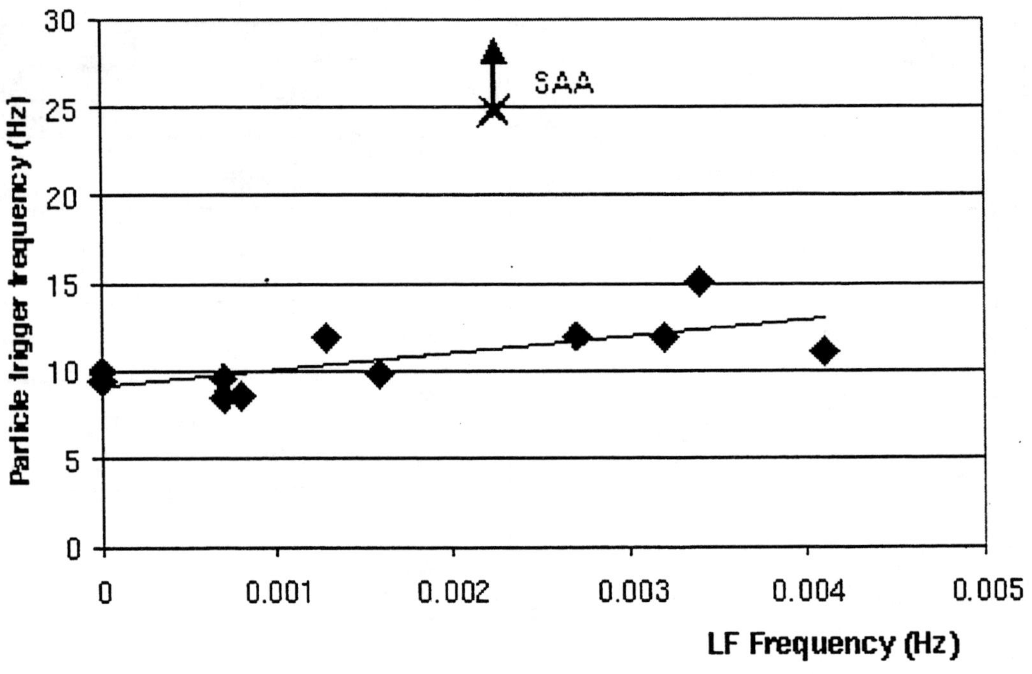

Figure 4: *Particle flux as a function of observed Light Flashes. SAA observed flux is well above the fluxes relative to other latitudes. The arrow indicates that the real SAA flux is higher than the measured one.*

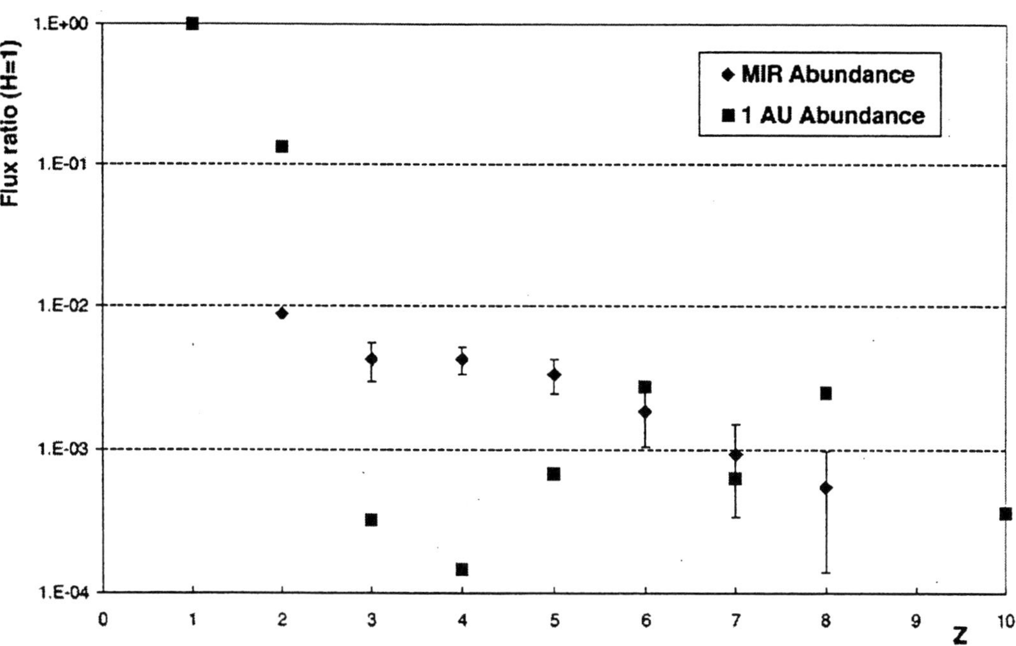

Figure 5: *MIR space station nuclear abundance with SilEye*

of this spectrum will be done in the future with longer sessions, to determine also the equivalent dose due to different nuclear species absorbed by astronauts in Space Station environment. In fact, high Z particles even being only the 1% of the total ionizing particle flux in cosmic radiation, contribute, due to their high quality factor, up to 25% of the total equivalent dose absorbed by Space Station crews [11].

Data analysis of SilEye-2 Flash data is now starting (at this time several sessions with astronauts have been performed). The construction [12] of a larger apparatus that combines the use of a large silicon detectors and an electroencephalograph, to directly correlate LF and particle crossing the head with brain activity, is under development (project ALTEA).

References

[1] W.Z.Osborne, L.S.Pinsky, J,V.Baily "Apollo light flash investigations", Biomedical results of Apollo. NASA SP-368, 355, 1975

[2] T.F.Budinger, C.A. Tobias, R.H.Huesman et al, " Light flash observations. Experiment mA-106", Apollo-Soyuz test project summary science report, NASA SP-412, 193.

[3] R.A.Hoffman, L.S.Pinsky,W.Z.Osborne, J.V.Baily "Visual light flash observation on Skylab 4" Biomedical results from Skylab, NASA SP-377, 127, (1977). 260, (1970).

[4] J.Fremlin, New Scientist 47, 42, (1970).

[5] G. Horneck, Nucl. Tracks Radiat. Meas., Vol.20, 1, 185, (1992).

[6] A. Galper, A. Morselli, P. Picozza et al., Proc. VI Eur. Symp. on Life Sciences Research in Space 17-21 June, Trondheim, Norway, (1996).

[7] A. Morselli, P. Picozza et al., XXIV ICRC, OG 10.2.8, Vol.5, (1997).

[8] V. Bidoli, A. Morselli, P. Picozza et al., NIM A 399, 477, (1997).

[9] A. Bakaldin, A. Morselli, P. Picozza et al., Astr. Phys., 8/1-2, 109, (1997).

[10] J.A. Simpson, Phys. Rev. Letters 51, 1010, (1983).

[11] G. Reitz et al. Radiat. Meas. Vol. 26, Nr.6, 679, (1996)

[12] M. Casolino, A. Morselli, P. Picozza et al., Il Nuovo Cimento Vol.19 D, N.10 (1997).

Detection of Multi-Joule Cosmic Rays Using Silicon/Solar Panel Detectors

D. B. Kieda

Department of Physics, University of Utah, Salt Lake City, Utah 84112-0830
kieda@krusty.physics.utah.edu

ABSTRACT

We describe the capability of a large array of non-imaging Čerenkov detectors for the observation of EeV and ZeV cosmic rays ($10^{18} - 10^{21} eV$). The detector employs an optical technique to observe the Čerenkov light emitted from extensive air showers both in daylight and at nighttime with inexpensive solid state (silicon) detectors. Monte Carlo simulations indicate that an array of 0.4 m^2 area Cz-Si solar panels with 1.5 km spacing will have a triggering threshold of approximately 30 EeV; an array using 1 cm^2 area PIN diodes will have a similar threshold. The typical energy resolution for this type of array is 10-25%, typical X_{max} resolution is 20-35 g cm^{-2}. The Silicon Detector Air Čerenkov Array has a factor of 5-10 improved on-time efficiency compared to the traditional Nitrogen Fluorescence (Fly's Eye) technique, and has improved point source sensitivity compared to traditional ground arrays of scintillator panels.

Cosmic Rays—Detection Techniques, Extremely High Energy, Atmospheric Čerenkov Detection

1. Introduction

The recent observation of cosmic rays with energies above 0.1 ZeV ($= 10^{20}$ eV = 16 Joules) (Bird 1994, Bird 1995, Hayashida 1994, Efimov 1991) and the apparent lack of a GZK cutoff in the cosmic ray spectrum (Takeda 1998) has challenged conventional theories of cosmic ray origin(Elbert 1995). Limitations on the canonical acceleration mechanisms have forced serious consideration of exotic production sites (Sigl 1994, Bhattacharjee 1994) and non-standard mechanisms (Chung 1998, Kephart 1996, Weiler 1997, Farrar 1996, Farrar 1998, Berezinsky 1997) to explain the observed cosmic flux. Many of these theories predict a large fraction of the multi-Joule cosmic rays will not originate isotropically, but instead will begin to cluster around several point sources, possibly AGN's (Biermann 1987). A cosmic ray detector with a very large aperture ($> 10^4$ km^2 sr) will be needed to measure the characteristics of these energetic cosmic rays, including clustering around known astrophysical objects, and changes in primary composition with increasing energy. These characteristics will provide essential information for elucidating the origin of these extremely high-energy cosmic rays.

A new, low cost technique is possible for the observation of cosmic rays in this energy range. The technique uses large area non-imaging optical detectors to directly observe the Čerenkov light emitted by extensive air showers (EAS) generated by cosmic ray primaries with energies greater than 10 EeV. The detection technique is based upon several concepts:

1. The pulsed Čerenkov light emitted by EAS generated by cosmic rays above 10 EeV energies is brighter than fluctuations in the daytime background light level over time scales of the pulse width.

2. The best directional reconstruction of the EAS is determined by a 'shadowing' technique; i.e. reconstructing the direction based upon relative Čerenkov light amplitudes measured by multiple solar panels at a single location oriented in different directions.

3. The Lateral distribution of Čerenkov light can be used to measure the Depth of Maximum Shower Development (X_{max}), which is can be related to the primary interaction cross section and hence the primary composition.

This technique should have energy and direction resolution similar to or better than traditional observation techniques, and should have X_{max} resolution similar to the existing Fly's Eye detector. A detailed summary of the expected resolution of this technique is provided elsewhere(Kieda 1995a, Kieda 1995b, Kieda 1996, Sorel 1997)

In this paper, we briefly describe the characteristics of Čerenkov light that make this technique possible, and the magnitudes of the expected background noise with respect to the Čerenkov signal. We describe the implications of the noise levels for detector threshold, spacing, and on-time. We describe calculated capabilities of a large array of Silicon Air Čerenkov detectors, especially energy and X_{max} resolution. Finally, we discuss point source sensitivity limitations due to background event 'spilldown' from lower energy cosmic rays. A general cost estimate of the project is presented.

2.   Čerenkov Light Intensity, Background Light Levels, and Inherent Noise

The general features of the Čerenkov light emission from ZeV cosmic Rays has been calculated using a modified MOCCA Monte Carlos simulation(Kieda 1995a). For a sea-level detector, the Čerenkov light has an intensity of $10^8 - 10^9$ photons m^{-2} at distances of 1 km from the shower core, and has an emission time duration of several hundred nanoseconds. The spectral emission is strongly peaked between 300-400 nm for vertical showers, and 500-600 nm for very steep showers (60-75 degree zenith angles). The lateral distribution of Čerenkov light extends over several kilometers, and the shape of this distribution is correlated with X_{max}.

A generic silicon sensor will have an area A, a wavelength dependent quantum efficiency $qe(\lambda)$. If $\frac{dN_{ch}(\lambda)}{d\lambda}$ is the wavelength dependent Čerenkov photon flux, the photoelectron signal S generated by the sensor is

$$S = (A) \int \frac{dN_{ch}(\lambda)}{d\lambda} qe(\lambda) d\lambda \tag{1}$$

The Čerenkov signal from these showers is extraordinary, especially compared to daytime sky background light levels. The noise due to fluctuations in the sky background light can be calculated using the standard solar spectrum of the sky background multiplied by the spectral response $qe(\lambda)$ of the silicon detector, integrated over an interval similar to the longest detectale Čerenkov light pulse width (5 μsec). The noise level is then scaled for various sky conditions (twilight, noon, full moon, etc). Figure 1 illustrates the Signal-to-Noise ratio for a 0.4 m^2 solar panel detector under various observing conditions, and for various primary energy cosmic ray protons. One observes that the Čerenkov signal from a 60 degree zenith angle proton at 0.1 ZeV is similar to fluctuations in the daytime direct noon sun!

Detector On-Time

In practice, one finds that a Signal-to Noise ratio (S/N) > 10 at 1 km distances from the shower core is required for adequate reconstruction of the primary energy, direction, and X_{max}. Using the known amounts of no-moon, full moon, twilight , and daytime sun conditions for a 45 degree latitude Northern hemisphere site, we can predict the percentage of time that the S/N ratio exceeds this minimum requirement. Figure 2 illustrates the possible duty cycle as a function of primary energy and zenith angle for a large solar panel (0.8 m^2) detector which is 1 km from the core of the shower. One notes that for 10^{20} eV cosmic rays, 60% duty cycle is easily achieved over almost all zenith angle trajectories. At 10^{21} eV, one approaches nearly 100% duty cycle for zenith angles between 40 to 70 degrees. The plot here assumes perfect weather, and should be compared to the canonical maximum 14% duty cycle typical of the Fly's Eye Technique and most air Čerenkov experiments.

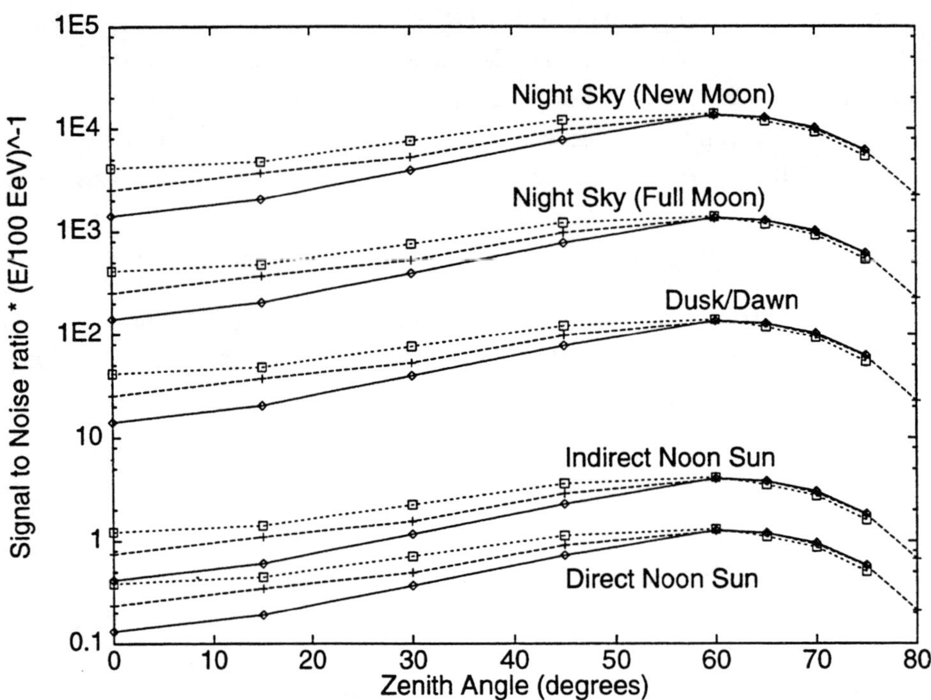

Fig. 1.— Signal-to-Noise for Air Čerenkov detection of Cosmic rays using Silicon Detectors as a function of Primary Energy, Zenith Angle, and Viewing Conditions. Horizontal Axis: Cosmic Ray Trajectory Zenith Angle (degrees); Vertical Axis: Signal-to-Noise ratio divided by primary energy (in units of 100 EeV). Dotted Line (squares): 10 EeV; Dashed line (crosses) 100 EeV; Solid Line (diamonds) 1000 EeV

Other types of silicon detectors show similar on-time efficiencies. For 1 cm^2 area PIN diodes (without Winston cones), one decreases the on time efficiency by about 20% for each curve to account for the reduced S/N for the devices. This is mainly due to their smaller area. A PIN diode with a small Winston cone might increase the signal by a factor of 10, giving increased duty cycle similar similar to the solar panel duty cycle.

Total Noise in Silicon Detectors

The above described scenario assumes a perfect detector, i.e. no electronic noise in the measurement of the Čerenkov photons with the large area detectors. For Silicon detectors, one cannot ignore the inherent noise N_{inh} due to the finite capacitance of the P-N or PIN junction region. The total noise N in a real silicon

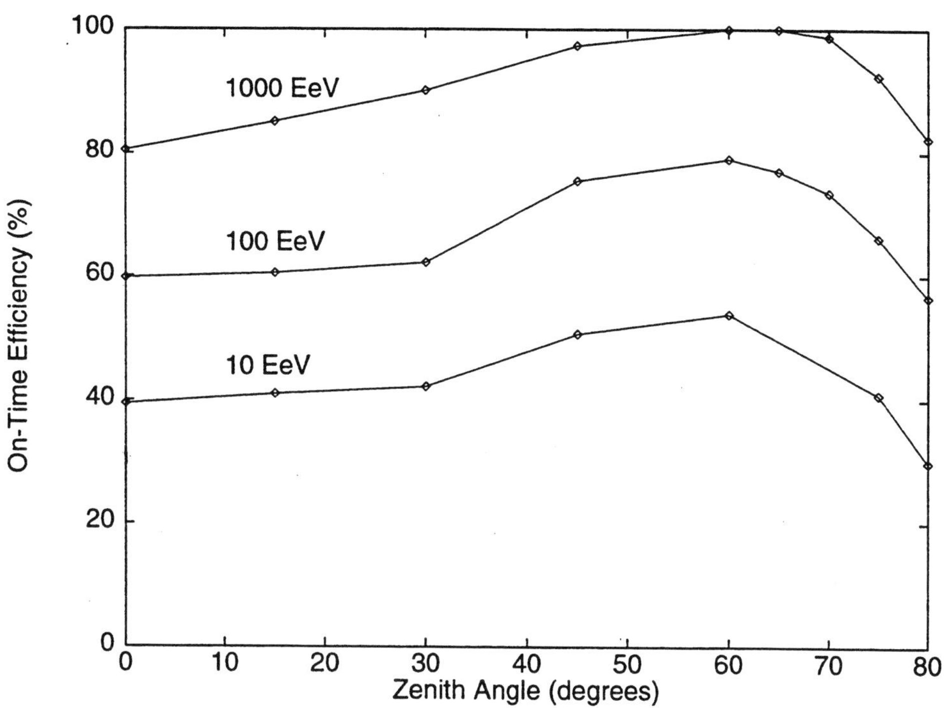

Fig. 2.— On-Time Efficiency for a 0.8 m^2 area solar panel detector located one km from the shower core as a function of primary energy and zenith angle. On-Time efficiency is defined as percentage time that the detector has a minimum S/N ratio of 10. Horizontal Axis: Zenith Angle (degrees) ; Vertical Axis: On-Time efficiency (%).

detector is therefore the combination of the statistical fluctuations in the sky background light level described above (N_{sky}), the statistical fluctuation in the Čerenkov photon flux (N_{signal}), and N_{inh}.

$$N^2 = N_{sky}^2 + N_{signal}^2 + N_{inh}^2 \qquad (2)$$

The Signal-to Noise ratio (S/N) including the effects of the statistical, sky, and inherent noise levels has been calculated as (Kieda 1997a, Kieda 1997b):

$$S/N = \frac{N_{ch} <QE>_{ch}}{\sqrt{\frac{dN_{sky}}{dt} <QE>_{sky} (\Delta t)}} \cdot \frac{\sqrt{(A)}}{\sqrt{1 + \frac{9 \times 10^{10} \sqrt{\varepsilon/(2\mu_h \rho (V_0 + V_b))}}{(\Delta t)\frac{dN_{sky}}{dt} <QE>_{sky}}}} \qquad (3)$$

Here $\varepsilon = 1.054 \times 10^{-10}$ is the permittivity of the silicon material, A is the junction area (in m^2) $\mu_h = 500$ cm^2 V^{-1} sec^{-1} is the hole mobility in silicon, ρ is the bulk resistivity, $V_0 = 1V$ is the contact potential, and V_b is the externally applied junction (reverse) bias voltage. N_{ch} is the integrated number of Čerenkov photons over all wavelengths per unit area, $< QE >_{ch}$ is the average quantum efficiency of the sensor to the Čerenkov spectrum, dN_{sky}/dt is the background sky light level per unit time per unit area, integrated over all wavelengths, and $< QE >_{sky}$ is the average quantum efficiency of the sensor for viewing the sky spectrum.

3.   Triggering threshold

For widely spaced detectors, with small area, and at low cosmic ray energy, the electronic noise level N_{inh} will dominate the total noise of the detector panel in Equation 3, and hence will limit the triggering threshold of the panel.

Using equation 3, and the Monte Carlo simulated number of Čerenkov photons per unit area $N_{ch}(E, d)$ as a function of primary energy E and distance d from the shower core, we can invert this expression to determine the triggering threshold as a function of detector area A and array spacing d given a minimum required S/N ratio of 10 (Kieda 1997b). The calculation an be performed for different types of Silicon detectors, including both CZ-SI (solar panel) and PIN diodes. We insert the measured night sky background photon rate $dN_{sky}/dt = 6.5 \times 10^6$ photons m^{-2} μsec^{-1}, a signal integration time of 5 μsec, and also assume $< QE >_{ch} \approx < QE >_{sky} = 0.6$.

Figure 3 illustrates the array triggering threshold as a function of detector spacing and detector area for a CZ-SI solar panel detector for vertical proton showers. Typical CZ-SI solar panels available now have areas of 0.4 m^2; implying a detection threshold of 30-50 EeV. Showers with 60 °inclinations have a factor of 2 energy lower threshold than vertical showers.

Figure 4 illustrates the vertical shower triggering threshold plot for a PIN diode sensor. For a large area PIN diode (1 cm^2) one can achieve a 30-50 EeV threshold with 1.5 km spacing between detector elements. For closer spacing (1 km) the threshold drops to 10 EeV; 1 EeV thresholds may be achievable with 300 m detector spacing.

4.   Air Čerenkov Detector Design and Performance

Based upon the considerations of the previous section , we have designed and simulated a strawman air Čerenkov detector. Events are generated by seeding MOCCA-simulated showers onto a simulated 3000 km^2 area array of detectors located at 100 m above sea level. Spacing between detectors is 1.5 km; each

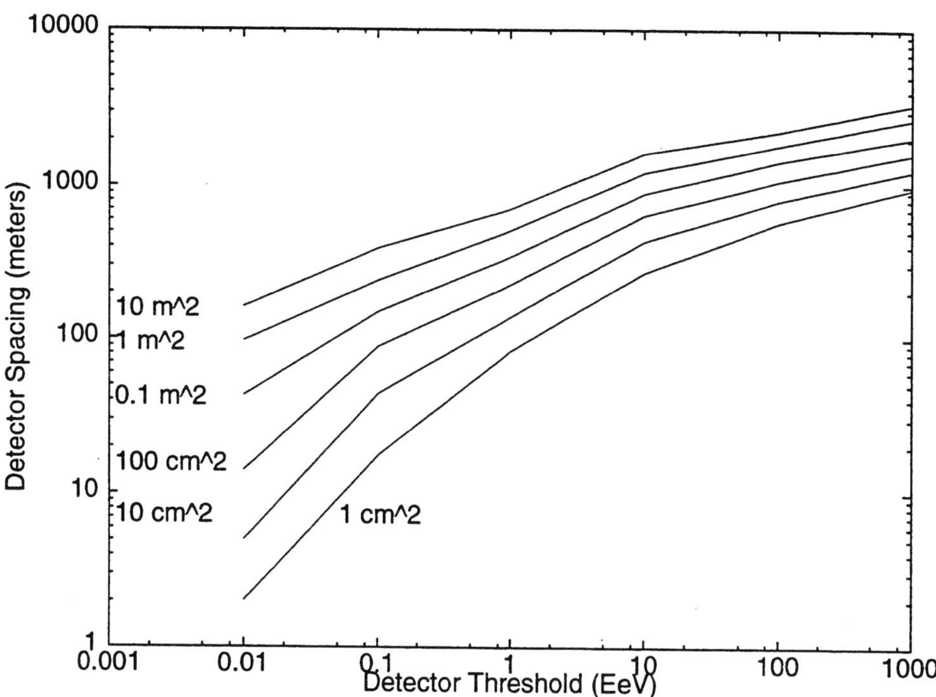

Fig. 3.— Minimum detector spacing as a function of energy threshold and detector area for the Silicon Detector Čerenkov Array using Silicon CZ-SI solar panels. Curves are labeled according to the detector area. Horizontal Axis: Detector Energy Threshold (EeV); Vertical Axis: Maximum detector spacing (meters).

detector consists of six 0.4 m^2 CZ-Si solar panel detector elements pointed in various directions to optimize the reconstructed direction accuracy (Suson 1996) . Events are analyzed by reconstructing the local Čerenkov light direction at each station using a 'shadow fitting' algorithm. The impact location for the core of the shower is calculated from the mean station position, weighted by the fitted station amplitudes derived in the first stage of reconstruction. A second reconstruction is performed to determine the direction of the original cosmic ray from the individual station directions. The lateral distribution of the Čerenkov photons is reconstructed using the fitted amplitudes, core position, and primary direction. The fitted lateral distribution function is used to calculate $\rho(1200)$ the Čerenkov photon density at a perpendicular distance of 1.2 km from the shower axis, and $\rho(500)/\rho(3000)$. $\rho(500)/\rho(3000)$ is a useful quantity because it is directly correlated with the shower X_{max}.

The resolution of the solar panel detector for the above quantities will depend mainly upon the Signal-to-Noise ratio (S/N) in a panel (Figure 1). The typical

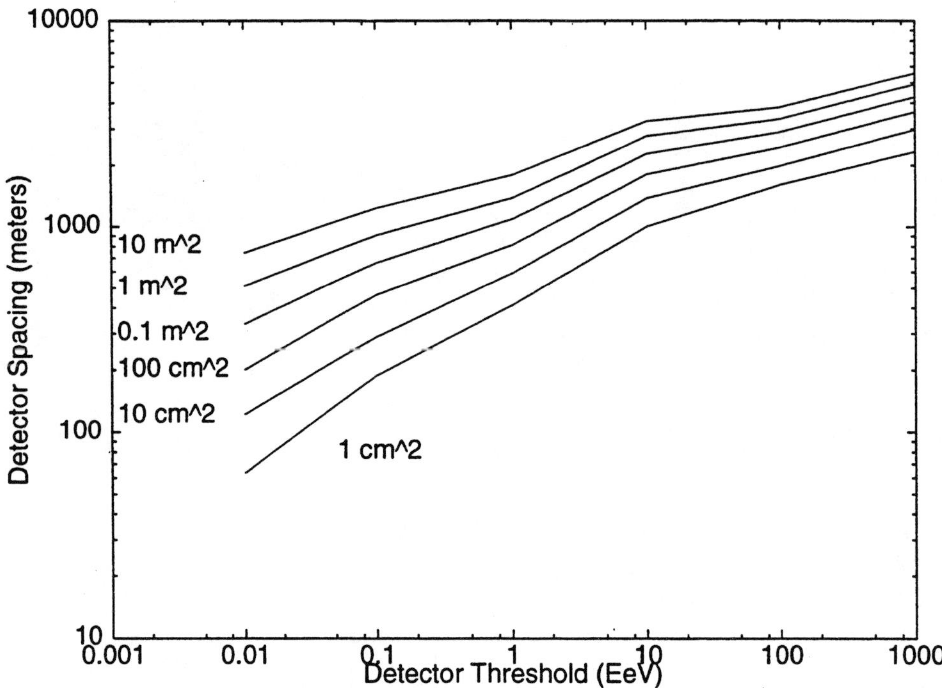

Fig. 4.— Minimum detector spacing as a function of energy threshold and detector area for the Silicon Detector Čerenkov Array using Silicon PIN diodes. Curves are labeled according to the detector area. Horizontal Axis: Detector Energy Threshold (EeV); Vertical Axis: Maximum detector spacing (meters).

reconstructed angular resolution for the original primary cosmic ray direction is approximately 1-2° for near vertical showers, and 0.3-0.5° for steep zenith angle shower (S/N =100).

Figure 5 illustrates the expected energy resolution as a function of zenith angle. For vertical showers with S/N = 10, the energy resolution at 100EeV is about 25%. It is unlikely the vertical energy resolution could ever get much better than 40% at 1 ZeV, due to the fact that vertical EAS at these energies can have X_{max} which approach the ground level! The best energy resolution ($< 10\%$) occurs for steep (60°) zenith angle showers.

In Figure 6 we show the X_{max} resolution of the simulated array as a function of S/N ratio. The typical X_{max} resolution is 20-35 g cm^{-2}, with the best X_{max} resolution at steep (60°) zenith angles.

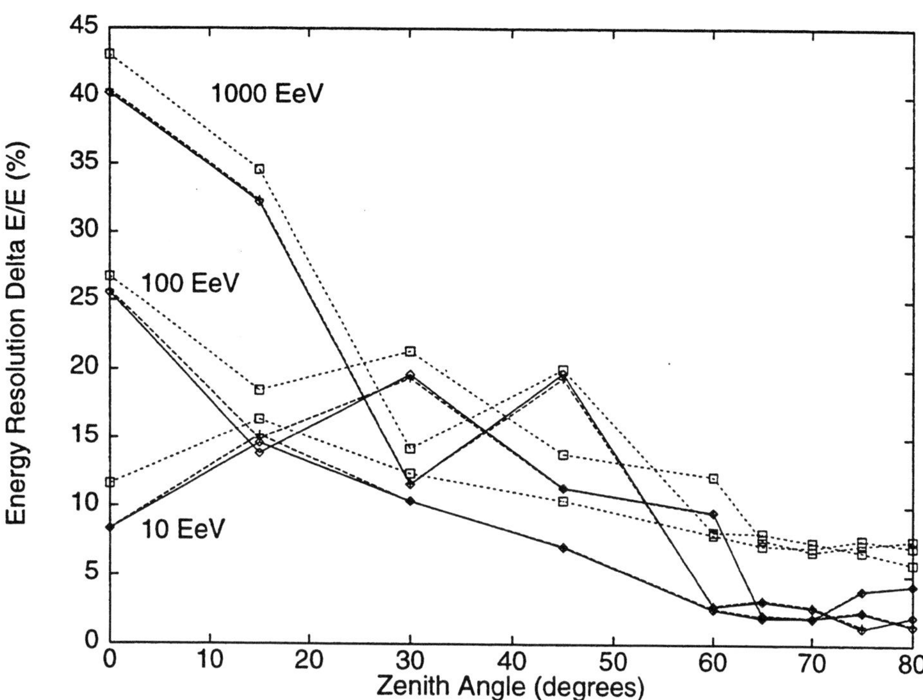

Fig. 5.— Calculated Energy Resolution as a function of Zenith angle, Energy, and Signal-to-Noise ratio. Horizontal Axis: Cosmic Ray Trajectory Zenith Angle (degrees); Vertical Axis: Energy resolution ($\Delta E/E$); Dotted Line (squares): S/N = 10; Dashed line (crosses) S/N=100; Solid Line (diamonds) S/N=1000

5.   Malmquist Bias Effects on Point Source Identification.

If EHE cosmic rays indeed originate from a finite number of discrete point sources, the Silicon Detector Air Čerenkov Array would have strong advantages for observation of these point sources over existing detectors. Traditional scintillator arrays, such as the AGASA array, have increasingly poorer energy resolution with increasing shower zenith angle. Consequently, the detector must either accept showers in a narrow field of view near the vertical, thereby substantially reducing their exposure to the point source, or must accept these showers at larger zenith angles, and accept the consequences of a large background rate due to 'spill-down' of mismeasured lower energy showers into the higher energy shower bins.

A Monte Carlo Simulation can illustrate the problems caused by the spilldown of lower energy cosmic rays into higher energy bins due to zenith-angle dependent energy resolution. The simulation first randomly seeds a set of 20 point sources of protons isotropically on the celestial sphere. The locations of these sources are

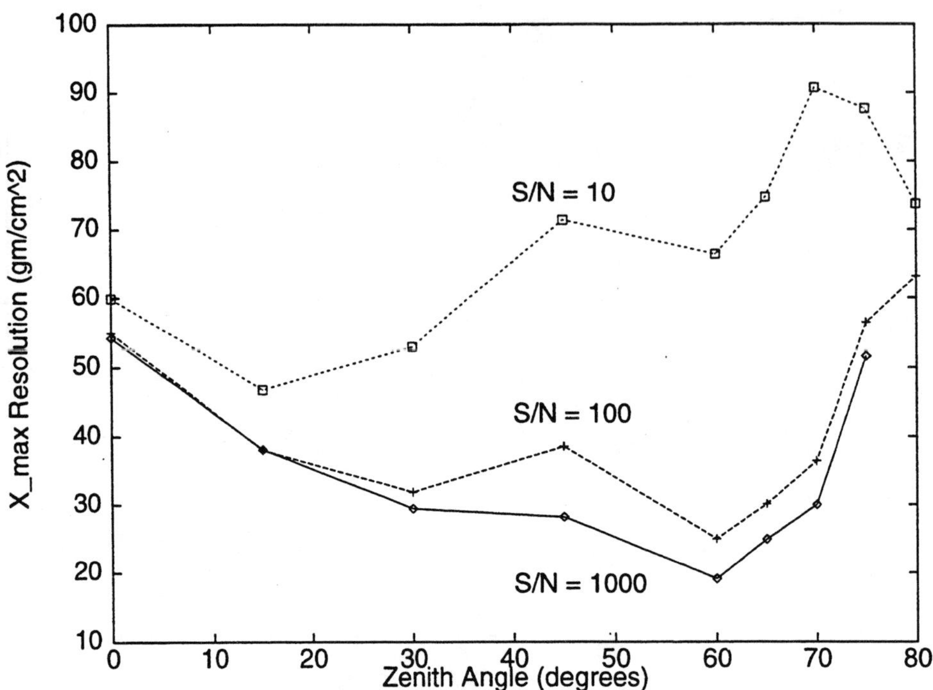

Fig. 6.— Expected X_{max} resolution of the Silicon Detector Čerenkov Array as a Function of Signal-to-Noise and Zenith Angle. Horizontal Axis: Cosmic Ray Trajectory Zenith Angle (degrees) Vertical Axis: X_{max} resolution (g cm^{-2}) Dotted Line (squares): S/N = 10; Dashed line (crosses) S/N=100; Solid Line (diamonds) S/N=1000

represented by a solid circle on a sky map plot (Figure 7). The individual point source intensities are chosen according to a $dN/dI \propto I^{-3/2}$ distribution, consistent with a homogeneous, equal intensity source distribution.

The sources a assumed to be extragalactic. The simulation deflects the measured proton directions from the true source direction depending on the rigidity of the protons in the Galactic magnetic field. Typical estimates of this deflection are approximately 10 degrees for a 4×10^{19} eV proton, and 1 degree for a 4×10^{20} eV proton.

The simulation assumes a 5 year exposure with a total ground area of 3000 km^2 in each hemisphere detector.

Figure 8 shows the sky distribution of events for the traditional scintillator array, with zenith-angle dependent energy resolution defined by the AGASA array. Figure 9 shows the same simulated events as Figure 8, but using the energy and

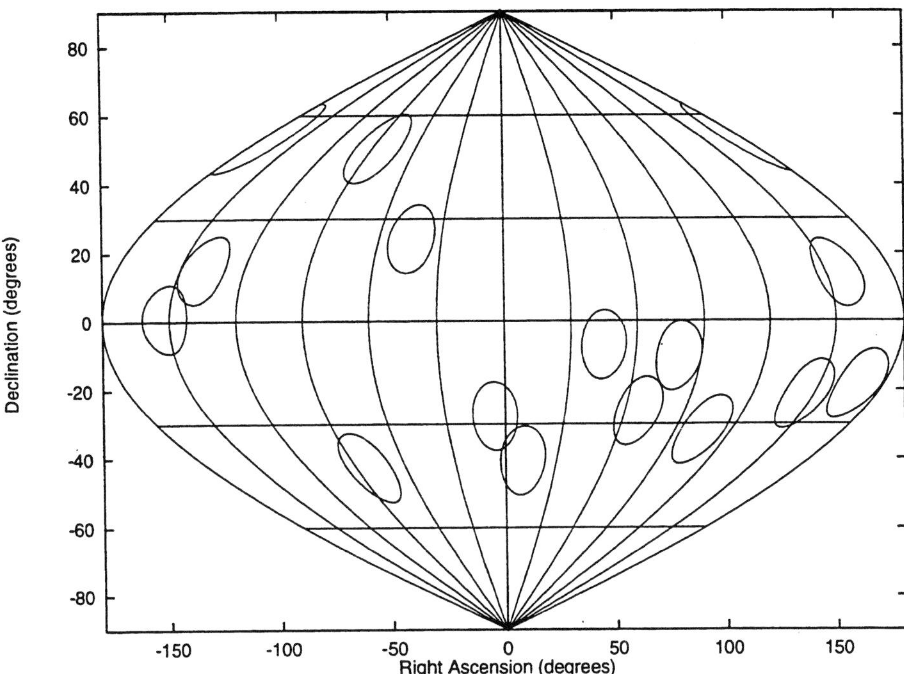

Fig. 7.— Seeded Location of Simulated Sources above 40 EeV. Seeded sources are at the center of the circular regions. Horizontal Axis: Right Ascension(degrees) . Vertical Axis: Declination (Degrees)

direction resolution of the Silicon Detector Air Čerenkov Array, as described above. The general background level of the Čerenkov array is substantially lower than the surface scintillator array, corresponding to an effective increase in sensitivity for point sources by a factor of 10. Sources whose locations were not statistically significant for the scintillator array become strongly visible for the Silicon Detector Air Čerenkov Array.

6. Cost

The expected cost of the Silicon Detector Air Čerenkov Detector has been calculated using similar accounting procedures to the Auger project. The estimated cost for a pair of arrays the size of the Auger project would be about \$20-25 Million. Much of this cost is non-detector related; it includes central facility costs, road building, etc. A detector with a detection area about a factor of 20 larger than the Auger Project should cost about \$100 Million US. The main difficulty may be trying to find a location to put such large detector (about 100,000 km^2

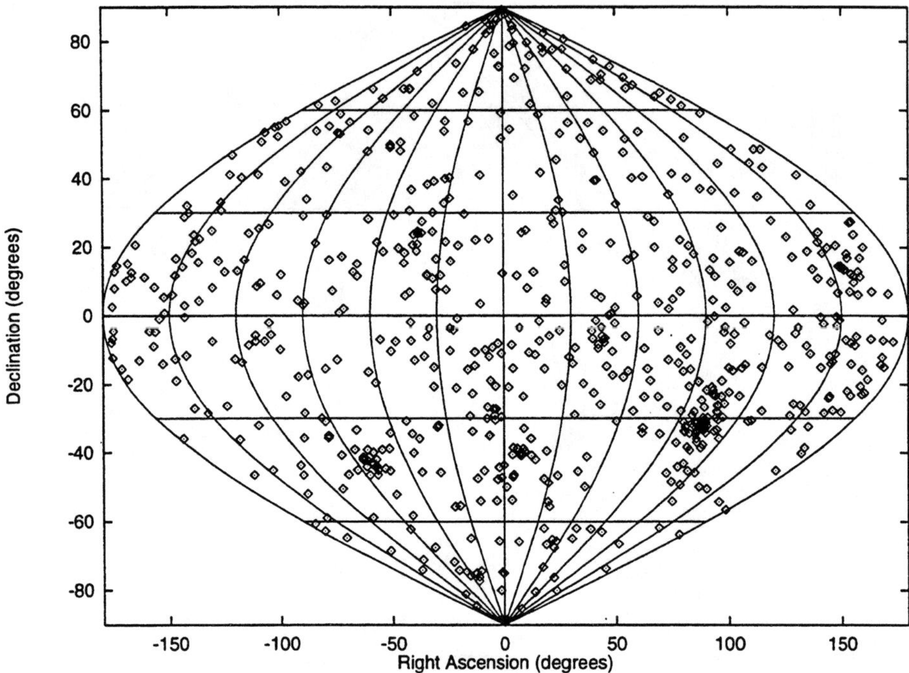

Fig. 8.— Reconstructed event locations for Simulated Sources above 40 EeV,
including Malmquist bias, for a scintillator particle detector array. Horizontal Axis:
Right Ascension(degrees) . Vertical Axis: Declination (Degrees)

area!). A detector of this size would have a reasonable chance of detecting cosmic
rays with energies up to nearly 10^{22} eV (10 ZeV), assuming the primary energy
spectrum above 10^{20} eV does not suffer a change in the spectral index.

7. Present Status

At the present time, there are three groups working on the construction of
a Silicon Detector Air Čerenkov Detector: a group at the University of Utah (D.
Kieda), a group at the AGASA Array in Tokyo (M. Takeda), and a group at the
University of Bologna (G. Giacomelli). All three groups are developing prototypes
for deployment at various existing particle detector arrays, for simultaneous
observations with the particle detectors . We look forward to seeing positive results
from these experiments over the next few years.

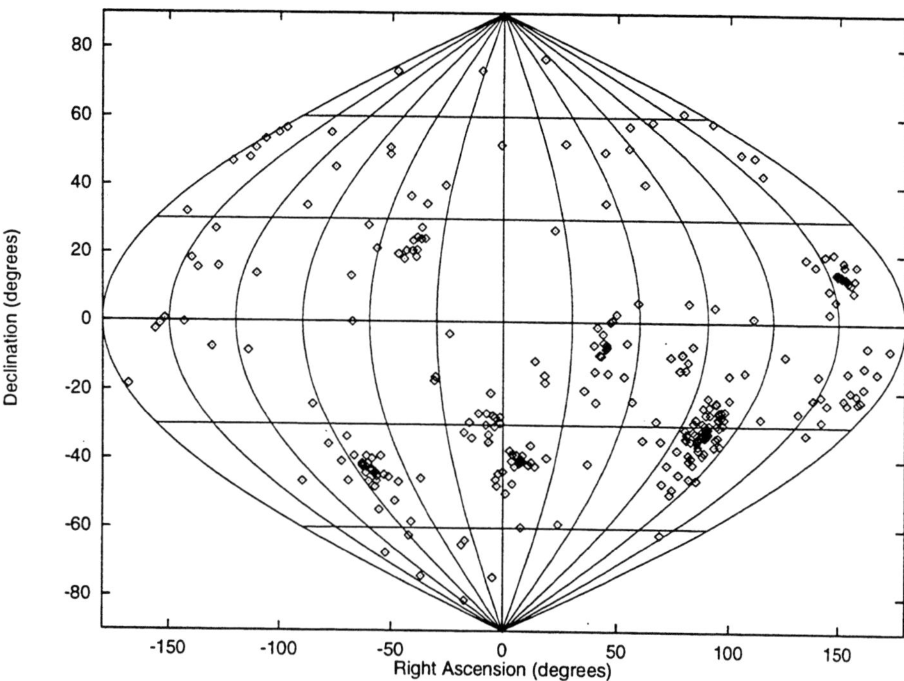

Fig. 9.— Reconstructed event locations for Simulated Sources above 40 EeV, including Malmquist bias, for the Silicon Detector Air Čerenkov Array. Horizontal Axis: Right Ascension(degrees) . Vertical Axis: Declination (Degrees)

8.   Acknowledgments

The author gratefully acknowledges the assistance of the HiRes and the CASA-MIA collaborations. This work is supported through institutional support by the University of Utah, and by the National Science Foundation Grant # PHY 9527421.

REFERENCES

Berezinsky, V, Kachelrieβ, M. and Vilenkin, A. , 1997, Phys. Rev. Lett. 79, 4302.

Bhattacharjee, P. and Sigl, G., 1994, Phys. Rev. D 51, 4079

Bird, D. *et al.*, 1994, ApJ 424, 497

Bird, D. *et al.*, 1995, ApJ 441, 144

Chung, D. , Farrar, G. and Kolb, E., 1998 , Phys. Rev. D 57 4606.

Efimov, N. *et al.*, 1991, in: *Proc. Intnl. Wrkshp. on Astrophysical Aspects of the Most Energetic Cosmic Rays*, Kofu, Japan, M. Nagano and F. Takahara, eds., (World Scientific, Singapore) 20

Elbert, J. W. and Sommers, P. , 1995 ApJ 441, 151.

Farrar, G. , 1996, Phys. Rev. Lett. 76, 4111.

Farrar, G. and Biermann, P., 1998, Phys. Rev. Lett. to be published.

Hayashida, N. *et al.*, 1994, Phys. Rev. Lett. 73, 3491

Kephart, T., and Weiler, T., 1996, Ap. Phys. 4, 3, 271.

Kieda, D.B., 1995, Ap. Phys. 4, 133

Kieda, D.B., March 23, 1995, "A Direct Measurement of Noise Levels in Large $(0.8m^2)$ CZ-Si Solar Panels, and Simple Calculation of Triggering and Event Rates in a 6 Panel Prototype Detector" HiRes Fly's Eye Note, University of Utah

Kieda, D. B., 1996, "Air Čerenkov Detector for the Auger Project," in *The Pierre Auger Observatory Design Report, 2nd Edition*, FNAL

Kieda, D. B., 1997, Proc. Intl. Symp. on Extremely High Energy Cosmic Rays, (Tanashi, Tokyo Sept 1996) (M. Nagano, ed., ICRR, Japan) , 320.

Kieda, D. B., 1997, Proc. 25th ICRC, Durban, South Africa, 5, 197.

Sigl, G., Schramm, D. N., and Bhattacharjee, P., 1994, Ap. Phys. 2, 401.

Sorel, M., 1997, Ph.D Thesis, University of Bologna

Suson, D.J. and Kieda, D.B, 1996, *Nucl. Inst. Meth. Phys.* **A 374**, 381

Takeda, M. *et al.*, 1998, Phys. Rev. Lett. 81, 1163,

Weiler, T. 1997, hep-ph/9710431, to appear in Ap. Phys. (1998).

Cosmic Ray Nuclei Measurements in Space with NINA Experiments

A. Morselli, R.Sparvoli on the behalf of NINA Collaboration
Dept. of Physics, II Univ. of Rome "Tor Vergata" and INFN, Italy

Abstract

In June 1998 the telescope NINA has been launched in space on board of the Russian satellite Resource-01 n.4. The main scientific objective of the mission is the study of the Anomalous, Galactic and Solar components of the Cosmic Rays in the energy interval 10-200 MeV/n.

1 Introduction: the detector

The active part of the NINA instrument is a telescope composed of 16 X-Y planes, giving information on the energy of the crossing particle and its incident angle. The sensitive element consists of two n-type silicon detectors, 60×60 mm^2, each divided in 16 strips and attached to a supporting ceramic frame passing under the lateral strips 1 and 16. The couple of detectors is mounted with the strips right-angled, in order to measure the X and Y coordinates of the detected particle. The thickness of the first two detectors is 150 ± 15 μm; all the others, instead, are 380 ± 15 μm thick, so to have 11.7 silicon mm in total [2].

The lateral strips (n.1 and n.16) are used for the anticoincidence system (AC) and are read together by the same electronic channel, except for the plane 1 where they are physically disconnected. The planes are vertically stacked; interplanar distance is 1.4 cm, except for the first and second planes separated by 8.5 cm, for a better measurement of the particle incident angle. The 16 planes are modular, so that mechanically and electronically they are interchangeable. Below the 16 active modules, other 4 cards are placed, dedicated to the trigger electronics, silicon power supply, analog-digital conversion, FIFO.

The geometrical factor is 8.3 cm^{2}sr for low energy particles which stop in the second plane, and it decreases with growing energy.

The whole structure is surrounded by an aluminum vessel, a cylinder of 284 mm diameter and 480 mm height. The vessel is 2 mm thick, apart from a little window in correspondence to the first silicon plane where it is reduced to 300 μm; this choice has been made so to decrease the amount of passive material the particles cross before reaching the sensitive detector.

The signal produced by the incoming particles in the silicon strips has to be amplified and shaped before performing the conversion from analog to digital. Each plane of the telescope has two 16 channels preamplifiers.

The analogic signal is digitized by a 12 bit (4096 channels) ADC. The ADC overflow channel corresponds to 2800 mip, where *1 mip* in 380 μm of silicon is equivalent to 30400 electrons or 105 keV; the resolution per channel is about 0.68 mip/ch, or 0.07 MeV/ch.

2   Operating conditions

NINA can work in different operating conditions, switched automatically or via telecommand, and affecting the trigger system. In particular:

1. Two thresholds for the energy deposits in the single silicon layers have been implemented: a *Low Threshold* (L.T.), corresponding to 2.5 mips, and a *High Threshold* (H.T.), corresponding to 25 mips.

   In the first two layers, in order to compensate the smaller thickness, the High Threshold corresponds to 0.4 of the one previously defined.

2. The strips 1 and 16 of every silicon layer are used as Lateral Anticoincidence system, and are physically connected to the same output.

3. The planes 15 and 16 can be used as Bottom Anticoincidence. The usual operating mode adopts the plane 16 but, in case of failure, the plane 15 can be selected by telecommand.

The basic operating trigger of the acquisition system is the following:

$$TRG\,M1 = D_{1x} \times D_{1y} \times ((D_{2x} + D_{2y}) + (D_{3x} + D_{3y})) \times$$
$$\times \overline{D_{16x}} \times \overline{D_{16y}} \times \overline{LAT},$$

where D_{ij} is the above-threshold signal coming from the plane i, view j (j=x,y), and LAT is the logic OR of all signals coming from the strips defining the Lateral Anticoincidence. The logic OR of planes 2 and 3 provides redundancy in case of a failure of plane 2.

Particle	Z	E_{min} (MeV/n) TRG M1	E_{max} (MeV/n)	E_{min} (MeV/n) TRG M2
^1H	1	10.0	48.0	18
^2H	1	6.5	32.0	12.5
^3H	1	5.0	25.3	9.7
^3He	2	11.0	55.7	21.3
^4He	2	9.25	47.2	18.0
^6Li	3	11.5	59.3	22.7
^7Li	3	10.6	54.4	20.7
^7Be	4	14.6	75.1	28.6
^9Be	4	12.7	65.2	24.8
^{10}B	5	16.0	79.0	31.0
^{11}B	5	14.6	74.6	29.1
^{12}C	6	17.5	87.5	33.3
^{14}N	7	18.6	95.0	36.4
^{16}O	8	20.0	103.1	39.4
^{19}F	9	21.0	106.8	40.5
^{20}Ne	10	23.0	117.0	44.5
^{28}Si	14	27.5	141.8	53.6
^{40}Ca	20	38.5	174.5	65.3
^{56}Fe	26	58.2	194.6	72.5

Table 1: *Energy windows for the most abundant particles in NINA detector. Column 3: E_{min} using TRG M1; Column 5: E_{min} using TRG M2.*

This trigger can be used with Low or a High Threshold, defining two different intervals of atomic numbers for nuclei. In particular TRG M1, together with High Threshold, cuts from the trigger all protons and a very small percentage of heliums (figure 1). Finally, TRG M1 allows a good reconstruction of the particle trajectory. For particular data taking demands, or in case of failure of the first plane, it is possible to switch, via telecommand, to a second trigger logic:

$$TRG\ M2 = (D_{2x} + D_{2y}) \times (D_{3x} + D_{3y}) \times (D_{4x} + D_{4y}) \times (D_{5x} + D_{5y}) \times$$
$$\times \overline{D_{16x}} \times \overline{D_{16y}} \times \overline{LAT}.$$

This trigger increases the acceptance angle, although with a worsening of the angular resolution. The combination of TRG M2 and High Threshold excludes most of the protons from the trigger (which are in much higher percentage and fill the storage memory very quickly), but keeps all He nuclei.

The acceptance window for Low Threshold acquisition and both TRG M1 and TRG M2 is shown in table 1. The spectrum of particles extends from hydrogen to iron, and the energy window in the interval 10 - 200 MeV/n.

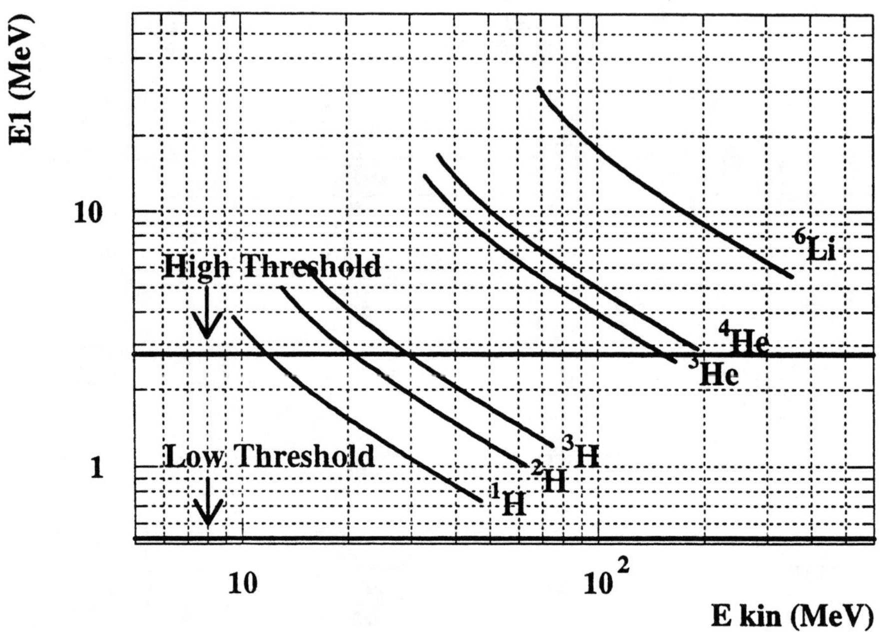

Figure 1: Effects of the High and Low Threshold on the acquisition of ^1H, ^2H, ^3H, ^3He, ^4He e ^6Li contained in the detector, in case of TRG M1.

3   Acquisition modes

The maximum amount of data that NINA can send to ground from the satellite varies between 2 and 4 Mbyte/day, out of a total mass memory of 14 Mbytes. Therefore, a limit to the acquisition capability of the instrument has been provided, organizing a system which, at high rate conditions, enables the detector to register events with less detail.

The flux of particles changes notably along the orbit, so that three data acquisition and storage modes have been foreseen, depending on the counting rate:

Full-Format mode (counting rate up to 10 Hz). This mode, in which the whole event topology is recorded, is the normal working configuration outside the Earth Radiation Belts. It allows to measure the energy released by the particle in each silicon detector (Bragg curve). Within the precision allowed by the strip pitch, we can also reconstruct the particle trajectory, identify its range with good precision and check for multiple tracks and for particles escaping the telescope, due to the geometrical inefficiencies of the anticoincidence system.

E_1 - E_{tot} **mode** (counting rate 10-100 Hz). At high fluxes, to make an optimal use of the mass memory storage, the total energies released in the first plane (E_1) and in the whole detector (E_{tot}) by the crossing particle are calculated. Moreover, to select mainly single and not escaping tracks, a Second Level trigger restricts the event acceptance only to particles crossing the 4 central strips of the first 2 planes of the telescope, and leaving a single cluster of fired strips (multiple tracks subtraction).

Once these conditions have been fulfilled, only the E_1 and E_{tot} information of the event are transmitted.

Rate Meter mode (counting rate above 100 Hz). If the trigger rate rises above 100 Hz, only the counting rates of some planes, chosen at different depths of the telescope, are read and stored every 6 minutes. Two signal integration systems have been implemented, so to have information both at high and low rates.

The scientific information in Rate Meter mode is obviously strictly reduced, but still allows to have a global check of the system and an estimation of the rate of particles inside the Radiation Belts or in presence of solar flares.

The switching among the different acquisition modes is instantaneous. In order to avoid oscillations between the modes, which could bring to a fragmented scientific information, every passage between one state and the next can happen only at minimum intervals of 60 seconds.

4   Energy calibration

The detector has been calibrated by comparing the energy deposits of the collected families of nuclei, from proton to carbon, expressed in ADC channels, with the corresponding simulated ones expressed in MeV. Montecarlo simulation programs had been previously calibrated by using monocromatic proton and helium beams at the PSI Laboratory of Willigen (Zurich - Switzerland).

Figure 2 shows some Bragg curves for carbon experimental data. The tracks chosen for the calibration were straight, clean, without double or missing signals, and right in the central strips of the detector.

The average ratio between MeV energy and ADC channels obtained from the data is the following:

$$R \ = \ (0.067 \pm 0.002) \ \frac{MeV}{ADC \ Ch} \ . \tag{1}$$

Figure 3 shows the E_1 *vs* E_{tot} curves of real particles, coming from the fragmentation of the original carbon beam by means of the polyethylene target, with energies expressed in MeV. All the products of fragmentation are visible, and the energy behavior is in perfect agreement with the expected simulated data [1].

5   Mass analysis

The *Full-Format* acquisition mode allows the complete registration of the track of the particle with all its energy deposits in the strips and therefore, in this condition, the best nuclear and isotope discrimination can be performed.

Figure 3 shows how the nuclear families coming from the fragmentation of the ^{12}C beam are well separated by the simple E_1 *vs* E_{tot} technique. Nevertheless, we can better identify groups of equal Z number by using a well tuned product of the E_1 and E_{tot} deposits. The product $E_1 \times E_{tot}$ is an effective tool for identifying groups of equal atomic number Z, as can be seen in figure 4.

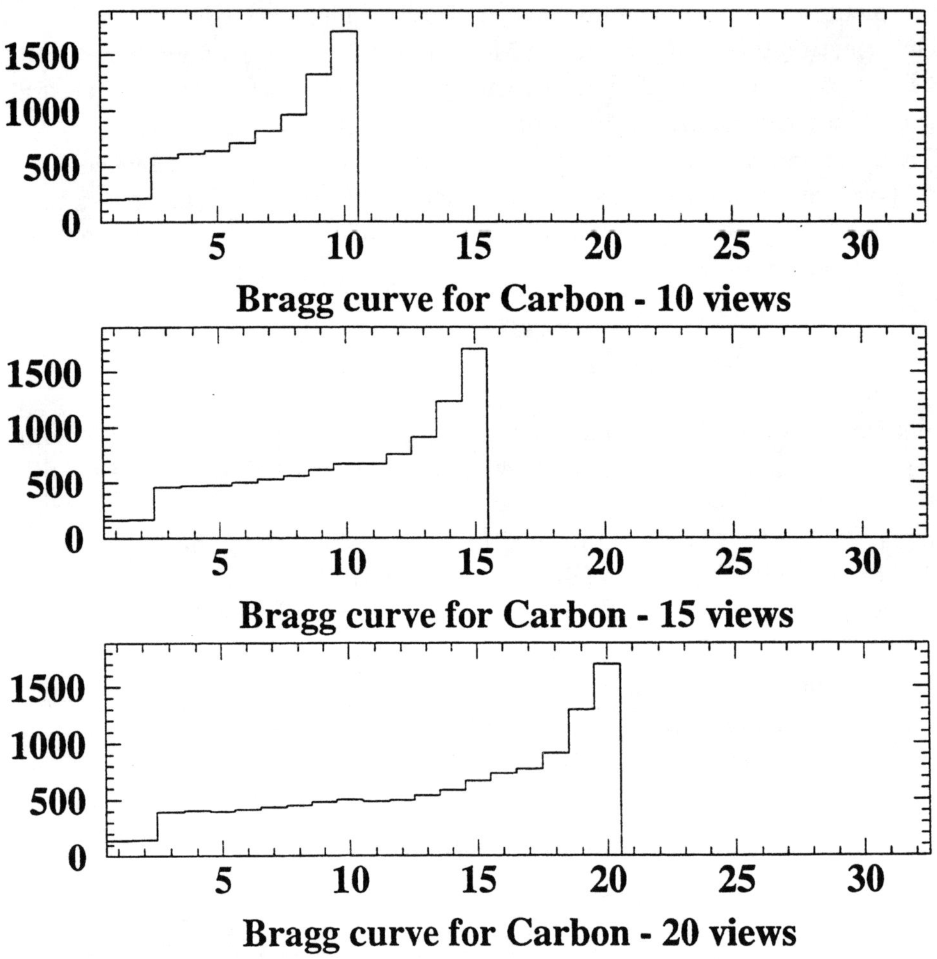

Figure 2: Bragg curves for real carbons at different ranges. X axis: n. of views in the detector $(1 < n < 32)$; Y axis: energy deposit in ADC channels.

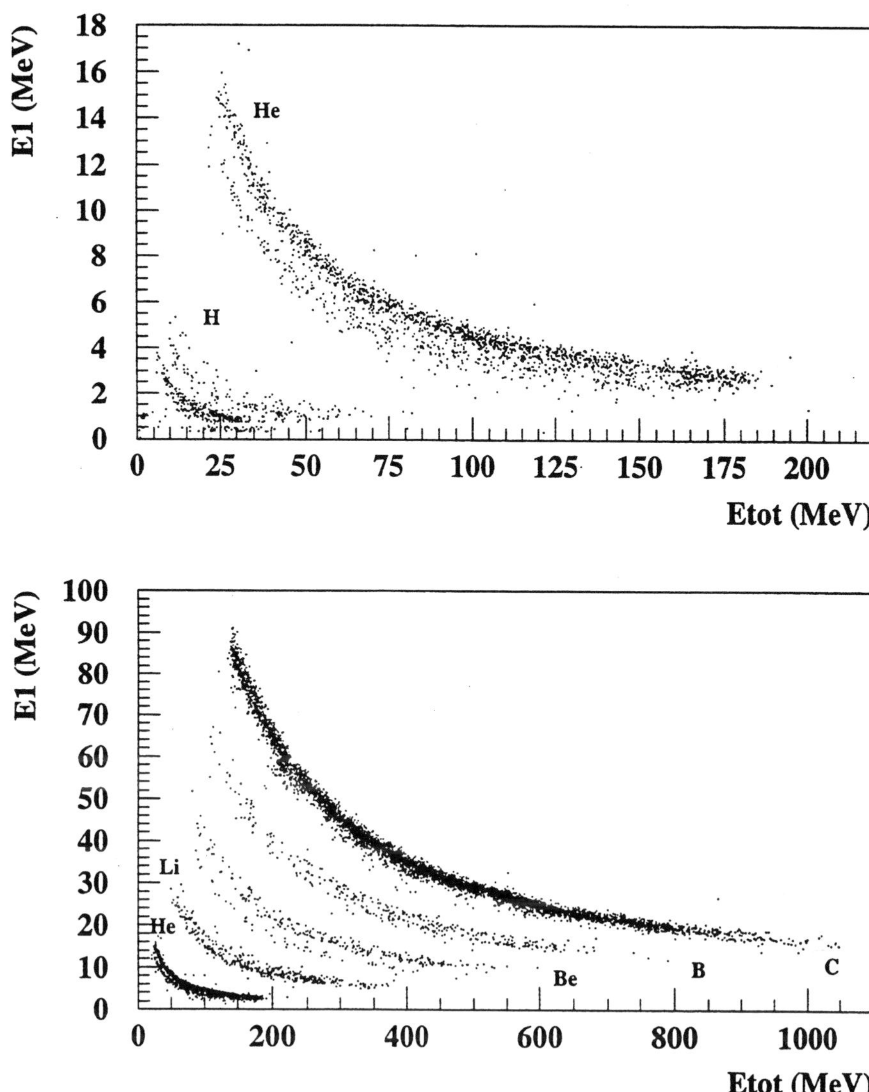

Figure 3: Distribution of the energy released in the first plane (E_1) versus the energy released in the whole detector (E_{tot}) for real particles coming from fragmentation of ^{12}C, produced at GSI.

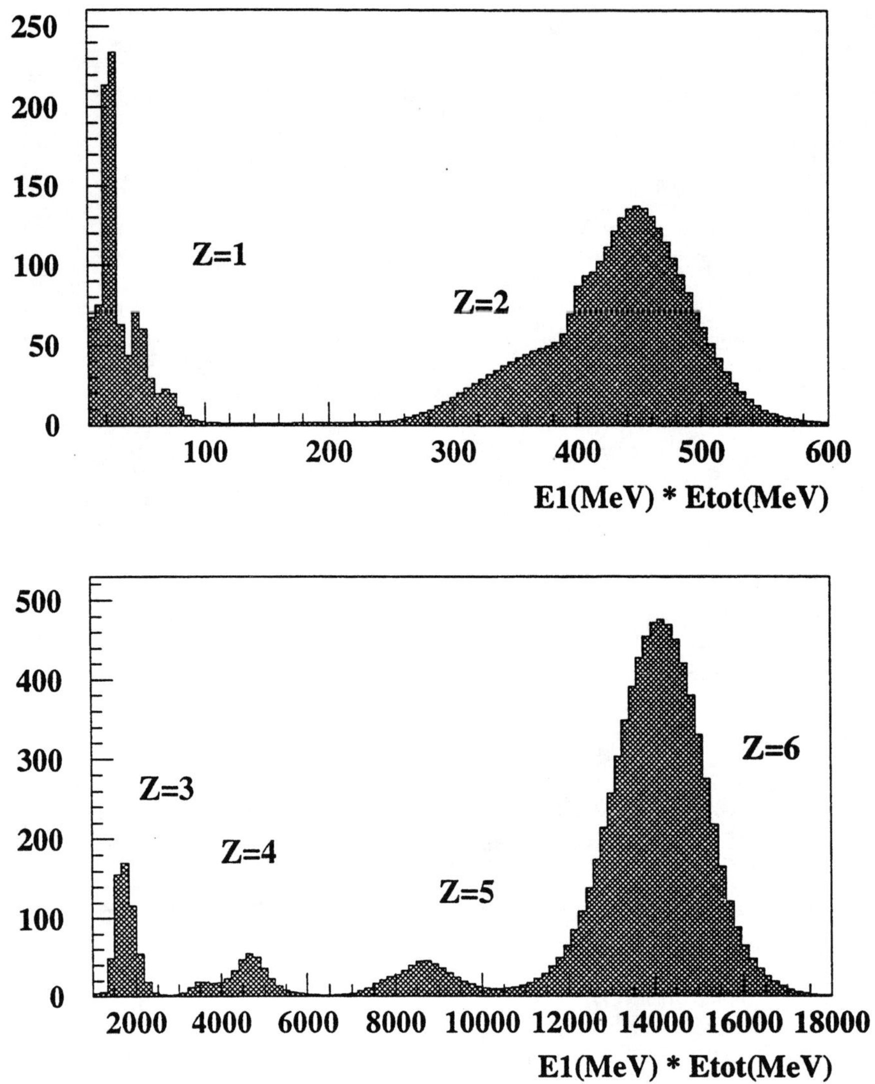

Figure 4: Distribution of the energy released in the first plane (E_1) multiplied by the energy released in the whole detector (E_{tot}) for real particles coming from fragmentation of ^{12}C, produced at GSI.

Once the charge has been identified, the masses M of the different isotopes can be reconstructed by means of the equation:

$$M = \left(\frac{a(E^b - (E - \Delta E)^b)}{Z^2 \Delta x} \right)^{\frac{1}{b-1}} \, , \tag{2}$$

where ΔE is the energy lost by the particle in an optimized thickness Δx measured starting from the first plane. This algorithm of mass reconstruction is well known in literature [3] and adopted in many experiments [4].

The parameter a is a constant of the medium and the parameter b has a value between 1.5 and 1.8 in NINA's energy range. A precise evaluation of such parameters for each particle has been obtained by a fit of the following expression:

$$R = a \frac{M}{Z^2} \left(\frac{E}{M} \right)^b \, , \tag{3}$$

where R and E are respectively the measured range and kinetic energy of known particles of mass M and charge Z. The procedure of fitting has been done for GSI and simulated data, and the two sets of a and b parameters have shown a perfect agreement.

The incident direction of the particle must be taken into account. For a particle impinging with angles θ_x and θ_y and hitting n (n_x and n_y) views, the expression for Δx will be:

$$\begin{aligned}
\Delta x &= (150\mu m + (n_x^{\Delta x} - 1)380\mu m)/cos\theta_x + \\
&\quad + (150\mu m + (n_y^{\Delta x} - 1)380\mu m)/cos\theta_y \, ,
\end{aligned} \tag{4}$$

where $n_x^{\Delta x}$ and $n_y^{\Delta x}$ are respectively the number of X and Y views composing Δx. 150 μm is the contribution of the first two thin silicon detectors and the rest is the contribution of all the normal 380 μm thick silicon views.

Z	Isotope, M(MeV)	$\overline{M}$ (MeV)	σ (MeV)	σ (amu)
1	^1H, 938	939	67	0.072
1	^2H, 1875	1834	153	0.164
1	^3H, 2814	2761	191	0.205
2	^3He, 2814	2828	146	0.157
2	^4He, 3727	3742	155	0.167
3	^6Li, 5603	5589	154	0.165
3	^7Li, 6535	6545	208	0.223
4	^7Be, 6536	6504	214	0.230
4	^9Be, 8394	8409	237	0.254
4	^{10}Be, 9328	9359	206	0.222
5	^{10}B, 9327	9278	282	0.303
5	^{11}B, 10255	10170	290	0.311
6	^{12}C, 11178	11239	279	0.300

Table 2: Medium averages M and sigmas σ from the Gaussian fits of the masses reconstructed by eq. 2, for real events collected at GSI.

Starting from $Z=1$ till $Z=6$, we reconstructed all isotope masses, as shown in figures 5 and 6. For every mass, a Gaussian fit has been performed, evaluating in such a way the average value of the reconstructed mass M and its corresponding sigma σ. The results of this analysis are resumed in table 2, for all isotopes studied. As it can be seen by the pictures and the table, the sigmas of the reconstructed masses become wider increasing the nuclear charge Z.

Results up to carbon give sigmas of about 0.3 amu. Such results confirm the good capability of this instrument to perform isotope analysis.

References

[1] A. Bakaldin, et al. - *Astroparticle Physics 8 (1997), 109.*

[2] R. Bellotti, et al. - *Astroparticle Physics 7 (1997), 219.*

[3] F. S. Goulding - *NIM 162 (1979), 609.*

[4] D. N. Baker, et al. - *IEEE Trans. on Geoscience and Remote Sensing 31-3 (1993), 531.*

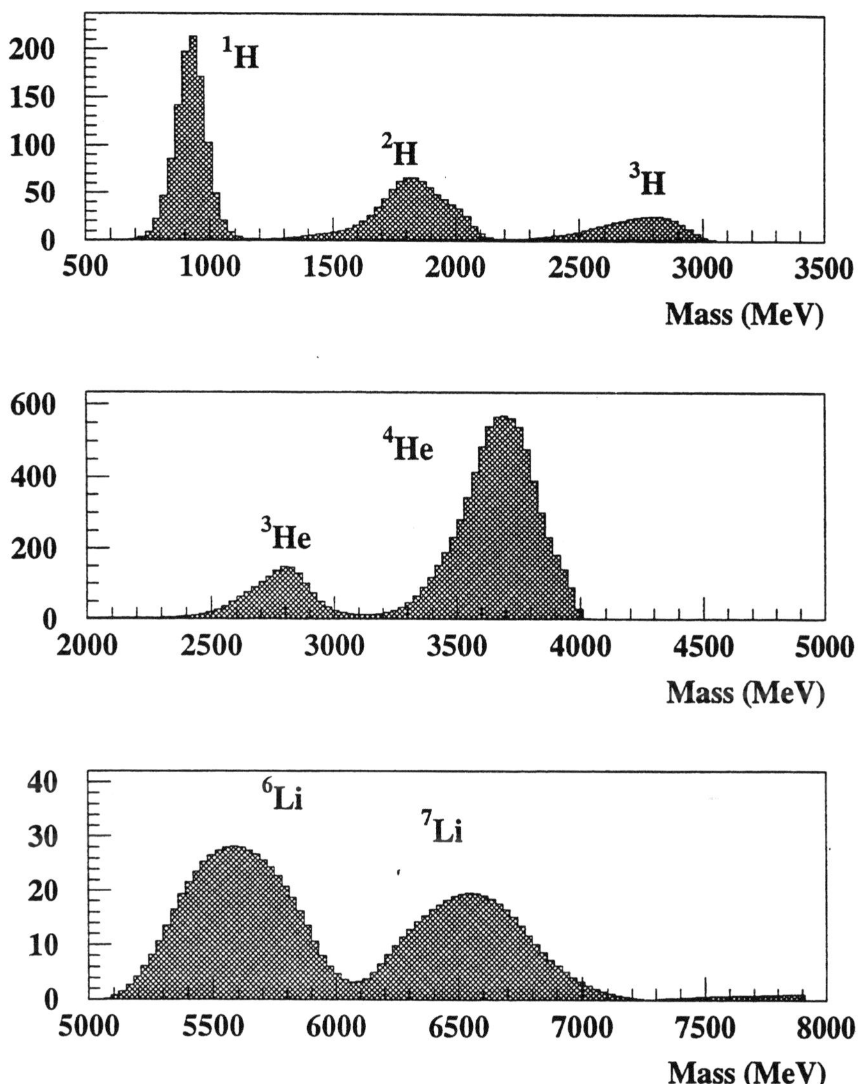

Figure 5: Distributions of the masses reconstructed by eq. 2 for real events of $Z = 1$, $Z = 2$ e $Z = 3$. N. of events on Y axis.

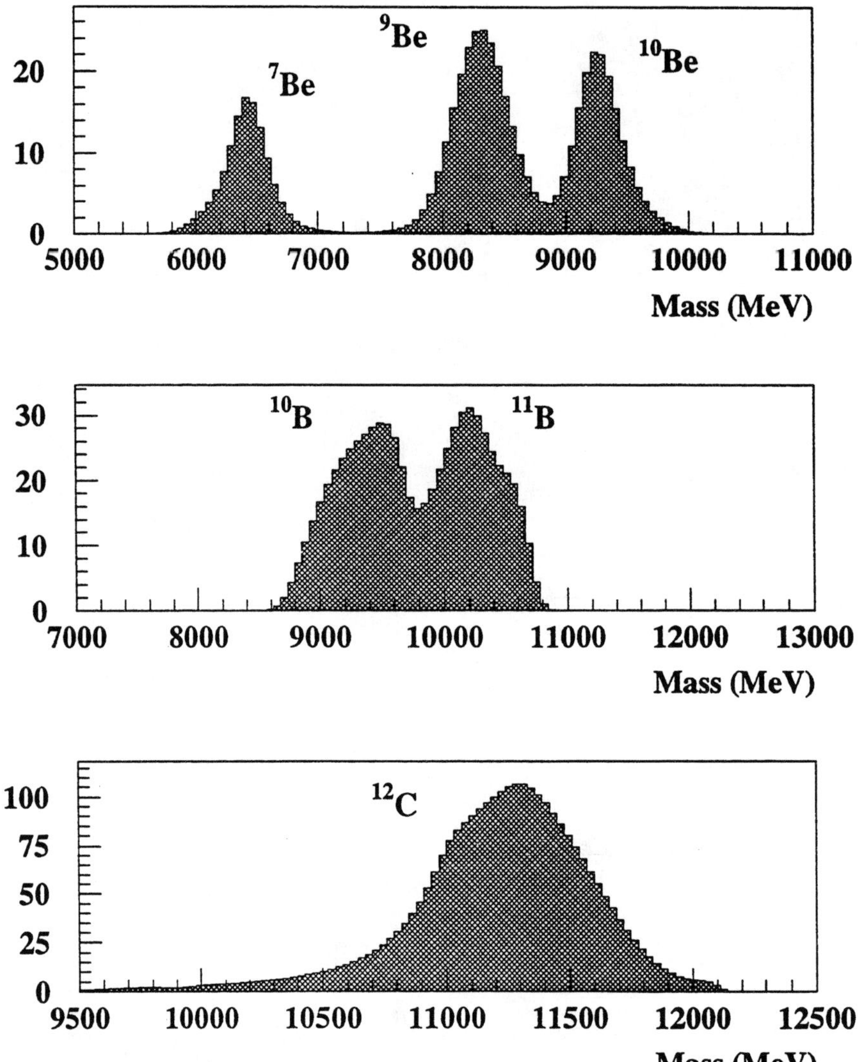

Figure 6: Distributions of the masses reconstructed by eq. 2 for real events of $Z = 4$, $Z = 5$ e $Z = 6$. N. of events on Y axis.

The PAMELA apparatus for the search of antimatter in cosmic rays

A. Morselli, P.Picozza on the behalf of PAMELA Collaboration
Dept. of Physics, II Univ. of Rome "Tor Vergata" and INFN, Italy

Abstract

The PAMELA experiment will be installed on-board of the RESURS-5 ARTIKA satellite to be launched in the 2002 for a mission at least three years long. The satellite orbit is polar, sun-synchronous and 700 km high. The observational objectives of the PAMELA instrument are the measurement of the spectra of antiprotons, positrons and nuclei in a wide range of energies, the search for primordial antimatter and the study the cosmic ray fluxes over half solar cycle. PAMELA will be able to measure magnetic rigidities (momentum /charge) up to its Maximum Detectable Rigidity (MDR) of 800 GV/c. Data gathered with the PAMELA instrument will deal with a wide range of fundamental issues and the peculiarity of its orbit will also allow to study several items in astrophysics, Solar-physics and Earth-physics.

PAMELA is an satellite-borne magnet spectrometer built by the WiZard-PAMELA collaboration [1]. The list of the people and the Institution involved in the collaboration together with the on-line status of the project is available at *http://www.roma2.infn.it/infn/aldo/pamela.html.*
The Pamela telescope, shown in figure 1, consists of the following elements: a magnet + tracker system, an imaging calorimeter, a Transition-Radiation-Counter (TRD), scintillation counter hodoscopes for Time-of-Flight and Trigger, an anticoincidence scintillation counter. The magnet + tracker system consist of 5 permanent magnets, each 8 cm high, interleaving 6 detection planes of the silicon microstrip tracker. The whole closed in a ferromagnetic screen and surrounded on its sides by a system of anticoincidence scintillation counters. The resolution of the tracking system is about 4 μm. The magnetic field inside the magnet will be $\sim$ 0.4 T, so the Maximum Detectable Rigidity (MDR) will be around 800 GV/c.

1   PAMELA Scientific Objectives

The observational objectives of the PAMELA instrument are the measurement of the spectra of antiprotons, positrons and nuclei in a wide range of energies, the search for primordial antimatter and the study the cosmic ray fluxes over half a solar cycle. PAMELA will be able to measure the magnetic rigidities (momentum/charge) up to its Maximum Detectable Rigidity of $\sim$ 800 GV/c. Data gathered with the PAMELA instrument will deal with a wide range of fundamental issues. These include:

- the role of Grand Unified Theories in Cosmology in relation to antimatter and dark matter.
- the understanding of the acceleration and propagation of cosmic rays.
- the role of solar, terrestrial and heliosperic relationships to energetic particle propagation in the heliosphere.

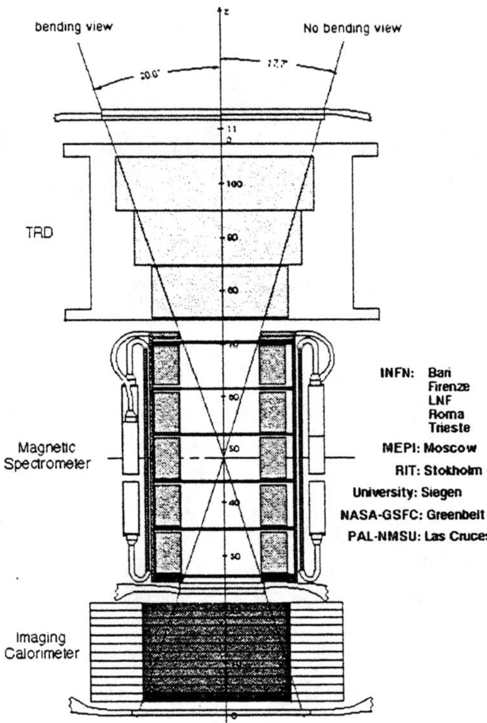

Figure 1: *Schematic of the PAMELA baseline instrument.*

The PAMELA observations will extend the results of balloon-borne experiments over an unexplored range of energies with unprecedented statistics and will complement information gathered from Great Space Observatories and ground-based cosmic-ray experiments. These observational objectives can be schematically listed in the following points:

• Measurement of the energy antiproton spectrum in a large energy range: from 100 MeV up to 150 GeV (present limits 0.4 - 20 GeV);
• Measurement of the energy positron spectrum in a large energy range: from 100 MeV up to 200 GeV (present limits 0.7 - 30 GeV);
• Search for anti-nuclei with a sensitivity of $6 \ 10^{-8}$ in the anti-helium/helium ratio (present limit about 10^{-5});
• Measurement of the electron energy spectrum up to 1000 GeV;
• Continuous monitoring of the cosmic rays solar modulation during and after the 23rd maximum of solar activity;
• Studies of the time and energy distributions of the energetic particles emitted in solar flares.

Figure 2 summarized the experimental and theoretical situation in the study of the antiproton flux compared with the proton flux. The range that will be covered by PAMELA is also shown.

The low energy antiproton and positron measurements and the last two objectives are peculiar of the PAMELA experiment because the satellite travels

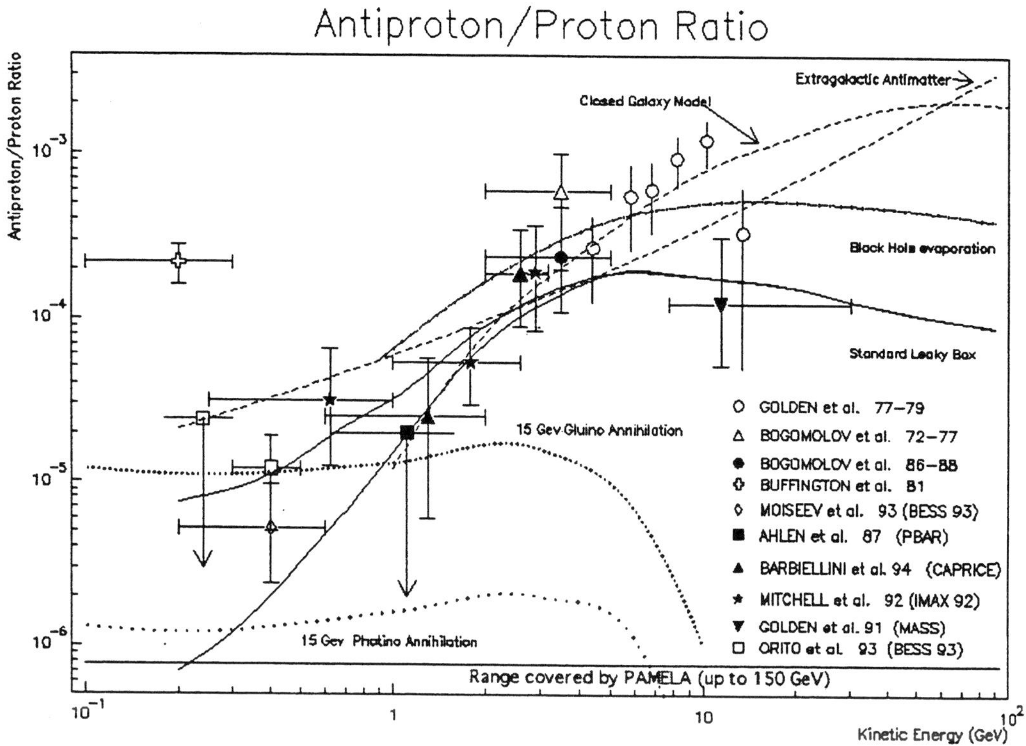

Figure 2: *The current experimental situation for the antiproton/proton ratio together with the expectation of the different models*

in a polar orbit. It spends a large fraction of its time in the high latitude and Polar Regions, where the cut-off due to the terrestrial magnetic field is negligible.

The scientific relevance of these objectives is enhanced by the length of the mission, that is planned to last not less than three years, but could be prolonged for many other years because of the orbit altitude and the maximization of the electric power due to its sun-synchronism.

In the following sub-sections the above objectives are discussed in some detail.

1.1 Antiprotons

Antiprotons have been observed in the cosmic rays since 1979 by balloon-borne experiments [2]; prior to these measurements it was generally expected that all primary cosmic rays experienced the same basic history during their acceleration and propagation. Similarly it was assumed that all secondary were produced and stored in the same regions of the Galaxy. These assumptions were quite adequate to explain the observations of secondary $Z \geq 3$ nuclei in cosmic rays.

However, the data from [3] appear higher than expected if the antiprotons are produced by interactions of the cosmic rays with the interstellar medium (ISM). The overabundance of antiprotons has led to speculations of their origin ranging from models where they are produced in shrouded supernova to annihilation of Majorana fermions created during the Big Bang. Other models explain the excess in term of how cosmic rays may propagate in the Galaxy, or postulate some kind of more "exotic" processes like the evaporation of mini black holes or the annihilation of super-symmetric particles in the galactic halo. These speculations predict different energy spectra at high-energy [4].

Because of atmospheric backgrounds and limited flight times, balloon-borne experiments can only measure the energy spectrum of antiprotons up to about 20 GeV and the present data at high energy (5-15 GeV) are in conflict and not allow to give firm indication. In addition, the sensitivity of the balloon observations is limited by the difficulty in eliminating large fluxes of atmospheric secondary particles. PAMELA will be able to measure the energy spectrum of antiprotons up to 150 GeV.

Recent experiments [5,6], performed with magnetic spectrometers on balloon, have investigated the low-energy spectrum of antiprotons in the few hundred MeV ranges. These measurements are consistent with secondary origin of low energy antiprotons but the large statistical uncertainties require further investigations to clarify if other sources exist. PAMELA will be able to measure the antiproton spectrum down to 100 MeV.

1.2 Positrons and Electrons.

Positrons and electrons are unique among cosmic rays because they are the lightest charged leptons. Due to their low mass, high-energy electrons and positrons undergo interactions with the ISM, which result in severe energy losses at high energies. While most of the observed electrons are believed to be of primary origin, the origin of positrons is yet to be established. Positrons are even harder to observe than antiprotons due to the high flux of protons (more than 1000 times higher). All observations to date have suffered from the risk of subtracting significant background. The majority of data shows an excess of positrons above the flux expected by the simple leaky-box model and may even indicate a rise in the positron/electron ratio at energies greater than 15 GeV. These direct observations, combined with the observation of a positron annihilation emission line from the galactic disk and the high antiproton fluxes, give rise to questions such as:

1. Are there positrons in the cosmic rays that are not produced as secondary?
2. Is there a relationship between the antiproton and positron excesses?
3. If positrons are indeed all secondary particles, at what point do the radiative losses become important?

Observation of positron over a very large energy range should yield new insights into galactic processes. In particular, the signature of WIMP particle existence in Dark Matter could be found in the high-energy spectrum of positron. As for antiprotons, PAMELA will aim to measure accurately the spectra of positrons from low (cut-off) energies up to the highest energies attainable (about 200 GeV). Furthermore, together with the measurement of the spectra of electrons (up to about 1000 GeV) PAMELA can provide information on the:

1. Acceleration of electrons and the distribution of acceleration sites;
2. Cosmic-ray lifetime, and the physical conditions in the containment volume;
3. Magnitude of re-acceleration by interstellar shock waves.

1.3 Search for antimatter.

Detection of antimatter of primary origin in cosmic rays would be a discovery of fundamental significance. Cosmic-ray searches that have been made so far have yielded only upper limits of one part in 10^{-4} for heavy nuclei (Z>2) and one part in 10^{-5} for helium [7,8]. The detection of anti-nuclei in cosmic rays would provide direct evidence of the existence of antimatter in the universe. Baryons and photons were produced in the Big Bang in equal amount, but from observation of the 2.7 K cosmic background radiation and the present matter density of the universe we know that only about one baryon remains for ever 10^9 photons. The current theory suggests that the remaining matter is the remnant of the almost complete annihilation of matter and antimatter at some early epoch, which stopped only when there was no more antimatter to annihilate. Starting from a matter/antimatter symmetric Universe, the required conditions for a following asymmetric evolution are the CP violation, the baryon number non-conservation and a non equilibrium environment. On the basis of gamma-ray observations, the coexistence of condensed matter and antimatter on scales smaller than that of clusters of galaxies has been virtually ruled out. However, no observations presently exclude the possibility that the domain size for establishing the sign of CP violation is as large as a cluster or super-cluster of galaxies. For example, there could be equality in the number of super-clusters and anti-super-clusters. Similarly, there is nothing that excludes the possibility that a small fraction of the cosmic rays observed at Earth reach our Galaxy from nearby

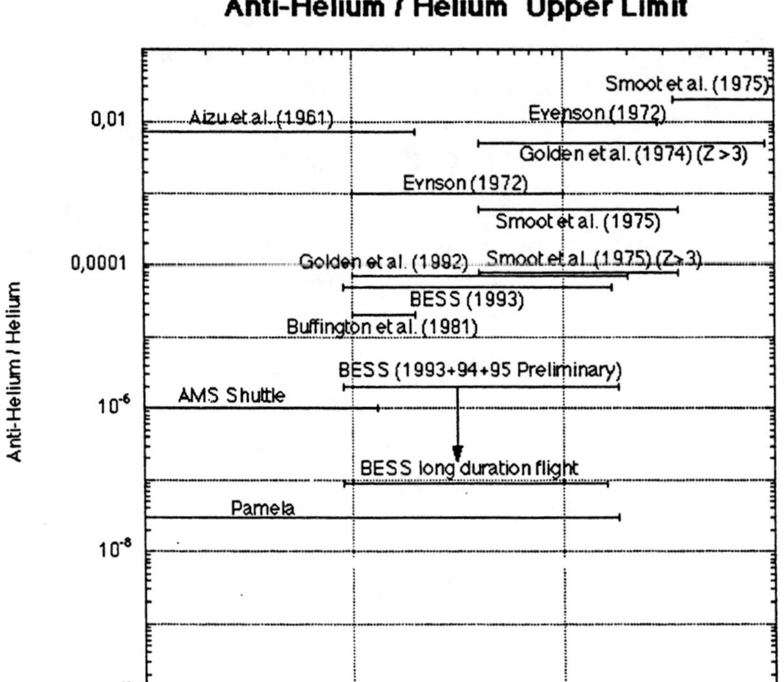

Figure 3: *The current experimental situation for the antihelium/helium ratio together with the expectation for PAMELA*

super-clusters. PAMELA will search for anti-nuclei with sensitivity on the anti-helium/helium ratio of some units in 10^{-8}.

1.4 Additional objectives

The continuous determination of the direction, latitude and longitude of the primary electrons, positrons, protons, antiprotons and light nuclei during several years and over a large energy range (100 MeV up to several tens GeV) will provide a great opportunity to investigate other scientific issues. The PAMELA experiment will be able to address these additional objectives besides the primary ones above described. Indeed, during its orbiting around the Earth, the satellite will encounter all that kind of events that are related to the solar activity and to the terrestrial geomagnetic effects.

The analysis of all data gathered over its mission will provide a significant complement to the measurements performed so far with dedicated experiments of different concepts. The additional objectives of PAMELA are the following:

1. Modulation of galactic cosmic rays in the heliosphere;
2. Solar flare particle spectra;
3. Distribution and acceleration of solar cosmic rays (SCR's) in the internal heliosphere;
4. Magnetosphere and magnetic field of the Earth;
5. Stationary and disturbed fluxes of high energy particles in the Earth's magnetosphere;
6. Anomalous component of cosmic rays.

The PAMELA mission will carry out its observations during and after the maximum of the 23rd solar activity cycle. The modulation effect on electrons, positrons, protons and nuclei during this period will be investigated in order to find any dependence on charge sign, energy (rigidity) and mass of the particles. Latitude and longitude distributions of the observed fluxes will be analysed as a function of the solar activity to look for possible correlation.

Most of the satellite measurements on the composition and spectrum of solar energetic particles from the flares are limited to energies below a few tens of MeV. Experiments carried by GOES-7, SAMPEX and in the next few months NINA telescope, could determine the spectrum of heavy nuclei to about 100 MeV/n for a few flares. Using PAMELA it is possible to measure the energy spectrum of nuclei from helium to at least oxygen from about 100 MeV/n up to an energy where the solar energetic particles can be distinguished from the galactic cosmic rays. Solar flares with generation of similar particles are rare but, as of 2002, when the solar activity is at its maximum, about ten of such events per year will be available. The composition and the spectral shape of nuclear components could vary from flare to flare and it would be very exciting to relate these variations to the physical conditions in the flare sites.

For the first time, there will be an opportunity to observe at the same time high energy solar particles with different charges and different masses. The observation of such particles will not only allow to choose between different SCR's acceleration processes on the Sun, but also to gather information on the distribution and acceleration processes of the particles in the internal part of the heliosphere. The observation of SCR's, the determination of energy and temporal distributions of various components will allow to carry out the internal tomography of the heliosphere: the shock wave propagation and the fluxes of solar wind drivers (coronal mass ejection) will be investigated.

The experimental data obtained by PAMELA both in conditions of quiet and disturbed Earth's magnetosphere will allow to study the variations of current systems and their influence on the trajectory of cosmic rays and on the geomagnetic rigidity thresholds.

The altitude of the PAMELA orbit is about 700 km. For almost one third of its orbits, the instrument will cross the internal part of the Earth radiation belt (the so called Brazilian anomaly region) giving the opportunity of observing both primary cosmic rays and the high energy particles of the Earth radiation belt itself.

The anomalous component of cosmic rays consists of partially (or single) ionised interstellar neutral atoms accelerated at the termination shock and penetrating inside the heliosphere. Their energy can vary from 10 MeV up to several hundred MeV which is sufficient to reach the vicinity of the Earth and to be observed by satellites at high latitudes or outside the magnetosphere. Due to the inclination of the PAMELA's satellite orbit (98 degrees) there is an opportunity to carry out measurements of the anomalous component of cosmic rays both at high latitudes and in the radiation belt of the Earth.

2   The PAMELA telescope.

The PAMELA telescope design is well defined and completed. In the past two years prototypes of detectors and sub-systems of the telescope have been built and tested, measuring their performances. Therefore, the expected behaviour of the whole PAMELA experiment is well known and the required techniques already defined and tested.

The concept of the PAMELA telescope is the same on which the proposed WIZARD experiment was based:

1. A magnetic spectrometer to determine the sign of the electric charge with a very high confidence degree, and to measure the momentum of the particles up to the highest energies for which a useful flux of rare particles (as antiprotons and positrons) can be collected;

2. An imaging calorimeter that can give, besides the measurement of the energy released by the interacting particle (and indeed of the extra energy released in an annihilation event), the pattern of the interaction of the particle inside the calorimeter, in order to identify the particle itself. This last point, observing the annihilation pattern of antiprotons and possible anti-nuclei, allows confirmation of their identity on an "event by event" basis;

3. A velocity measurement system to help the calorimeter in the identification of

the nature of the particle;

The PAMELA experiment can make use of all the experience accumulated in the last years by its proponents, collecting in one experiment advanced versions of detectors developed for previous balloon flights, allowing a high selectivity and redundancy for many of the physics goals to which PAMELA may contribute. The redundancy in the particle identification detectors allows to recognize the kind of particle on event by event basis, which is a fundamental task in the search of rare events and in the case of low fluxes, as antiparticles and antimatter in cosmic rays.

The technical choices had to be made taking into account the mass and power limits of the payload, dictated by the launch opportunity. They are the following:

Permanent magnet system and micro-strip silicon sensors for the magnetic spectrometer. The magnetic system is composed by five permanent magnet of Nd-Fe-B, each 8 cm high, that provide a field inside the tracking volume of about 0.4 T. There are six planes of silicon micro-strip detectors and the tests made in the past years on PAMELA detector prototypes resulted in a spatial resolution of 2.9 μm. Therefore, by considering a conservative approach, the spatial resolution requirement of the PAMELA tracker is fixed to 4 μm. The chosen cross section of the magnet gap is 13x16 cm^2, matching the acceptance of the spectrometer with the maximum reachable energy;

Silicon strip sensors for the imaging calorimeter, interleaved with tungsten plates as absorber; this choice minimizes the volume of the calorimeter, and maximizes indeed its geometrical acceptance; the high granularity assumed in PAMELA allows a very good separation between electromagnetic showers and interacting or not interacting hadrons; the chosen depth of 16 radiation length allows a good resolution in the measurement of the energy of electromagnetic particles, further extending the energy spectrum measurements of electrons and positrons;

A Transition Radiation Detector (TRD) for distinguishing electromagnetic particles from hadrons up to very high energy (about 1000 GeV); the TRD is based on small diameter straw tubes arranged in double layer planes interleaved by carbon fibre radiator. The use of the straw tubes allows to perform many energy loss measurements along the particle trajectory bringing the selection capability of the instrument down to less than 1 GeV, and to track all particles before their entrance in the magnetic spectrometer, cleaning the sample of the particles accepted at its entrance;

Several (6) scintillation counter hodoscopes (each 7 mm thick) for the construction of the triggers and for the TOF measurements; the use of several hodoscopes allows independent TOF measurements, improving the precision and the safety; the hodoscope structure maximizes the yield of the collected light and the safety of the system; both the hodoscopization and the number of hodoscopes allow a good flexibility in constructing the first level triggers;

Finally a set of scintillation counters covering the top edge and the sides of the magnetic spectrometer and the bottom part of the calorimeter completes the telescope, for a further labelling of contaminating events.

The main characteristics of the PAMELA experiment are:

1. A Maximum Detectable Rigidity of 800 GV/c;

2. An acceptance (geometrical factor) of 20.5 cm^2 sr;

3. A maximum energy reachable in the antiproton spectrum measurement of 150 GeV or greater, depending on the content of antiprotons in cosmic rays;

4. A maximum energy reachable in the positron spectrum measurement of 200 GeV;

5. A maximum energy reachable in the electron spectrum measurement of 1000 GeV;

6. A sensitivity for the anti-helium search of 6 10^{-8} in the anti-helium/helium ratio;

7. A total volume of 90x90x110 cm^3;

8. A total mass of 395 kg;

9. A number of read-out channels of 43498;

10. A power consumption of 267 W.

2.1   The PAMELA rates and data flow.

The PAMELA experiment will be put in polar orbit; therefore the total number of collected events will be high in spite of the small Geometric Factor (GF): the magnetic cut-off due to the Earth magnetic field nearly cancels on the poles, while at medium and low latitudes the bulk of cosmic rays is cut out by the Earth magnetic field, that acts for them like a mirror.

The total rate of particle collected by the PAMELA telescope, averaged on the polar orbit and on the solar activity cycle, is 3.3 event/s, i.e. about 3 10^5 event/day (mainly protons), ranging from 2.3 event/s to 4.2 event/s going from the period of maximum solar activity to its minimum. Averaging on the solar activity of the first three years foreseen for the PAMELA flight (2002-2005) we expect to collect in 10^8 s the following approximate numbers for particles, antiparticles and some nuclei:

protons	$3\ 10^8$	anti-protons	$3\ 10^4$
electrons	$3\ 10^6$	positrons	$1\ 10^5$
He nuclei	$4\ 10^7$	Be nuclei	$4\ 10^4$
C nuclei	$4\ 10^5$	anti-nuclei limit	$6\ 10^{-8}$ (90% C.L.)

The expected information flow from PAMELA is of the order of 2.5 kB/trigger; with not more than 5 10^5 trigger/day, this results in a maximum amount of information of 1.25 GB/day.

2.2   The mission profile and the spacecraft RESURS-5 ARKTIKA.

The PAMELA experiment will be installed on the up-ward side of the RESURS-5 ARKTIKA satellite, that will be continuously oriented down-ward to the Earth during all its mission, in order to fulfil a program of Earth surface observation. Furthermore the satellite will travel in a quasi-circular, about 700-km high, polar orbit. This is an optimal situation for the observation of cosmic rays:

1. the up-ward orientation of the PAMELA telescope on board of the satellite is the required direction to observe cosmic rays without interference with the Earth and keeping far away from the telescope acceptance the showers produced by quasi horizontal cosmic rays on the terrestrial atmosphere; these showers are responsible of the strong increase of the background at low zenith angles;

2. as above mentioned, the polar orbit maximizes the cosmic ray collection rate and also minimises the geomagnetic cut-off in a significant portion of the satellite trajectory (important for all scientific issues related to the solar activity and to the terrestrial geomagnetic effects);

3. finally, the large height of the orbit insures a long permanence of the satellite in space due to the very small effect of the atmosphere. The stabilisation of the satellite is obtained by magnetic devices to make the best use of this situation without being dependent from the possible shortness of fuel.

The satellite is the upgraded version of the RESURS 01 satellites that the All-Russian Science-Research Institute of Electromechanical (VNIIEM) of Moscow is regularly delivering in a solar synchronous polar orbit for a long duration program of Earth surface observation.

References

1. R.L.Golden et al., Nuovo Cimento, 105 B (1990), 191
2. R.L.Golden et al., Phys. Rev. Letters 43 (1979) 1196; E.A.Bogomolov et al., 16th ICRC , Kyoto 1979, 1, 330; A.Buffington et al., Ap. J.248 (1981) 1179.
3. R.L.Golden at al., Ap. Lett., 24 (1984) 75.
4. S.A.Stephens and R.L.Golden: Astron. Astrophys. 202 (1988) 1.
5. Orito S. et al., 24th ICRC (Rome), 3 (1995) 76.
6. Labrador A.W. et al., 24th ICRC (Rome), 3 (1995) 64.
7. G.F.Smoot et al., Phys. Rev. Lett., 35 (1975) 258.
8. Ormes J.F. et al., 24th ICRC (Rome), 3 (1995) 92.

III. Galactic cosmic rays: acceleration and propagation

Nonlinear phenomena in diffusive shock acceleration

M. A. Malkov

Max-Planck Institut für Kernphysik, D-69029, Heidelberg, Germany

ABSTRACT

A popular view that cosmic rays (CRs) are efficiently accelerated in astrophysical shocks implies a strong coupling of accelerated particles with the gas flow. Such an accelerating shock is, in fact, a complex dynamical system with subtle interrelations between physical processes taking place at smallest (thermal subshock) and largest scales, relevant to the CRs at highest energy. We discuss conditions under which a strong shock may channel a significant part of its energy into CRs. The main emphasis is on the bifurcation of the solutions for acceleration efficiency in terms of the rate at which particles are drawn from thermal plasma (injection rate), their maximum energy and the Mach number of the shock. The bifurcation curve for the acceleration efficiency depending on the injection rate is of an S-type, so that there exists a critical injection rate below which only a relatively inefficient acceleration is possible and an injection rate above which the acceleration is extremely efficient. This may invalidate the one-dimensional treatment of shock acceleration since different parts of the shock front will propagate at different speeds due to the strong variation in the reaction from accelerated particles. Also, a vast extent of the turbulent CR precursor may result in an enhanced heating of thermal plasma. These two latter aspects must lower both the acceleration efficiency and the maximum energy of CRs.

1. Introduction

The first order Fermi or diffusive shock acceleration is probably the most viable process to explain the cosmic ray (CR) origin. The acceleration sites are thought to be strong shocks such as supernova remnant (SNR) shocks. The mechanism is particularly attractive because it produces in a straightforward way a power-law momentum distribution $f \propto p^{-q}$ of an index q determined by merely the shock compression i.e., virtually by the Mach number M, $q = 4/\left(1 - M^{-2}\right)$. Since the best candidates are sufficiently strong shocks, $M^2 \gg 1$, the index $q \approx 4$ i.e., it is practically the same for all these shocks and, in fact, very close to what is inferred from the observations of the CR background.

It should be remembered, however, that the derivation of the above spectrum is strictly valid only for shocks whose internal structure is much smaller than the particle mean free path. At first glance this restriction seems harmless enough particularly for energetic particles because their mean free path cannot be shorter than their gyroradius since they are scattered by long hydromagnetic waves (collisionless plasma) while the shock structure is determined by thermal and slightly suprathermal particles and short waves driven by them. Thus, this approximation must be increasingly good for the high energy particles. An immediate caveat, however is that they may blow up the shock transition, should they be accelerated in large numbers. Indeed, they escape upstream to the distance $\sim \kappa(p)/u$ where $\kappa(p)$ is their momentum dependent diffusion coefficient and u is the velocity of counterstreaming gas, whereas their mean free path is $\sim \kappa(p)/v$, where $v \simeq c$ is their velocity. Consequently, since $v \gg u$, an opposite approximation should be employed in this case that is, one in which the shock transition is much broader than the mean free path. Moreover, the total width of the shock is $\sim \kappa(p_1)/u_1$ where p_1 is the maximum momentum and u_1 is the flow speed far upstream. Therefore, if the shock is modified strongly, i.e., the total compression r is much larger than the compression of the gaseous subshock r_s, particles with lower momenta $(\kappa(p) < \kappa(p_1))$ do not feel the total shock compression at all. This means that the above formula for the power-law index q that depends only on the shock compression r, $q = 3r/(r-1)$ is not valid anymore and its usage is rather misguiding. We determine the particle spectrum and the velocity field within the shock for the strongly nonlinear acceleration regime in Sec. 3.

Whether or not this second, non-linear acceleration regime dominates over the first, test particle or linear one depends not only on the relative number of CRs (which is always fairly small) but equally on their maximum momentum. Indeed, since for strong shocks the test particle index $q \simeq 4$, the CR pressure has a clear tendency to diverge with maximum momentum that is the acceleration efficiency which can be defined as the ratio of CR pressure to the shock ram pressure, $P_c/\rho_1 u_1^2$ may be very high (≈ 1) also for a small number density of CRs. Furthermore, the very existence of the ultrarelativistic component in the gas drives its specific ratio to its limiting value $\gamma = 4/3$ instead of $\gamma = 5/3$ in a nonrelativistic gas. The shock compression should then be closer to 7 rather than to the conventional 4 and the CR pressure may really diverge with p_1. Moreover, CRs partly escape from the shock through the upper energy cut-off since there are no sufficiently long magnetohydrodynamic waves to resonate with them. This increases the total compression even stronger (Eichler 1984) so that the pressure diverges indeed and, as we shall see, the CR pressure scales as $P_c \propto n_c \sqrt{p_1}$. Clearly, the overall energy requirements impose a strong constraint on the above combination of n_c and p_1 since $P_c \propto n_c \sqrt{p_1} < \rho_1 u_1^2$. Traditionally, it is implied that this energetic constraint must be interpreted in favor of p_1, should it conflict with the last relation. That is, if there is no external energy cut-off caused by geometry or by the above-mentioned lack of resonant waves, the backreaction of the CR pressure on the shock structure reduces n_c while p_1 keeps growing with time. In Sec. 5 we demonstrate that this is indeed so if there is no heating of the upstream plasma by CRs. If, however, the turbulent heating of the precursor is sufficiently strong the above conflict must be resolved by restricting the maximum momentum p_1 as well or, else, the acceleration becomes inefficient. This may indicate at the possibility of an intrinsic energy cut-off in efficiently accelerating quasi-stationary shocks.

2.　Energetic particle transport at a shock front

When a thermal proton is overrun by a collisionless shock it has always a chance to become entangled in a chain of interactions with the shock and particularly with the magneto-hydrodynamic (MHD) turbulence associated with it. Such a particle, in contrast to the majority left behind after their thermalization at the shock front, can be considered as a seed for the first order Fermi acceleration. This process that should bring up the particles to such a high energy that they may repeatedly catch up the escaping shock is referred to as injection (Lee 1982, Malkov & Völk 1995 [MV95], Malkov 1998 [M98], Scholer, Kucharek & Trattner 1998). Once this has happened, particle kinetics greatly simplifies and the leading, pitch angle independent part of the momentum distribution $f(t, x, p)$ can be described by the so called diffusion-convection equation (see e.g., Drury 1983)

$$\frac{\partial f}{\partial t} + U\frac{\partial f}{\partial x} - \frac{\partial}{\partial x}\kappa\frac{\partial f}{\partial x} = \frac{1}{3}\frac{\partial U}{\partial x}p\frac{\partial f}{\partial p} \tag{1}$$

Here the coordinate x is directed along the shock normal, $\kappa(p, x)$ is the spatial diffusion coefficient originating from the pitch angle scattering (wave-particle collisions) and $U(x)$ is the bulk plasma speed.

3.　Nonlinear Theory. An Elementary Approach

3.1.　Basic equations

In a test particle (linear) theory the high energy distribution f is decoupled from the plasma flow given by the velocity field $U(x)$ in equation (1). If the number density of CRs n_c is not vanishingly small, and the particle spectrum is extended to sufficiently high energies, the pressure exerted by these particles on the inflowing gas cannot be neglected while determining the shock structure. As a result, the latter acquires the form shown in Figure 1. The effect of the slowing down of the upstream flow may be very strong even for small n_c due to the following positive feedback: an increase of the CR pressure hardens the spectrum due to stronger compression which further increases the pressure. As a result the system jumps from a nearly test particle solution with small P_c to a solution in which almost all the flow ram pressure is converted into the CR pressure. To describe this situation we must complement equation (1) with an equation for the flow profile $U(x)$. This can be obtained from the conservation of mass and momentum. Assuming the steady state, introducing for convenience $u(x) = -U(x)$ and $g(x, p) = p^3 f(x, p)$ the system of equations to determine both the particle distribution and the flow profile $u(x)$ takes the form

$$\frac{\partial}{\partial x}\left(ug + \kappa(p)\frac{\partial g}{\partial x}\right) = \frac{1}{3}\frac{du}{dx}p\frac{\partial g}{\partial p}, \tag{2}$$

$$\rho u = \rho_1 u_1, \tag{3}$$

$$P_c + \rho u^2 = \rho_1 u_1^2, \quad x > 0 \tag{4}$$

Here the number density of CRs is normalized to $4\pi g dp/p$, the particle momentum p to mc, $\rho(x)$ is the mass density, $\rho_1 = \rho(\infty)$, P_c is the CR pressure

$$P_c(x) = \frac{4\pi}{3}mc^2 \int_{p_0}^{p_1} \frac{pdp}{\sqrt{p^2+1}}g(p, x) \tag{5}$$

The upper limit p_1 stands for a boundary in the momentum space (cut-off) beyond which particles are assumed to leave the system instantaneously ($g \equiv 0$, $p > p_1$). Note, that any separable $x-$ dependence of particle diffusivity $\kappa(p, x) = \kappa(p)\mathcal{K}(x)$ can be removed from equation (2) by the transformation $x \to \int dx/\mathcal{K}(x)$.

In the downstream medium, $x < 0$, the only bounded solution is $g = G(p) \equiv g(p, x = 0)$. As indicated, equation (4) is written in the region $x > 0$ where we have neglected the contribution of a cold gas (i.e., particles with $0 < p < p_0$) confining our consideration to sufficiently strong shocks with $M^2 \equiv \rho_1 u_1^2/\gamma P_{g1} \gg (u_1/u_0)^\gamma$, where γ is the specific heat ratio of the gas and P_{g1} is its kinetic pressure at $x = \infty$ (see Malkov 1997a, hereafter M97a, for a detailed discussion of this approximation). The thermal gas pressure is of course still critical at the subshock which is described by a regular Rankine-Hugoniot (RH) condition

$$r_s \equiv \frac{u_0}{u_2} = \frac{\gamma+1}{\gamma-1+2M_0^{-2}} \tag{6}$$

where M_0 is the Mach number of the flow in front of the subshock. The last equation is coupled with equations (2-4) through the gas deceleration and heating rates in the precursor. In the case of a purely adiabatic heating

$$M_0 = MR^{-(\gamma+1)/2} \tag{7}$$

with $R = u_1/u_0$. In what follows we set $\gamma = 5/3$.

3.2. An exact solution

Introducing the flow potential ϕ, such as $u = d\phi/dx$ we seek the solution of equation (2) in the form [1]

$$g = g_0(p) \exp\left\{-\frac{1+\beta}{\kappa(p)}\phi(x)\right\}, \quad x > 0 \tag{8}$$

where

$$\beta(p) \equiv -(1/3)d\ln g_0/d\ln p \tag{9}$$

Considering u as $u(\phi)$ and substituting equation (8) in equation (2) we obtain

$$du/d\phi = \lambda u/\phi \tag{10}$$

$$p\frac{d\beta}{dp} = (1+\beta)\left(\frac{d\ln\kappa}{d\ln p} - \frac{3}{\lambda}\beta\right) \tag{11}$$

where λ is a separation constant. Equation (10) may be readily integrated and yields for the flow potential

$$\phi(x) = \phi_0^{-\lambda/(1-\lambda)}\left[(1-\lambda)u_0 x + \phi_0\right]^{1/(1-\lambda)} \tag{12}$$

where $\phi_0 = \phi(0)$ is another constant (will be given in the next subsection). It is straightforward to verify that the following expression is the first integral of system (9,11)

$$g_0(p)\kappa^\lambda(1+\beta)^{-\lambda} = const \tag{13}$$

Denoting $\kappa_0 \equiv \kappa(p_0)$ and $\beta_0 \equiv \beta(p_0)$, for g_0 we then have

$$g_0(p) = g_0(p_0)\left(\frac{p}{p_0}\right)^3 \left[1 + 3\frac{\beta_0 + 1}{\lambda\kappa_0}p_0^{-3/\lambda}\int_{p_0}^{p}\kappa(p')p'^{3/\lambda-1}dp'\right]^{-\lambda} \tag{14}$$

For $p \gtrsim p_0$, more precisely for $(\kappa/\kappa_0)(p/p_0)^{3/\lambda} \gg 1$, the spectral slope is determined by merely $\kappa(p)$ and λ, i.e., it "forgets" its behavior at $p \simeq p_0$:

$$g_0(p) \propto \kappa^{-\lambda}(p) \quad \text{and} \quad \beta \simeq (\lambda/3)d\ln\kappa/d\ln p \tag{15}$$

As we shall see, the parameter λ depends on the scaling of $\kappa(p)$ as well, and the most surprising consequence of this dependence is that the resulting slope of $g_0(p)$ is, in fact, independent of $\kappa(p)$.

The region $p \lesssim p_0$ cannot be described within the present approach which produces two integration constants, the magnitude $g_0(p_0)$ and the slope $\beta(p_0) \equiv \beta_0$

---

[1]Some rationales behind this substitution are given in M97a

of particle distribution in the solution (14). They serve as external parameters provided by the "injection" theory (MV95, M98) that operates on an anisotropic at the shock front distribution function to which equation (2) is irrelevant. It should be noted, however, that a consistent asymptotic theory must be able to obtain the parameter β_0, also within the present approach to ensure a smooth matching of the spectrum at $p \sim p_0$ (see Sec.5).

3.3.   Approximate self-consistent solution

What we obtained so far is a one parameter (λ) family of exact solutions to equation (2) that require a rather special form of the flow profiles $u(\phi)$. One parameter is, generally speaking, not enough to satisfy a functional relation. Therefore, it is by no means guaranteed that this solution satisfies the pressure balance (4). A "miracle", however, is that it does in a fairly large part of the shock transition, of course, when the parameter λ is chosen properly. To demonstrate this we substitute solution (8) into equations (4,5). Using equation (3), condition (4) rewrites

$$u(\phi) + \mu \int_{s_1}^{s_0} \frac{ds}{\beta(s)} s^{\lambda-1} e^{-\phi s} \frac{p^2(s)}{\sqrt{1+p^2(s)}} = u_1. \tag{16}$$

We have introduced a new variable s in place of p

$$s = (1+\beta)/\kappa \tag{17}$$

and the limits $s_{0,1} = s(p_{0,1})$. We have also used the first integral (13), $g_0 \propto s^\lambda$. The parameter $\mu = (\lambda/3)\eta u_1 s_0^{-\lambda}$, where the injection rate η is defined as

$$\eta = \frac{4\pi}{3} \frac{mc^2}{\rho_1 u_1^2} g_0(p_0) \tag{18}$$

and the function $p(s)$ in equation (16) should be determined from equation (17). The most plausible $\kappa(p)$ dependence is believed to be of a Bohm-type, $\kappa(p) = Kp^2(1+p^2)^{-1/2}$, i.e., the mean free path of a particle is proportional to its Larmor radius (here K is a reference diffusivity). Then, equation (16) rewrites

$$u(\phi) = u_1 - \frac{\mu}{K} \int_{s_1}^{s_0} \left(1 + \frac{1}{\beta(s)}\right) s^{\lambda-2} e^{-\phi s} ds \tag{19}$$

According to equations (15,17), $\beta(s)$ is a very simple function, taking in the most part of its domain nearly constant and relatively close values, $\beta \simeq \lambda/3$ for $Ks \ll 1$ and $\beta \simeq 2\lambda/3$ for $Ks \gg 1$. It varies monotonically between these limiting values where $Ks \sim 1$. Differentiating equation (19) with respect to ϕ, assuming $0 < \lambda < 1$ and considering first the region $1/s_0 \ll \phi \ll 1/s_1$ we may obviously replace the lower limit by zero and the upper one by infinity. From equation (19) we then obtain

$$\frac{du}{d\phi} \simeq \frac{\mu}{K\phi^\lambda} \int_0^\infty \left[1 + 1/\beta(\tau/\phi)\right] \tau^{\lambda-1} e^{-\tau} d\tau$$

$$= \frac{\mu \Gamma(\lambda)}{K\phi^\lambda} \left[1 + 1/\beta(\bar{\tau}/\phi)\right] \tag{20}$$

where Γ is the gamma function and $\beta(\tau/\phi)$ is replaced in the last integral by its mean value at $\bar{\tau}/\phi$ with $\bar{\tau} \sim 1$. As we have already seen the function $\beta(\bar{\tau}/\phi)$ varies slowly and it is close to $\lambda/3$ for $\phi > K$ and to $2\lambda/3$ for $\phi < K$. Therefore, the ϕ dependence of $du/d\phi$ is determined by the factor $\phi^{-\lambda}$ and is indeed consistent with equation (10), i.e. with $u \propto \phi^\lambda$ provided that $\lambda = 1/2$. Equation (20) becomes invalid for $\phi \gtrsim 1/s_1$, since the lower limit in equation (19) cannot be replaces by zero in this case and the function $du/d\phi$ cuts off as (see equation (19))

$$\frac{du}{d\phi} \simeq \frac{\mu}{K\sqrt{s_1\phi}} \left(1 + \frac{1}{\beta(s_1)}\right) e^{-s_1\phi} \tag{21}$$

This results directly from the momentum cut-off at $p = p_1$. The last asymptotics obviously matches the formula (20) at $\phi \sim 1/s_1$ but equation (10) is no longer consistent with it. At the same time, this is a periphery of the shock transition where $u(\phi)$ exponentially approaches its limit u_1 at $x \sim l \equiv \kappa(p_1)/u_1$.

To illustrate the asymptotic nature and limitations of the present solution we choose a not quite favorable (moderate) ratio of $p_1/p_0 \approx 3 \cdot 10^4$. Clearly, for higher p_1 our approximation works better since the region of the scale invariant behavior $u - u_0 \propto x$ expands linearly with p_1. Note that for typical supernova shock conditions one usually expects $p_1/p_0 \sim 10^8$. To produce a very strong modification ($u_1/u_0 \sim \eta p_1 \gg 1$, see Sec. 5) we take the injection parameter η to be fairly (may be somewhat unrealistically) high but, in order to prevent extremely strong subshock reduction, we set the Mach number M to a very high value as well. This does not affect the flow and the spectrum in the precursor where $u(x) > u_0$. The case of moderate M and arbitrary η will be considered in Sec. 5

Figure 2 shows the solution (14) for particle spectrum with $\kappa(p)$ given above and corrected at higher momenta $p \sim p_1$ according to a more consistent treatment of this energy range in M97a; the spectrum is multiplied by a factor $\sqrt{1 + p/p_1}$, akin to the spatial region where $u(\phi)$ is given by equation (21). As we mentioned before, at the low energy end the solution liberates itself from the subshock control very rapidly. Already for $(p/p_0)^8 \gg 1$ the spectrum becomes universal, $G \propto p^{-1}$, independent of r_s. The next turn occurs at $p \sim 1$ where the spectrum transforms to $G \propto p^{-1/2}$, again independent of both the total and subshock compression.

Figure 3 shows the flow profile $u(x)$ (equation (19)) along with the particle distribution $g(p, x)$ at different values of p. One sees that the linear behavior of

$u(x)$ or, equivalently, $P_c(x)$ in the internal part of the shock transition results from the sum of exponentially decaying partial contributions to the CR pressure coming from different momenta.

4. Spectral universality of strong shocks

The downstream particle spectrum given by equations (13) with β from equation (15) being expressed in terms of kinetic energy E rather than momentum exhibits a fairly uniform behavior throughout the entire energy range, relativistic and nonrelativistic. In a standard normalization $F(E)dE$ this spectrum has the form

$$F \propto E^{-3/2} \sqrt{\frac{(E+1)(7E^2 + 14E + 8)}{(E+2)^3}} \tag{22}$$

where E is measured in mc^2. Clearly, the overall spectral index is close to 1.5 everywhere except the injection energy (if $r_s < 4$), the cut-off energy (both ignored in the last formula) and the region $E \sim 1$. We emphasize that this 1.5 value of the spectral index is not affected by any parameters involved in these calculations and is in *this* sense universal.

One may ask then what does the power-law index depend on? It should be primarily the momentum dependence of the CR diffusivity $\kappa(p)$ that we have specified in our treatment above. To examine this surmise we rescale κ as follows $\kappa' = \kappa^\alpha$, where $\alpha > 1/2$ and the rescaling of all primed variables below is induced by the above transformation of κ. In particular, the spectral slope β is now to be replaced by $\beta' = (\alpha\lambda'/3)d\ln\kappa/d\ln p$ (equation (15)). Recalculating $du/d\phi$ in equation (20) with these rescaled spectrum and CR diffusivity κ' we obtain $du/d\phi \propto \phi^{1/\alpha - 1 - \lambda'}$. Since the formula $u \propto \phi^{\lambda'}$ (equation (10)) holds, we deduce that the rescaled $\lambda' = 1/2\alpha \equiv \lambda/\alpha$. Consequently, the spectral slope β remains unchanged, $\beta' = \beta$.

We conclude that the spectral universality survives also the rescaling of $\kappa(p)$, in other words, *it is insensitive to the spectrum of the underlying MHD turbulence*. On the other hand, the velocity profile does depend on α. From the above analysis we obtain $u \propto x^{1/(2\alpha - 1)}$ which also explains the condition $\alpha > 1/2$. [2] Of course,

---

[2] Remarkably, a precisely opposite condition $\alpha < 1/2$ is required to produce a steady velocity jump *without* momentum cut-off and injection but with a secularly broadening CR precursor. This has been shown by Drury (1983). The fact that our strictly stationary solution that is based essentially on energy losses through the upper cut-off and particle injection at $p = p_0$ appears immediately beyond $\alpha = 1/2$ is perhaps more than a coincidence. Also, the time saving numerical solutions with $\alpha \leq 1/2$ might be inadequate for modeling the more realistic case of $\alpha = 1$. Some further interesting nonlinear studies of the case $\alpha = 0$ have been performed recently by Toptygin

there is a small deviation from this scaling at the distances x corresponding to the diffusion length of particles with $p \sim 1$ since β depends on ϕ in equation (20) at $\phi \sim K$.

Summarizing the last results, when $\kappa(p)$ rescales, so does the flow profile $u(x)$ but β remains invariant. It is not difficult to understand why this is so. As usual in the Fermi process *the spectral slope of course depends on the flow compression*. But, since the flow is modified, a particle with momentum p, bound diffusively to the shock front, samples not the total compression but only a compression accessible to it. The latter is determined by the relation $\phi(x) \propto \kappa^{\alpha}$ (equation (8)). As we have shown, $u(\phi) \propto \phi^{1/2\alpha}$. Therefore, the flow compression u/u_2, as seen by this particle, scales as $u/u_2 \propto \phi^{1/2\alpha} \propto \sqrt{\kappa(p)}$. As this is independent of α the index β must also be.

5.   Advanced theory. Reduction to Integral Equation

In the preceding section we guessed the form of the spectrum and satisfied iteratively the pressure balance (4) for all $x > 0$, equation (19). This was possible since the solution (8) remains a good approximation in the outer part ($u \simeq u_1$). One only needs to drop β in equation (8) which would not change the result significantly since β is numerically small ($\beta = 1/6$ for $p \gg 1$). [3] More importantly, the region where u approaches u_1 (as we have seen exponentially) is controlled by exponentially small amount of particles and is thus negligible in the RH jump conditions across the whole shock transition. The same arguments apply for the inefficient acceleration when $u_0 \lesssim u_1$. These remarks open a door for a more consistent approximate treatment of system (2-4, 6) in which all the four equations are solved simultaneously in a non-iterative fashion for all x and for both the efficient and inefficient acceleration regimes (M97a,b). A fundament of this approach is the exact internal solution (8). A central mathematical object here is what we called a spectral function which bears the information about both the particle spectrum and the flow profile. It appears naturally in the derivation of equation for the particle spectrum G by integrating equation (2) between $0-$ and $+\infty$ as a generalization of a standard test particle procedure (Eichler 1979). A

(1997).

[3]This is not true for the internal, scale-invariant part of the shock transition and the correct spectral index could not be calculated consistently without the β- term in equation (8) (see also the next footnote).

decisive step is to use the substitution (8). The result reads

$$-\frac{1}{3}\frac{\partial \ln G}{\partial \ln p} = \frac{1}{\bar{V}}\left(u_2 + \frac{1}{3}\frac{\partial \bar{V}}{\partial \ln p}\right) \tag{23}$$

Here the spectral function $\bar{V}$ is defined as follows

$$\bar{V}(p) = \int_{0-}^{\infty} e^{-\hat{s}(p)\Psi} du(\Psi) \tag{24}$$

with $\Psi = \phi - \phi_0 = \int_0^x u\,dx$

$$\hat{s}(p) = \frac{1}{\kappa(p)\bar{V}(p)}\left[u_2 + \bar{V}(p) + \frac{1}{3}\frac{\partial \bar{V}}{\partial \ln p}\right] \tag{25}$$

The functional dependence of the variable $\hat{s}$ on $\bar{V}$ is not very critical here and (24) should be regarded logically as an integral transform $u(\Psi) \mapsto \bar{V}(p)$ rather than an equation for $\bar{V}(p)$ given $u(\Psi)$. The function $\bar{V}(p)$ reflects explicitly a degree of shock modification. In an unmodified shock $\bar{V}(p) \equiv \Delta u \equiv u_0 - u_2$, since then $du/dx = 0$ in the upstream region; the spectral index is just the conventional $q = 3u_2/(u_1 - u_2)$ (see equation (23)). In general $\Delta u \le \bar{V}(p) \le u_1 - u_2$ and $\bar{V}(p) \to u_1 - u_2$ as $p \to \infty$. Even if the shock is appreciably modified, one may show that at small $p \gtrsim p_0$, $\bar{V}(p) \simeq \Delta u$. The spectral index then corresponds simply to the subshock compression ratio, and at lower momenta we have

$$q \simeq q_0 = \frac{3u_2}{u_0 - u_2}, \quad G(p) = Q_{\text{inj}}\left(\frac{p}{p_0}\right)^{-q_0} \tag{26}$$

The injection solution MV95, M98 produces essentially the same asymptotic result for $p \gtrsim p_0$, yielding thus the injection rate Q_{inj}. The solution $G(p)$ can be obtained then for all p from equation (23). To this end an independent equation for $\bar{V}(p)$ should be derived from equation (4). Using the solution (8) the latter can be rewritten as

$$\frac{d\Psi}{dx} + F(\Psi) = u_1 \tag{27}$$

where

$$F(\Psi) = \frac{4\pi}{3}\frac{mc^2}{\rho u_1}\int_{p_0}^{p_1}\frac{pG\,dp}{\sqrt{p^2 + 1}}\exp\left[-\frac{1 + \beta(p)}{\kappa(p)}\Psi\right] \tag{28}$$

The low energy asymptotics of G is fixed by equation (26) determining the CR input from the thermal plasma. Substituting equation (23) into (27) the latter may be manipulated into the following integral equation for the normalized spectral function (see M97a) $J(p) = \bar{V}(p)/\bar{V}(p_0)$

$$J(\tau) = \frac{\varsigma}{\varepsilon}\int_{\varepsilon}^{\varepsilon^{-1}}\frac{d\tau'}{\tau' + \tau}\frac{1}{\tau' J(\tau')}\exp\left[\Omega\Phi(\tau')\right] + 1. \tag{29}$$

We have used the notations

$$\tau = \frac{\kappa_0 s}{\varepsilon}\left(1 - \frac{1}{r_{\rm s}}\right), \quad \varepsilon^2 = \left(1 - \frac{1}{r_{\rm s}}\right)\frac{p_0}{p_1}\theta \ll 1, \tag{30}$$

$$\Omega = \frac{3}{\theta(r_{\rm s} - 1)} \quad \text{and} \quad \Phi(\tau) = \int_\varepsilon^\tau \frac{d\tau'}{\tau' J(\tau')} \tag{31}$$

The eigenvalue ζ is related to an injection rate ν through

$$\nu = \frac{\zeta}{R}\exp\left[\frac{3}{\theta(r_{\rm s} - 1)}\int_\varepsilon^{1/\varepsilon}\frac{d\tau}{\tau J(\tau)}\right] \tag{32}$$

The numerical factor $\theta \simeq 1.09$ and the injection rate is redefined here as

$$\nu \equiv \frac{4\pi}{3}\frac{mc^2}{\rho_1 u_1^2}p_0 Q_{\rm inj} \simeq \frac{p_0 r_{\rm s}}{r_{\rm s} - 1}\frac{m n_{\rm c} c^2}{\rho_1 u_1^2} \tag{33}$$

Note, that $\nu \sim p_0 \eta$ and the number density of CRs behind the shock is given by

$$n_{\rm c} = 4\pi \int_{p_0}^{p_1} G(p)dp/p \tag{34}$$

For simplicity, the diffusion coefficient κ is assumed to have a relativistic form $\kappa = \kappa_0 p/p_0 = \kappa_1 p/p_1$ for all p, since the relativistic particles are assumed to be dynamically much more important than the nonrelativistic ones ($p_1 \gg 1$). In order to close the system formed by equations (6) and (29) we need another equation to relate the three variables $(\nu, R, r_{\rm s})$ of which only one, say ν we consider as given. As an intermediate step we relate the precursor compression R and the spectral function J by inverting equation (24):

$$\frac{du}{d\Psi} = \frac{1}{2\pi i}\int_{-i\infty}^{i\infty} e^{s\Psi}\bar{V}(s)ds + \Delta u \delta(\Psi) \tag{35}$$

where δ is a delta function corresponding to the jump of u at the subshock. Integrating then equation (35) over Ψ between $\Psi = 0+$ and ∞, using analytic properties of $J(\tau)$ that has two branch points at $\tau = -1/\varepsilon, -\varepsilon$, we get

$$u_1 - u_0 = \frac{\Delta u}{2\pi i}\int_{-1/\varepsilon}^{-\varepsilon}\frac{d\tau}{\tau}[J(\tau + i0) - J(\tau - i0)] \tag{36}$$

We have taken $\bar{V}(p_0) \approx \Delta u$ (M97a). The integral around the cut $(-1/\varepsilon, -\varepsilon)$ may be evaluated with the help of equation (29) and the last equation rewrites

$$\frac{R - 1}{1 - r_{\rm s}^{-1}} = \frac{\zeta}{\varepsilon}U \tag{37}$$

where

$$U = \int_\varepsilon^{1/\varepsilon}\frac{d\tau}{\tau^2 J(\tau)}e^{\Omega\Phi(\tau)} \tag{38}$$

Equations (6), (29), and (37) form a closed system for describing nonlinear shock acceleration given the Mach number M, the cut-off momentum p_1 and the injection rate ν. They may have multiple solutions.

5.1. Inefficient or perturbative solution

Irrespective of the multiplicity, for sufficiently small injection rates ν, there must always be a solution that corresponds to a test particle (linear) acceleration regime in which $J \to 1$ as $\nu \to 0$. This solution can be written in terms of a Neuman series in ν or ζ as follows

$$J = 1 + \frac{\zeta}{\varepsilon^2} \ln \frac{\varepsilon^{-1} + \tau}{\varepsilon + \tau} + \mathcal{O}(\zeta^2) \tag{39}$$

where we have taken $r_s = 4$ and $\theta = 1$ for simplicity. The last solution is essentially perturbative and cannot describe multiple solutions of equation (29) that appear beyond some $\nu > 0$ (the series in ζ does not converge).

5.2. Bifurcation to efficient solution. Overall Rankine-Hugoniot Relations

If the Mach number M and the cut-off momentum p_1 are both sufficiently high, a pair of new solutions branch off at $\nu = \nu_1 > 0$. All the three solutions may be conveniently described by a single valued function $\nu = \nu(R)$ in the (R, ν) plane, Figure 4. We term the solution with $R > R_1$ efficient, and that with $R_2 < R < R_1$ – intermediate. It merges with the inefficient solution at the point $\nu = \nu_2, R = R_2$. For a fixed $\nu \in (\nu_1, \nu_2)$ all the three solutions have different values of R and, hence, different subshock compression ratio r_s.[4]

In a strongly nonlinear acceleration regime, the solution of equation (29) is dominated by the first term in the r.h.s. in contrast to the perturbative solution (39). Besides that the exponential factor in the r.h.s. may be replaced by unity since J turns out to be large enough. The remaining equation was solved in M97a in the limit $\varepsilon \ll 1$. The particle spectrum $G(p)$ and the velocity field $u(x)$ have

---

[4]Formally, the multiplicity can be removed by an appropriate variable transformation, $\nu \mapsto \nu' = f(\nu, R)$. Moreover this may happen naturally if the actual injection rate depends on R in a suitable way (see Figure 4). However, it is important to realize that this changes nothing in the inability of the perturbative approach ($R - 1 \ll 1$) to cope with the solutions on the branches 2 and 3 where $R \gg 1$, contrary to recent claims in the literature (see e.g., Berezhko 1996, Baring et al. 1999). For example, the perturbative expression for the spectral slope at $p = p_1 - 0$ suggested by Berezhko (1996), in addition to its illegitimate application to the case $R \gg 1$, still depends on a fitting parameter k (equations (32-34) of his paper) which cannot be specified accurately unless the solution for the case $R \gg 1$ is obtained. It is only *an assumption* that $k = 1$. Hence, the spectrum given by equation (34) of that paper, is also an assumption, not quite accurate particularly at $p = p_1$, as the solution in M97a showed. It should be also noted here that, even when the spectral slope is obtained acurately at $p = p_1 - 0$, it is not meaningful physically, since the real spectrum there is always determined by a concrete loss mechanism and/or temporal evolution.

been restored from J and correspond to those already presented in subsection 3.3 except a simplified form of $\kappa(p)$ and unimportant hardening of the spectrum just before the cut-off at p_1 (included, however, in Figure 2).

From equation (37) one can determine the jump condition across the shock appropriate for the efficient branch. The main result here is the following universal (Mach number independent) relation between the flow deceleration in the smooth and in the discontinuous parts of the shock transition

$$\frac{u_1 - u_0}{u_0 - u_2} = \frac{\pi}{\theta}\nu p_1/p_0 \tag{40}$$

One sees that the flow deceleration upstream is, in fact, an amplified subshock jump. In other words, the modification effect disappears whenever does the subshock. The strength of the latter may be obtained from the conventional Rankine-Hugoniot relation (6) provided that the precursor heating is identified, e.g., adiabatic, equation (7).

6. Dependence on parameters

The method of integral equation outlined in the preceding section allows one to describe the acceleration process on a universal basis in terms of the bifurcation analysis. In a steady state, or in a situation with a slowly advancing cut-off $p_1(t)$, a natural parameter space is two-dimensional and contains the Mach number M and the cut-off momentum p_1. A convenient dependent variable is the flow compression R that obviously signifies the efficiency of acceleration since $P_c(0)/\rho_1 u_1^2 \simeq 1 - 1/R$ provided that there is no strong heating in the precursor. In the present study, however, we add to this parameter space also the injection rate ν, since this latter, even though being in principle calculable, may vary depending on the model of the subshock dissipation used (M98). Thus we perform our bifurcation analysis here in the three-dimensional parameter space.

The analysis carried out in M97b on the basis of analytic solution of the system (6), (29), and (37) included only the adiabatic precursor heating of the form (7). Here, we present a numerical solution of this system of equations allowing also for an additional precursor heating. Such a heating may be caused by the dissipation of the MHD waves excited by the streaming instability of CRs (see McKenzie & Völk 1982) and/or by an acoustic instability driven by the pressure gradient of CRs first studied by Drury (1984) (see Webb, Zakharian & Zank 1997 for recent nonlinear studies). The first mechanism contains a smallness $\sim 1/M_A$, where M_A is the Alfvenic Mach number, in the conversion efficiency of the CR energy into MHD energy. Besides that it is difficult to assess due to our incomplete understanding of the cascading process of the MHD waves to short scales where dissipation should take place. The second mechanism seems to be more suitable for this purpose since the acoustic waves steepen into shocktrains and thus heat protons very efficiently. At the present stage, parametrisation is needed in either case and we modify equation (7) as follows

$$M_0 = MR^{-4/3} + \xi \cdot (1 - R^{-1})p_1/p_0 \tag{41}$$

Here the second term on the right hand side represents the non-adiabatic heating with the efficiency ξ powered by a (normalized) pressure contrast of CRs $(1 - R^{-1})$ and assumed to be proportional to the precursor length $l \sim \kappa(p_1)/u_1 \propto p_1$.

The numerical solution of the system (6,29,37) is represented in Figure 5a in the case of a purely adiabatic heating ($\xi = 0$). The multiplicity of the solution $R = R(\nu)$ is always present for sufficiently large values of M and p_1. The injection rate ν decreases monotonically with p_1 for all R which means that as p_1 slowly grows in an otherwise quasi-stationary acceleration, the injection requirements for the efficient acceleration become lower as it was mentioned in the introduction section.

The situation changes dramatically in the case of the turbulent heating ($\xi > 0$), Figure 5b. Starting from some critical value of $p_1 = p_* \propto 1/\xi$ the injection-compression diagram $\nu(R)$ suddenly lifts up increasing thus the injection requirements. For sufficiently high $p_1 > p_*$ these requirements become prohibitively high so that the acceleration should transit to an inefficient mode with a substantially reduced R.

These two sets of diagrams suggest the following scenario of the time-dependent acceleration. If it starts at a subcritical level, i.e., ν is below a plateau in Figures 5 ($R \gtrsim 1$, inefficient mode) a sudden jump to an efficient (supercritical) regime ($R \gg 1$ and ν is above the plateau) occurs when p_1 passes through a critical value determined by the current value of ν.

The further development depends dramatically on the extra heating. If $\xi = 0$ the acceleration may proceed ad infinitum provided that ν can still be maintained at a required level.[5] Clearly, these requirements are continuously relaxed while p_1 grows. The compression R will tend to its limiting value $M^{3/4}$ while the subshock strength will be asymptotically reduced, $r_s - 1 \sim M^{3/4}p_0/p_1\nu$ (M97a). This can also be seen from equation (40). The CR number density will decline asymptotically as well. Such a picture is at least in a qualitative agreement with widely accepted earlier ideas (see e.g., Drury 1983).

---

[5]Presumably, ν remains close to its initial value. Also note, that the number density of CRs does not depend on p_1 explicitly (the density integral is controlled by its lower limit).

If $\xi > 0$, a reversed transition to inefficient regime occurs at some $p_1 = p_*$ as discussed before. Interestingly, if the acceleration starts in the supercritical regime, this transition occurs first. To summarize, if injection is subcritical, R should rise sharply at some point. If ν is supercritical R should drop. This seems to be in agreement with numerical simulations by Berezhko, Yelshin & Ksenofontov (1996) (see Figures 5 and 7 of their paper).

We summarize the properties of the compression R as follows. It is always restricted by a condition of subshock preservation $R < M^{3/4}$ (see equations (6), (7) and (40)) or by an even stringent constraint in the case of a stronger heating. The limiting compression $R \approx M^{3/4}$ may be achieved only for $\nu p_1/p_0 \gg M^{3/4}$. For larger M (opposite condition) the compression R always saturates with M and scales as $R \sim \nu p_1/p_0$ on the efficient branch. [6] Clearly, it is very sensitive to ν in this case. On the lower, inefficient branch (see Figure 5) the function $R(\nu)$ is again relatively insensitive to the injection rate ν.

7. Discussion and Conclusion

We have considered the nonlinear shock acceleration using two approaches. First, we presented an exact scale-invariant strongly nonlinear solution pertinent to the internal part of a strong CR modified shock. The energy spectrum has a universal form which, except for the lowest and highest energies, is a power-law $F = CE^{-3/2}$ with a normalization factor C that changes only around $E = mc^2$. Interestingly enough that the nonrelativistic part of this spectrum is what *the test particle theory* predicts for a shock in a nonrelativistic gas while the relativistic part is precisely what it does for relativistic gases. The above spectrum is in a good agreement with the Monte Carlo simulation of highly compressive nonlinear shocks (Baring et al. 1999).

As a next step, we extended this description to the whole shock transition in an approximate manner with controllable accuracy. This allowed us to locate a parameter region where this solution changes dramatically. For fixed Mach number and maximum energy the crucial quantity that divides the parameter space into regions of efficient and inefficient acceleration is the injection rate. This is in agreement with the numerical results of Berezhko, Yelshin, & Ksenofontov (1996).

Clearly, our analysis highlights the role of injection. Although R indeed saturates at high injection rates, as emphasized by Berezhko, Yelshin, &

---

[6]This formally contradicts to the unbounded growth $R \propto M^{3/4}$ observed in simulations by Berezhko, Yelshin, & Ksenofontov (1996). Since in the simulations ν is assumed to scale as $\nu \propto R$, the condition $M^{3/4} \gg \nu p_1/p_0$ can never be fulfilled and the saturation does not occur.

                                                      M. A. Malkov

Ksenofontov (1996), this looks like the "insensitivity" of a Heaviside function with respect to its argument. The critical (discontinuity) region is identifiable with a quasi-plateau on $\nu - R$ diagram shown in Figures 5a,b. It is important to keep in mind that injection is a strongly self-regulating process. Any increase of its rate generates strong waves that trap particles downstream and inhibit injection (M98). Therefore, it is doubtful that it can be maintained at the saturation level without self-quenching.

A more realistic scenario seems to be one, analogous to a more general concept of *self-organized criticality* (SOC) (Bak, Tang & Wiesenfeld 1987, see also Diamond & Hahm 1995 for an example which is closer to the present subject). According to this scenario the actual injection rate should be maintained at its critical level (marginal stability) since any deviation from it causes a strong precursor instability (trough growing R) and heating with a subsequent backreaction on the subshock and return to the critical state. The solution ceases to be steady so that parameters like injection fluctuate around their critical values. In addition, the $\nu - R$ diagram is indicative of breaking the translational symmetry along the shock front, since even small variations in ν along it (e.g., due to fluctuations of the local magnetic field) will cause strong variation in R and thus the strong deformation of the shock front.

An important self-regulating ingredient in our consideration is the precursor heating. Its gross effect is clearly a reduction of acceleration efficiency which is not surprising. More interesting is that the heating reduces the maximum energy achievable in efficient acceleration regime. Under certain circumstances this can terminate the acceleration process after a limiting momentum p_* is reached. Indeed, the general idea of Fermi acceleration at strong shocks implies a bootstrap turbulent scattering of particles which should be inhibited in the low-efficiency mode. Clearly, after the disruption of the high-mode, the maximum energy may decrease due to particle escape from a laminar shock. This will naturally recover the high efficiency regime if the injection rate is still high enough. Note, that an intermittent rather than quasi-steady behavior has also been conjectured earlier on the grounds of simplified hydrodynamical studies of shock nonlinearity (e.g., Drury & Völk 1981, Drury, Markiewicz & Völk 1989). Whether this hysteretic behavior may dominate over SOC depends on the details of injection and heating which are still known rather poorly. Nevertheless, one conclusion can be drawn from the strong nonlinear response to these processes. That is, future models should include self-organizing impacts due to injection and heating adequately.

It is also necessary to realize that while the acceleration is in the low-mode, the spectrum may be described by the test particle theory which is very different from the strongly nonlinear solution presented in Sec.3.3.

This work was done within the Sonderforschungsbereich 328 of the Deutsche Forschungsgemeinschaft (DFG).

REFERENCES

Axford, W. I., Leer, E., & Skadron, G. 1977, Proc. 15th ICRC (Plovdiv) 11, 132

Bak, P., Tang, C & Wiesenfeld, K. 1987, Phys. Rev. Lett., 59, 381

Baring, M.G., Ellison, D.C. , Reynolds, S. P., Grenier, I. A., & Goret, P., astro-ph/9810158, to appear in ApJ

Bell, A. R. 1978, MNRAS, 182, 147

Berezhko, E. G. 1996, Astropart. Phys., 5, 367

Berezhko, E. G., Yelshin, V., & Ksenofontov, L. 1996, Sov. Phys. JETP, 82, 1

Blandford, R. D., & Ostriker, J. P. 1978, ApJ, 221, L29

Diamond, P. H., & Hahm, T. S. 1995, Phys. Plasmas, 2, 3640

Drury, L. O'C. 1983, Rep. Prog. Phys. 1983, 46, 973

Drury, L. O'C. 1984, Adv. Space Res., 4, 185

Drury, L. O'C., & Völk, H. J. 1981, ApJ, 248, 344 (DV)

Drury L.O'C., Markiewicz W.J., Völk H.J. 1989, A&A, 225, 179

Eichler, D. 1979, ApJ, 229, 419

Eichler, D. 1984, ApJ, 277, 429

Krymsky, G.F., 1977, Dokl. Akad. Nauk SSSR 234, 1306 (Engl. Transl. Sov. Phys.-Dokl. 23, 327).

Lee, M. A. 1982, J. Geophys. Res. 87, 5063

Malkov, M. A. ApJ, 1997a, 485, 638 (M97a)

Malkov, M. A. ApJ, 1997b, 491, 584 (M97b)

Malkov, M. A. Phys. Rev. E, 1998, 58, 4911 (M98)

Malkov, M. A., & Völk, H. J. 1995, A&A, 300, 605 (MV95)

McKenzie, J. E., & Völk, H. J. 1982, A&A, 116, 191

Scholer, M., Kucharek, H. & Trattner, K. J. 1998, Adv. Space Res., 21, 533

Toptygin, I. N. 1997, Proc. 25th ICRC (Durban), 4, 377

Webb, G. M., Zakharian, A., & Zank, G. P. 1997, Proc. 25th ICRC (Durban), 4, 369

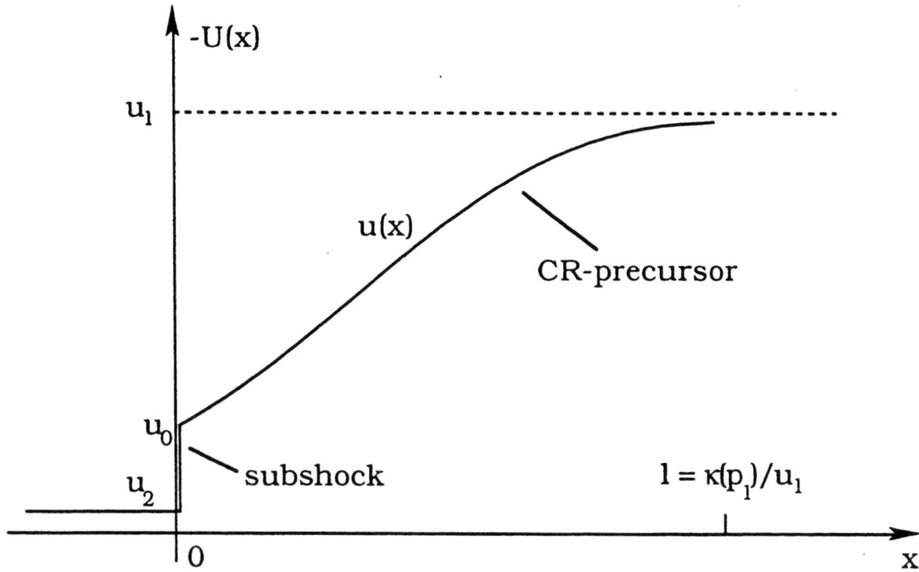

Fig. 1.— The flow structure in a strongly modified CR shock. The shock propagates to the right but the velocity field is shown (in shock frame) with the opposite sign.

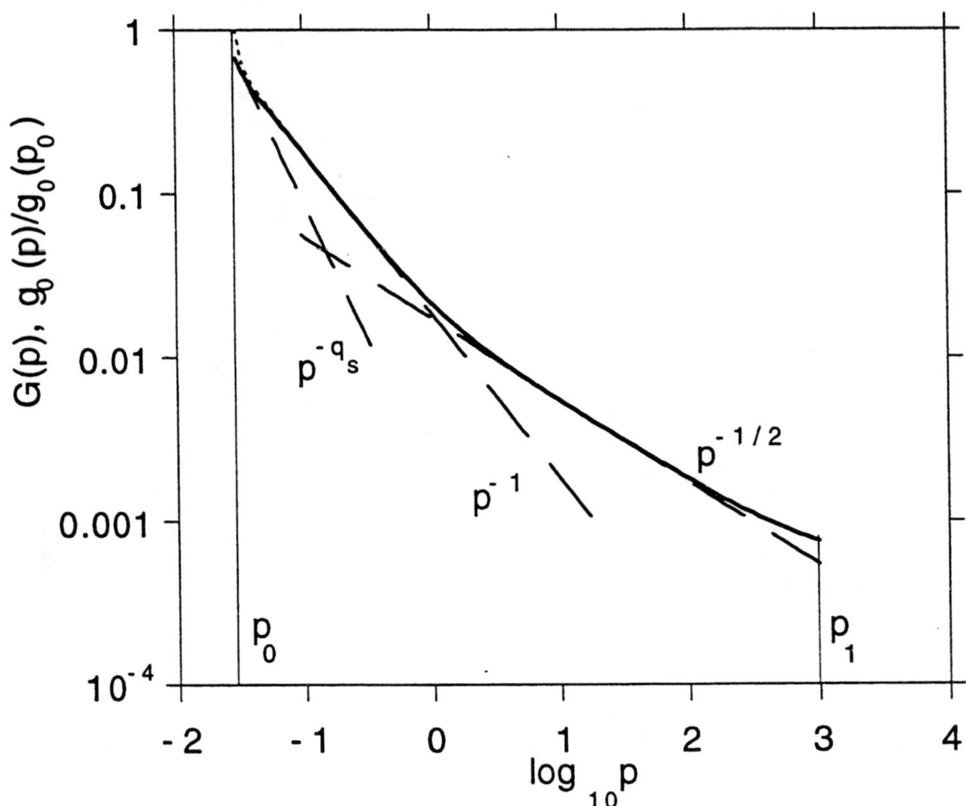

Fig. 2.— The post-shock spectrum $G = g_0(p)\exp[-s(p)\phi_0]$ (solid line) and $g_0(p)$ (dotted line, coincides with G, except at $p \approx p_0$) calculated using equation (14) for $p_0 = 0.03$; $p_1 = 1000$; $M = 9 \cdot 10^4$; $\eta = 0.84$ and normalized to dp/p. The subshock compression and other parameters involved in this calculation are found to be $r_s = 2.8$; $\mu = 0.0025$; $\beta_0 = 1.81$ and $\phi_0/K = 1.1 \cdot 10^{-4}$. The dashed lines show test particle power-law spectra with the indices $q = 3/(r-1)$ corresponding to shocks of compression $r = r_s$ ($q_s = 1.67$); $r = 4$ and $r = 7$, respectively.

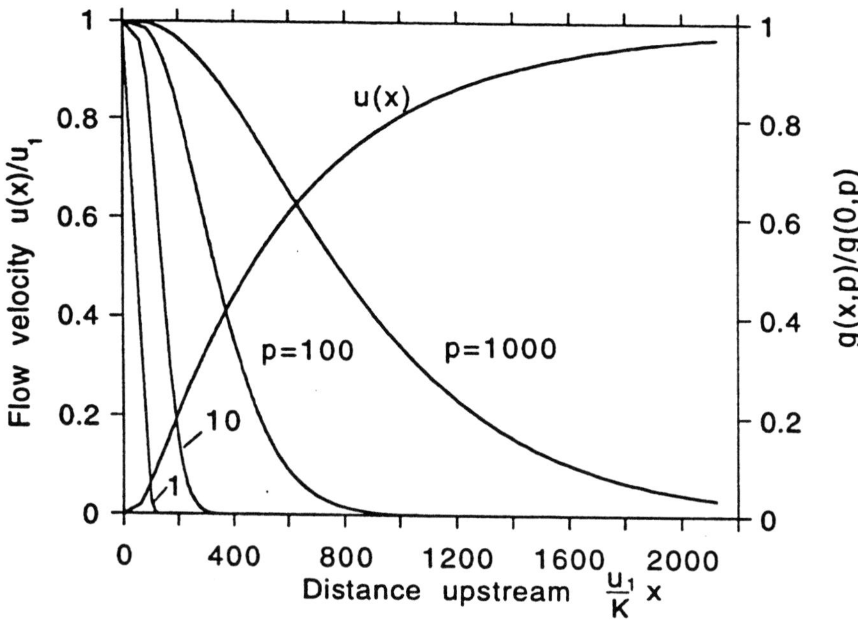

Fig. 3.— The flow profile $u(x)$ calculated using equation (19) and $d\phi/dx = u(\phi)$ for the same parameters as in Figure 1. The calculated precursor compression $u_1/u_0 \approx 2500$. The particle distribution $g(x,p)$ is drawn for different p and normalized for each p to its value at $x = 0$.

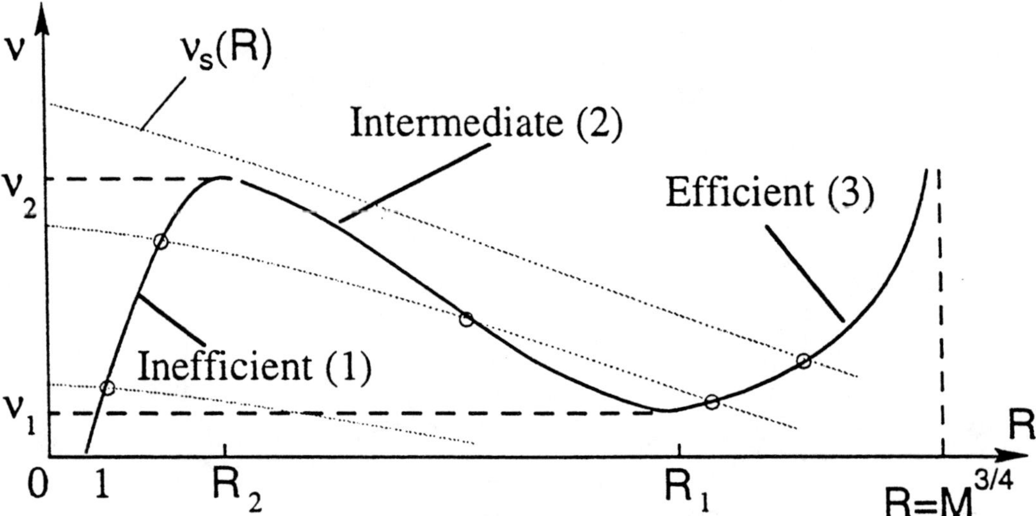

Fig. 4.— The nonlinear response R of an accelerating shock to the thermal injection of a rate ν represented in the form of a single valued function $\nu(R)$. Given $\nu \subset (\nu_1, \nu_2)$ there are three substantially different acceleration regimes. A few possible graphs of an actual subshock injection rate $\nu_s(R)$ are drawn with thin lines. The true steady state solutions should correspond to the intersection points marked by circles.

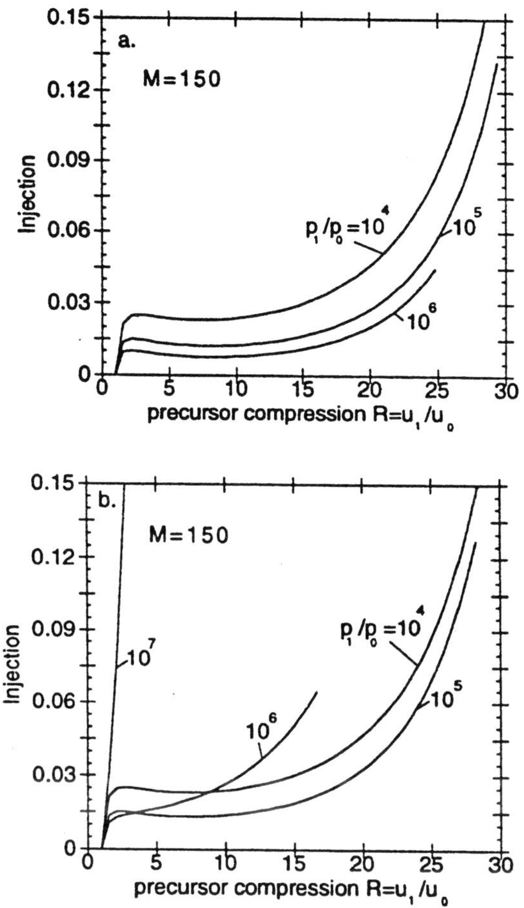

Fig. 5.— a.) A compression-injection diagram of a strong shock with only an adiabatic precursor heating ($\xi = 0$) given in the same format as in Figure 4. The Mach number $M = 150$. Different values of a cut-off parameter p_1/p_0 are indicated at each curve.

b.) The same as (a) but for $\xi \sim 10^{-8}$.

Galactic cosmic rays and gamma rays: a unified approach

Andrew W. Strong[1] and Igor V. Moskalenko[1,2]

[1]*Max-Planck-Institut für extraterrestrische Physik, Postfach 1603, D-85740 Garching, Germany*

[2]*Institute for Nuclear Physics, M.V.Lomonosov Moscow State University, 119 899 Moscow, Russia*

ABSTRACT

We are constructing a model which aims to reproduce observational data of many kinds related to cosmic-ray origin and propagation: direct measurements of nuclei, antiprotons, electrons and positrons, γ-rays, and synchrotron radiation. These data provide many independent constraints on any model. Propagation of primary and secondary nucleons, primary and secondary electrons and positrons are calculated self-consistently. Fragmentation and energy losses are computed using realistic distributions for the interstellar gas and radiation fields, and diffusive reacceleration is also incorporated. The models are adjusted to agree with the observed cosmic-ray B/C and ^{10}Be/^{9}Be ratios.

Our main results include evaluation of diffusion/convection and reacceleration models, estimates of the halo size, calculations of the interstellar positron and antiproton spectra, evaluation of alternative hypotheses of hard nucleon and hard electron interstellar spectra, and computation of the Galactic diffuse γ-ray emission.

1. Introduction

We have developed a model which aims to reproduce self-consistently observational data of many kinds related to cosmic-ray origin and propagation: direct measurements of nuclei, antiprotons, electrons and positrons, γ-rays, and synchrotron radiation. These data provide many independent constraints on any model and our approach is able to take advantage of this since it must be consistent with all types of observation.

A numerical method and corresponding computer code (GALPROP) for the calculation of Galactic cosmic-ray propagation in 3D has been developed. The basic spatial propagation mechanisms are (momentum-dependent) diffusion and convection, while in momentum space energy loss and diffusive reacceleration are treated. Fragmentation and energy losses are computed using realistic distributions for the interstellar gas and radiation fields. The code is sufficiently flexible that it can be extended to include new aspects as required. The basic procedure is first to obtain a set of propagation parameters which reproduce the cosmic ray B/C and ^{10}Be/^{9}Be ratios; the same propagation conditions are then applied to primary electrons. Gamma-ray and synchrotron emission are then evaluated with the same model.

Our approach is not intended to perform detailed source abundance calculations with a large network of reactions, which is still best done with the path-length distribution approach (see e.g. DuVernois, Simpson, & Thayer 1996 and references therein). Instead we use just the principal progenitors and weighted cross sections based on the observed cosmic-ray abundances (see Webber, Lee, & Gupta 1992). The B/C data is used since it is the most accurately measured ratio covering a wide energy range and having well established cross sections. A re-evaluation of the halo size is desirable since new ^{10}Be/^{9}Be data are available from Ulysses with better statistics than previously.

Preliminary results were presented in Strong & Moskalenko (1997) (hereafter Paper I) and full results for protons, Helium, positrons, and electrons in Moskalenko & Strong (1998a) (hereafter Paper II). Evaluation of the B/C and ^{10}Be/^{9}Be ratios, evaluation of diffusion/convection and reacceleration models, and setting of limits on the halo size, as well as full details of the numerical method and energy losses for nucleons and electrons are summarized in Strong & Moskalenko (1998) (hereafter Paper III). Evaluation of antiprotons in connection with diffuse Galactic γ-rays and interstellar nucleon spectrum are given in Moskalenko, Strong, & Reimer (1998) (hereafter Paper IV). For a recent discussion of diffuse Galactic continuum γ-rays and synchrotron emission in the context of this approach see Strong, Moskalenko, & Reimer (1998) (hereafter Paper V) and Moskalenko & Strong (1998d).

For interested users our model is available in the public domain on the World Wide Web (*http://www.gamma.mpe-garching.mpg.de/~aws/aws.html*)

2. Motivation

It was pointed out many years ago (see Ginzburg, Khazan, & Ptuskin 1980, Berezinskii et al. 1990) that the interpretation of radioactive cosmic-ray nuclei is model-dependent and in particular that halo models lead to a quite different physical picture from homogeneous models. The latter show simply a rather lower average matter density than the local Galactic hydrogen (e.g., Simpson & Garcia-Munoz 1988, Lukasiak et al. 1994a), but do not lead to a meaningful estimate of the size of the confinement region, and the corresponding cosmic-ray 'lifetime' is model-dependent. In such treatments the lifetime is combined with the grammage to yield an 'average density'. For example Lukasiak et al. (1994a) find an 'average density' of 0.28 cm^{-3} compared to the local interstellar value of about 1 cm^{-3}, indicating a z-extent of less than 1 kpc compared to the several kpc found in diffusive halo models. Our model includes spatial dimensions as a basic element, and so these issues are automatically addressed.

The possible rôle of convection was shown by Jokipii (1976), and Jones (1979) pointed out its effect on the energy-dependence of the secondary/primary ratio. Recent papers give estimates for the halo size and limits on convection based on existing calculations (e.g., Webber, Lee, & Gupta 1992, Webber & Soutoul 1998), and we attempt to improve on these models with a more detailed treatment.

Previous approaches to the spatial nucleon propagation problem have been mainly analytical: Jones (1979), Freedman et al. (1980), Berezinskii et al. (1990), Webber, Lee, & Gupta (1992), Bloemen et al. (1993), and Ptuskin & Soutoul (1998) treated diffusion/convection models in this way. Bloemen et al. (1993) used the 'grammage' formulation rather than the explicit isotope ratios, and their propagation equation implicitly assumes identical distributions of primary and secondary source functions. These papers did not attempt to fit the low-energy (< 1 GeV/nucleon) B/C data (which we will show leads to problems) and also did not consider reacceleration. It is clear than an analytical treatment quickly becomes limited as soon as more realistic models are desired, and this is the main justification for the numerical approach. The case of electrons and positrons is even more intractable analytically, although fairly general cases have been treated (Lerche & Schlickeiser 1982). Recently Porter & Protheroe (1997) made use of a Monte-Carlo method for electrons, with propagation in the z-direction only. This method would be very time-consuming for 2- or 3-D cases. Our method, using numerical solution of the propagation equation, is a practical alternative.

Reacceleration has previously been handled using leaky-box calculations (Letaw, Silberberg, & Tsao 1993, Seo & Ptuskin 1994, Heinbach & Simon 1995); this has the advantage of allowing a full reaction network to be used (far beyond what is possible in the present approach), but suffers from the usual limitations of leaky-box models, especially concerning radioactive nuclei, which were not included in these treatments. Our simplified reaction network is necessary because of the added spatial dimensions, but we believe it is fully sufficient for our purpose, since we are not attempting to derive a comprehensive isotopic composition. A more complex reaction scheme would not change our conclusions.

3. Description of the models

The models are three dimensional with cylindrical symmetry in the Galaxy, and the basic coordinates are (R, z, p), where R is Galactocentric radius, z is the distance from the Galactic plane, and p is the total particle momentum. The distance from the Sun to the Galactic centre is taken as $R_\odot = 8.5$ kpc. In the models the propagation region is bounded by $R = R_h$, $z = \pm z_h$ beyond which free escape is assumed. We take $R_h = 30$ kpc. The range $z_h = 1 - 20$ kpc is considered. For a given z_h the diffusion coefficient as a function of momentum is determined by B/C for the case of no reacceleration; if reacceleration is assumed then the reacceleration strength (related to the Alfvén speed) is constrained by the energy-dependence of B/C. The spatial diffusion coefficient for the case of no reacceleration is taken as $D_{xx} = \beta D_0 (\rho/\rho_0)^{\delta_1}$ below rigidity ρ_0, $\beta D_0(\rho/\rho_0)^{\delta_2}$ above rigidity ρ_0, where the factor $\beta\ (= v/c)$ is a natural consequence of a random-walk process. Since the introduction of a sharp break in D_{xx} is a contrived procedure which is adopted just to fit B/C at all energies, we also consider the case $\delta_1 = \delta_2$, i.e. no break, in order to investigate the possibility of reproducing the data in a physically simpler way. The convection velocity (in z-direction only) $V(z)$ is assumed to increase linearly with distance from the plane ($V > 0$ for $z > 0$, $V < 0$ for $z < 0$, and $dV/dz > 0$ for all z); this implies a constant adiabatic energy loss. The linear form for $V(z)$ is suggested by cosmic-ray driven MHD wind models (e.g., Zirakashvili et al. 1996).

We include diffusive reacceleration since some stochastic reacceleration is inevitable, and it provides a natural mechanism to reproduce the energy dependence of the B/C ratio without an *ad hoc* form for the diffusion coefficient (Letaw, Silberberg, & Tsao 1993, Seo & Ptuskin 1994, Heinbach & Simon 1995, Simon & Heinbach 1996). The spatial diffusion coefficient for the case of reacceleration assumes a Kolmogorov spectrum of weak MHD turbulence so $D_{xx} = \beta D_0(\rho/\rho_0)^\delta$ with $\delta = 1/3$ for all rigidities. For this case the momentum-space diffusion coefficient D_{pp} is related to the spatial coefficient using the formula given by Seo & Ptuskin (1994), and Berezinskii et al. (1990)

$$D_{pp}D_{xx} = \frac{4p^2 v_A{}^2}{3\delta(4 - \delta^2)(4 - \delta)w} \,, \tag{1}$$

where w characterises the level of turbulence, and is equal to the ratio of MHD wave energy density to magnetic field energy density. The free parameter in this relation is v_A^2/w, where v_A is the Alfvén speed; we take $w = 1$ (Seo & Ptuskin 1994).

The adopted distributions of atomic and molecular hydrogen and of ionized hydrogen are described in detail in Paper III; Fig. 1 shows the radial distribution

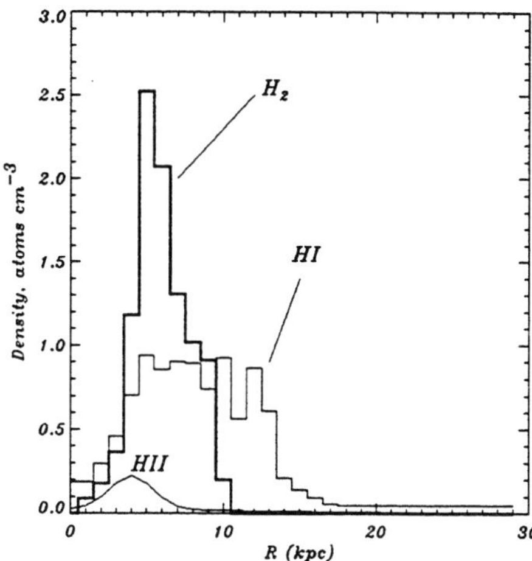

Fig. 1.— The adopted radial distribution of atomic (HI), molecular (H$_2$) and ionized (HII) hydrogen at $z = 0$.

of density in the Galactic plane. The He/H ratio of the interstellar gas is taken as 0.11 by number (see Paper III for a discussion).

The distribution of cosmic-ray sources is chosen to reproduce (after propagation) the cosmic-ray distribution determined by analysis of EGRET γ-ray data (Strong & Mattox 1996). The form used is

$$q(R, z) = q_0 \left(\frac{R}{R_\odot} \right)^\eta e^{-\xi \frac{R - R_\odot}{R_\odot} - \frac{|z|}{0.2 \text{ kpc}}} , \qquad (2)$$

where q_0 is a normalization constant, η and ξ are parameters; the R-dependence has the same parameterization as that used for SNR by Case & Bhattacharya (1996, 1998). We compute models with their SNR distribution, but also with different parameters to better fit the γ-ray gradient. We apply a cutoff in the source distribution at $R = 20$ kpc since it is unlikely that significant sources are present at such large radii. The z-dependence of q is nominal and reflects simply the assumed confinement of sources to the disk.

The primary propagation is computed first giving the primary distribution as a function of (R, z, p); then the secondary source function is obtained from the gas density and cross sections, and finally the secondary propagation is computed. The bremsstrahlung and inverse Compton γ-rays are computed self-consistently from the gas and radiation fields used for the propagation. The π^0-decay γ-rays are calculated explicitly from the proton and Helium spectra using Dermer's (1986) approach. The secondary nucleon and secondary $e^\pm$ source functions are computed from the propagated primary distribution and the gas distribution.

4. Propagation equation

The propagation equation we use is written in the form:

$$\frac{\partial \psi}{\partial t} = q(\vec{r},p) + \vec{\nabla} \cdot (D_{xx}\vec{\nabla}\psi - \vec{V}\psi) + \frac{\partial}{\partial p} p^2 D_{pp} \frac{\partial}{\partial p} \frac{1}{p^2}\psi - \frac{\partial}{\partial p}\left[\dot{p}\psi - \frac{p}{3}(\vec{\nabla}\cdot\vec{V})\psi\right] - \frac{1}{\tau_f}\psi - \frac{1}{\tau_r}\psi,$$

$$(3)$$

where $\psi = \psi(\vec{r},p,t)$ is the density per unit of total particle momentum, $\psi(p)dp = 4\pi p^2 f(\vec{p})$ in terms of phase-space density $f(\vec{p})$, $q(\vec{r},p)$ is the source term, D_{xx} is the spatial diffusion coefficient, $\vec{V}$ is the convection velocity, reacceleration is described as diffusion in momentum space and is determined by the coefficient D_{pp}, $\dot{p} \equiv dp/dt$ is the momentum loss rate, τ_f is the time scale for fragmentation, and τ_r is the time scale for the radioactive decay.

The numerical solution of the transport equation is based on a Crank-Nicholson (Press et al. 1992) implicit second-order scheme. The three spatial boundary conditions

$$\psi(R_h, z, p) = \psi(R, \pm z_h, p) = 0 \qquad (4)$$

are imposed on each iteration.

We use particle momentum as the kinematic variable since it greatly facilitates the inclusion of the diffusive reacceleration terms. The injection spectrum of primary nucleons is assumed to be a power law in momentum for the different species, $dq(p)/dp \propto p^{-\Gamma}$ for the injected *density* as expected for diffusive shock acceleration (e.g., Blandford & Ostriker 1980). This corresponds to an injected *flux* per kinetic energy interval $dF(E_k)/dE_k \propto p^{-\Gamma}$, a form often used; the value of Γ can vary with species. The injection spectrum for ^{12}C and ^{16}O was taken as $dq(p)/dp \propto p^{-2.35}$, for the case of no reacceleration, and $p^{-2.25}$ with reacceleration. These values are consistent with Engelmann et al. (1990) who give an injection index 2.23 ± 0.05. The same indices reproduce the observed proton and ^{4}He spectra (Paper II). For primary electrons, the injection spectrum can be adjusted to reproduce direct measurements or γ-ray and synchrotron data; all details are given in our series of papers (I–V).

For secondary nucleons, the source term is $q(\vec{r},p) = \beta c\,\psi_p(\vec{r},p)[\sigma_H^{ps}(p)n_H(\vec{r}) + \sigma_{He}^{ps}(p)n_{He}(\vec{r})]$, where $\sigma_H^{ps}(p)$, $\sigma_{He}^{ps}(p)$ are the production cross sections for the secondary from the progenitor on H and He targets, ψ_p is the progenitor density, and n_H, n_{He} are the interstellar hydrogen and Helium number densities.

To compute B/C and ^{10}Be/^{9}Be it is sufficient for our purposes to treat only one principal progenitor and compute weighted cross sections based on the observed cosmic-ray abundances, which we took from Lukasiak et al. (1994b). Explicitly, for a principal primary with abundance I_p, we use for the production cross section $\bar{\sigma}^{ps} = \sum_i \sigma^{is} I_i / I_p$, where σ^{is}, I_i are the cross sections and abundances of all species producing the given secondary. For the case of Boron, the Nitrogen progenitor is secondary but only accounts for $\approx 10\%$ of the total Boron production, so that the approximation of weighted cross sections is sufficient.

For the fragmentation cross sections we use the formula given by Letaw, Silberberg, & Tsao (1983). For the secondary production cross sections we use the Webber, Kish, & Schrier (1990) parameterizations in the form of code obtained from the Transport Collaboration (Guzik et al. 1997). For the important B/C ratio, we take the ^{12}C, ^{16}O $\rightarrow$ ^{10}B, ^{10}C, ^{11}B, ^{11}C cross sections from the fit to experimental data given by Heinbach & Simon (1995). For electrons and positrons the same propagation equation is valid when the appropriate energy loss terms (ionization, bremsstrahlung, inverse Compton, synchrotron) are used. The energy loss formulae for these loss mechanisms are given in Paper III.

5.   Evaluation of models

We consider the cases of diffusion+convection and diffusion+reacceleration, since these are the minimum combinations which can reproduce the key observations. In principle all three processes could be significant, and such a general model can be considered if independent astrophysical information or models, for example for a Galactic wind (e.g., Zirakashvili et al. 1996, Ptuskin et al. 1997), were to be used.

In our evaluations we use the B/C data summarized by Webber et al. (1996), from HEAO–3 and Voyager 1 and 2. The spectra were modulated to 500 MV appropriate to this data using the force-field approximation (Gleeson & Axford 1968). We also show B/C values from Ulysses (DuVernois, Simpson, & Thayer 1996) for comparison, but since this has large modulation (600–1080 MV) we do not base conclusions on these values. We use the measured ^{10}Be/^{9}Be ratio from Ulysses (Connell 1998) and from Voyager–1,2, IMP–7/8, ISEE–3 as summarized by Lukasiak et al. (1994a).

The source distribution adopted has $\eta = 0.5$, $\xi = 1.0$ in eq. (2) (apart from the cases with SNR source distribution). This form adequately reproduces the small observed γ-ray based gradient, for all z_h; a more detailed discussion is given in Section ??.

5.1.   Diffusion/convection models

The main parameters are z_h, D_0, δ_1, δ_2 and ρ_0 and dV/dz. We treat z_h as the main unknown quantity, and consider values 1–20 kpc. For a given z_h we show

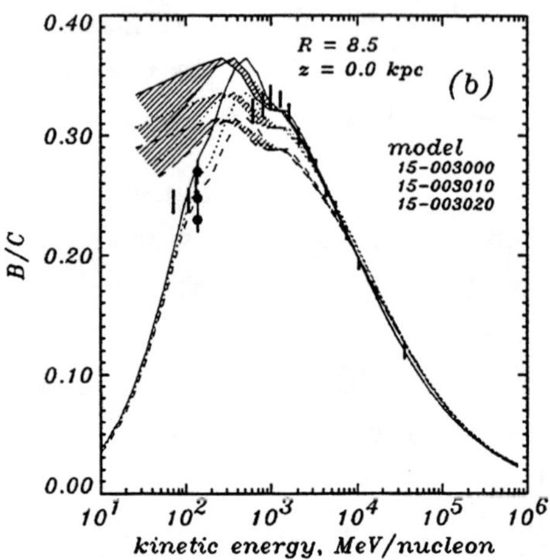

Fig. 2.— B/C ratio for diffusion/convection models without break in diffusion coefficient, for z_h = 3 kpc, dV/dz = 0 (solid line), 5 (dotted line), and 10 km s^{-1} kpc^{-1} (dashed line). Solid line: interstellar ratio, shaded area: modulated to 300–500 MV. Data: vertical bars: HEAO-3, Voyager (Webber et al. 1996), filled circles: Ulysses (DuVernois, Simpson, & Thayer 1996: Φ = 600, 840, 1080 MV).

B/C for a series of models with different dV/dz.

Fig. 2 shows the case of no break, $\delta_1 = \delta_2$; for each dV/dz, the remaining parameters D_0, δ_1 and ρ_0 are adjusted to fit the data as well as possible. It is clear that a *good* fit is *not* possible; the basic effect of convection is to reduce the variation of B/C with energy, and although this improves the fit at low energies the characteristic peaked shape of the measured B/C cannot be reproduced. Although modulation makes the comparison with the low energy Voyager data somewhat uncertain, Fig. 2 shows that the fit is unsatisfactory; the same is true even if we use a very low modulation parameter (300 MV) in an attempt to improve the fit. This modulation is near the minimum value for the entire Voyager 17 year period (cf. the average value of 500 MV; Webber et al. 1996). The failure to obtain a good fit is an important conclusion since it shows that the simple inclusion of convection cannot solve the problem of the low-energy falloff in B/C.

We can however force a fit to the data by allowing a break in $D_{xx}(p)$, i.e. $\delta_1 \neq \delta_2$. In the absence of convection, the falloff in B/C at low energies requires that the diffusion coefficient increases rapidly below $\rho_0 = 3$ GV ($\delta_1 \sim -0.6$) reversing the trend from higher energies ($\delta_2 \sim +0.6$). Inclusion of the convective term does not reduce the size of the *ad hoc* break in the diffusion coefficient, in fact

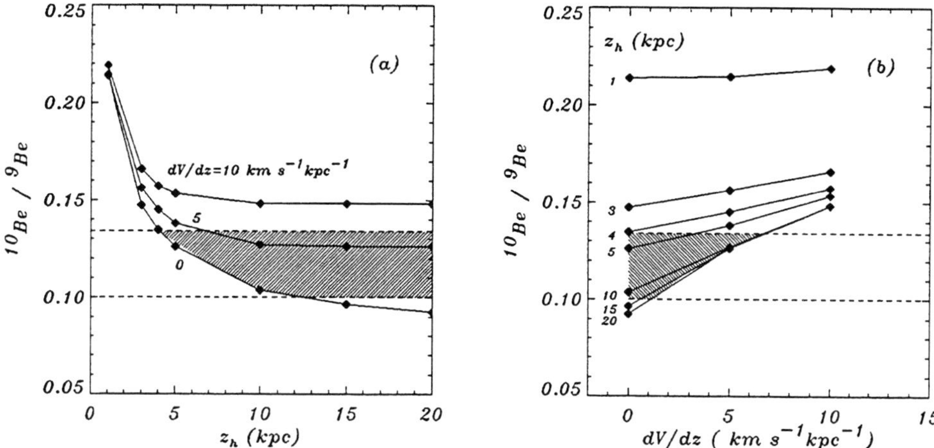

Fig. 3.— Predicted ^{10}Be/^{9}Be ratio as function of (a) z_h for $dV/dz = 0$, 5, 10 km s^{-1} kpc^{-1}, (b) dV/dz for $z_h = 1 - 20$ kpc at 525 MeV/nucleon corresponding to the mean interstellar value for the Ulysses data (Connell 1998); the Ulysses experimental limits are shown as horizontal dashed lines. The shaded regions show the parameter ranges allowed by the data.

it rather exacerbates the problem by requiring a larger break[1].

Fig. 3 summarizes the limits on z_h and dV/dz, using the ^{10}Be/^{9}Be ratio at the interstellar energy of 525 MeV/nucleon appropriate to the Ulysses data (Connell 1998). For $z_h < 4$ kpc, the predicted ratio is always too high, even for no convection; no convection is allowed for such z_h values since this increases ^{10}Be/^{9}Be still further. For $z_h \geq 4$ kpc agreement with ^{10}Be/^{9}Be is possible provided $0 < dV/dz < 7$ km s^{-1} kpc^{-1}. We conclude from Fig. 3a that in the absence of convection 4 kpc $< z_h < 12$ kpc, and if convection is allowed the lower limit remains but no upper limit can be set. It is interesting that an upper as well as a lower limit on z_h is obtained in the case of no convection, although ^{10}Be/^{9}Be approaches asymptotically a constant value for large halo sizes and becomes insensitive to the halo dimension. From Fig. 3b, $dV/dz < 7$ km s^{-1} kpc^{-1} and this figure places upper limits on the convection parameter for each halo size. These limits are rather strict, and a finite wind velocity is only allowed in any case for $z_h > 4$ kpc. Note that these results are not very sensitive to modulation since

---

[1]Note that the dependence of interaction rate on particle velocity itself is not sufficient to cause the full observed low-energy falloff. In leaky-box treatments the low-energy behaviour is modelled by adopting a constant path-length below a few GeV/nucleon, without attempting to justify this physically. A convective term is often invoked, but our treatment shows that this alone is not sufficient.

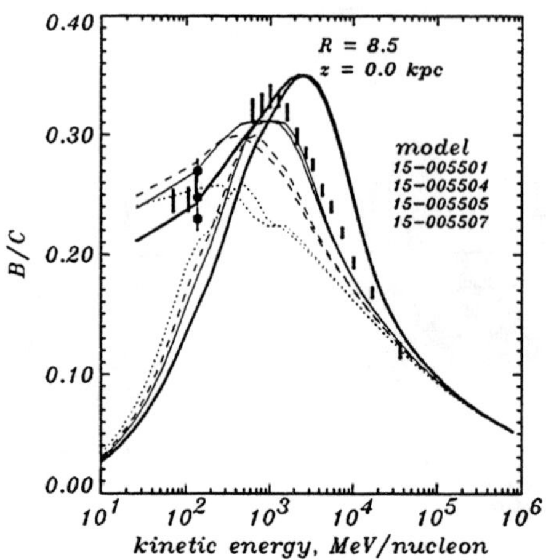

Fig. 4.— B/C ratio for diffusive reacceleration models with $z_h = 5$ kpc, $v_A = 0$ (dotted), 15 (dashed), 20 (thin solid), 30 km s^{-1} (thick solid). In each case the interstellar ratio and the ratio modulated to 500 MV is shown. Data: as Fig. 2.

the predicted ^{10}Be/^{9}Be is fairly constant from 100 to 1000 MeV/nucleon.

5.2. Diffusive reacceleration models

The main parameters are z_h, D_0 and v_A. Again we treat z_h as the main unknown quantity. The evaluation is simpler than for convection models since the number of free parameters is smaller. Fig. 4 illustrates the effect on B/C of varying v_A, from $v_A = 0$ (no reacceleration) to $v_A = 30$ km s^{-1}, for $z_h = 5$ kpc. This shows how the initial form becomes modified to produce the characteristic peaked shape. Reacceleration models thus lead naturally to the observed peaked form of B/C, as pointed out by several previous authors (e.g., Letaw, Silberberg, & Tsao 1993, Seo & Ptuskin 1994, Heinbach & Simon 1995).

Fig. 5 shows ^{10}Be/^{9}Be for the same models, (a) as a function of energy for various z_h, (b) as a function of z_h at 525 MeV/nucleon corresponding to the Ulysses measurement. Comparing with the Ulysses data point, we conclude that 4 kpc $< z_h < 12$ kpc. Again the result is not very sensitive to modulation since the predicted ^{10}Be/^{9}Be is fairly constant from 100 to 1000 MeV/nucleon.

Energy losses attenuate the flux of stable nuclei much more than radioactive nuclei, and hence lead to an increase in ^{10}Be/^{9}Be. Clearly if losses are ignored the predicted ratio will be too low and the derived value of z_h will be too small since z_h will have to be reduced to fit the observations.

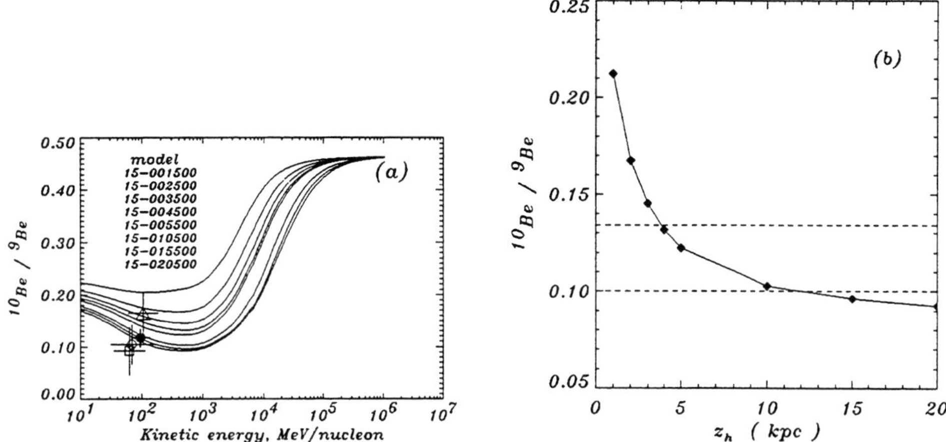

Fig. 5.— ^{10}Be/^{9}Be ratio for diffusive reacceleration models: (a) as function of energy for (from top to bottom) $z_h = 1$, 2, 3, 4, 5, 10, 15 and 20 kpc; (b) as function of z_h at 525 MeV/nucleon corresponding to the mean interstellar value for the Ulysses data (Connell 1998); the Ulysses experimental limits are shown as horizontal dashed lines. Data points from Lukasiak et al. (1994a) (Voyager-1,2: square, IMP-7/8: open circle, ISEE-3: triangle) and Connell (1998) (Ulysses): filled circle.

Our results on the halo size can be compared with those of other studies: $z_h \geq 7.8$ kpc (Freedman et al. 1980), $z_h \leq 3$ kpc (Bloemen et al. 1993), and $z_h \leq 4$ kpc (Webber, Lee, & Gupta 1992). Lukasiak et al. (1994a) found 1.9 kpc $< z_h < 3.6$ kpc (for no convection) based on Voyager Be data and using the Webber, Lee, & Gupta (1992) models. We believe our new limits to be an improvement, first because of the improved Be data from Ulysses, second because of our treatment of energy losses (see Section **??**) and generally more realistic astrophysical details in our model. Recently, Webber & Soutoul (1998), Ptuskin & Soutoul (1998) have obtained $z_h = 2 - 4$ kpc and 4.9^{+4}_{-2} kpc, respectively, in agreement with our results.

6. Cosmic-ray gradients

An important constraint on any model of cosmic-ray propagation is provided by γ-ray data which give information on the radial distribution of cosmic rays in the Galaxy. For a given source distribution, a large halo will give a smaller cosmic-ray gradient. It is generally believed that supernova remnants (SNR) are the main sources of cosmic rays (see Webber 1997 for a recent review), but unfortunately the distribution of SNR is poorly known due to selection effects. Nevertheless it

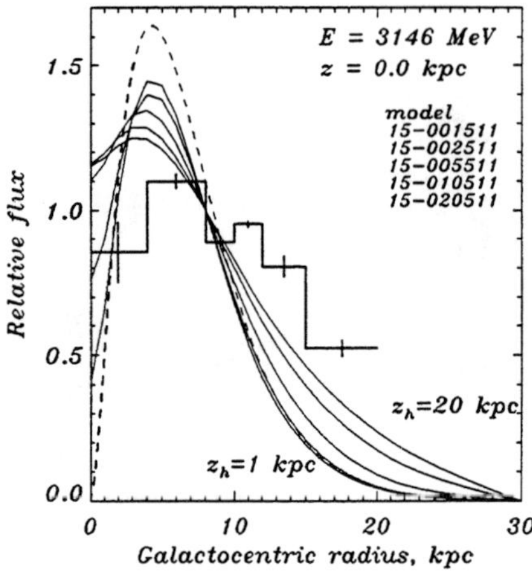

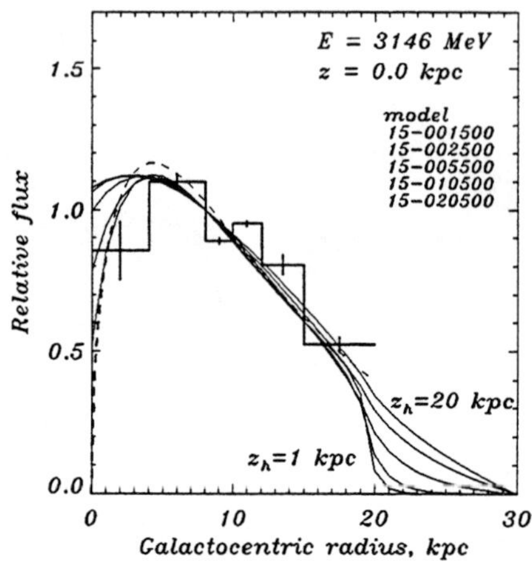

Fig. 6.— *Left panel*: Radial distribution of 3 GeV protons at $z = 0$, for diffusive reacceleration model with halo sizes $z_h = 1, 3, 5, 10, 15$, and 20 kpc (solid curves). The source distribution is that for SNR given by Case & Bhattacharya (1996), shown as a dashed line. The cosmic-ray distribution deduced from EGRET >100 MeV γ-rays (Strong & Mattox 1996) is shown as the histogram. *Right panel*: Radial distribution of 3 GeV protons at $z = 0$, for diffusive reacceleration model with various halo sizes $z_h = 1, 3, 5, 10, 15$, and 20 kpc (solid curves). The source distribution used is shown as a dashed line. It was adopted to reproduce the cosmic-ray distribution deduced from EGRET >100 MeV γ-rays (Strong & Mattox 1996) which is shown as the histogram.

is interesting to compare quantitatively the effects of halo size on the gradient for a plausible SNR source distribution. For illustration we use the SNR distribution from Case & Bhattacharya (1996), which is peaked at $R = 4 - 5$ kpc and has a steep falloff towards larger R.

Fig. 6 (left panel) shows the effect of halo size on the resulting radial distribution of 3 GeV cosmic-ray protons, for the reacceleration model. For comparison we show the cosmic-ray distribution deduced by model-fitting to EGRET γ-ray data (> 100 MeV) from Strong & Mattox (1996), which is dominated by the π^0-decay component generated by GeV nucleons; the analysis by Hunter et al. (1997), based on a different approach, gives a similar result. The predicted cosmic-ray distribution using the SNR source function is too steep even for large halo sizes; in fact the halo size has a relatively small effect on the distribution. Other related distributions such as pulsars (Taylor, Manchester & Lyne 1993, Johnston 1994) have an even steeper falloff. Only for $z_h = 20$ kpc does the gradient approach that observed, and in this case the combination of a large halo and a slightly less steep SNR distribution could give a satisfactory fit. For diffusion/convection models the situation is similar, with more convection tending to make the gradient follow more closely the sources. A larger halo ($z_h \gg 20$ kpc), apart from being excluded by the ^{10}Be analysis presented here, would in fact not improve the situation much since Fig. 6 shows that the gradient approaches an asymptotic shape which hardly changes beyond a certain halo size. This is a consequence of the nature of the diffusive process, which even for an unlimited propagation region still retains the signature of the source distribution.

Based on these results we have to conclude, in the context of the present models, that the distribution of sources is not that expected from the (highly uncertain: see Green 1991) distribution of SNR. This conclusion is similar to that previously found by others (Webber, Lee, & Gupta 1992, Bloemen et al. 1993). In view of the difficulty of deriving the SNR distribution this is perhaps not a serious shortcoming; if SNR are indeed cosmic-ray sources then it is possible that the γ-ray analysis gives the best estimate of their Galactic distribution. Therefore in our standard model we have obtained the source distribution empirically by requiring consistency with the high energy γ-ray results.

Alternatively it is possible that the diffusion is not isotropic but occurs preferentially in the radial direction, so smoothing the source distribution more effectively. This possibility will be addressed in future work.

Fig. 6 (right panel) shows the source distribution adopted in the present work, and the resulting 3 GeV proton distribution, again compared to that deduced from γ-rays. The gradients are now consistent, especially considering that some systematic effects, due for example unresolved γ-ray sources, are present in the γ-ray based distribution.

7. Interstellar positrons and antiproton spectra

The positron and antiproton fluxes reflect the proton and Helium spectra throughout the Galaxy and thus provide an essential check on propagation models and also on the interpretation of diffuse γ-ray emission (Paper IV). Secondary positrons and antiprotrons in Galactic cosmic rays are produced in collisions of cosmic-ray particles with interstellar matter[2]. These are an important diagnostic for models of cosmic-ray propagation and provide information complementary to that provided by secondary nuclei. However, unlike secondary nuclei, antiprotons reflect primarily the propagation history of the protons, the main cosmic-ray

---

[2]Secondary origin of cosmic-ray antiprotons is basically accepted, though some other exotic contributors such as, e.g., neutralino annihilation (Bottino et al. 1998) are also discussed.

component.

In our model the proton and Helium spectra are computed as a function of (R, z, p) by the propagation code. The injection spectrum is adjusted to give a good fit to the locally measured spectrum, normalizing at 10 GeV/nucleon.

For the injection spectra of protons, we find $\Gamma = 2.15$ reproduces the observed spectra in the case of no reacceleration, and $\Gamma = 2.25$ with reacceleration. We find it is necessary to use slightly steeper (0.2 in the index) injection spectra for Helium nuclei in order to fit the observed spectra in the 1–100 GeV range of interest for positron production. The spectra fit up to about 100 GeV beyond which the Helium spectrum without reacceleration becomes too steep and the proton spectrum with reacceleration too flat; these deviations are of no consequence for the positron and antiproton calculations.

Our calculations of the interstellar antiproton spectra and $\bar{p}/p$ ratio for these spectra are shown in Fig. 7. The computed antiproton spectrum is divided by the same interstellar proton spectrum, and the ratio is modulated to 750 MV. The corresponding ratios are shown on the right panel. We have performed the same calculations for models with and without reacceleration and the results differ only in details. As seen, our result agrees well with the calculations of Simon et al. (1998), showing that our treatment of the production cross-sections is adequate (for the details of the cross sections see Paper IV).

Fig. 8 shows the computed secondary positron spectra for the cases without and with reacceleration. Our predictions are compared with recent absolute measurements above a few GeV where solar modulation is small (Barwick et al. 1998), and the agreement is satisfactory in both cases; this comparison has the advantage of being independent of the electron spectrum, unlike the positron/electron ratio which was the main focus of Paper II.

8. Probes of the interstellar nucleon spectrum

Diffuse Galactic γ-ray observations have been interpreted as requiring a harder average nucleon spectrum in interstellar space than that observed directly (Hunter et al. 1997, Gralewicz et al. 1997, Mori 1997, Moskalenko & Strong 1998b,c, see also Section ??). A sensitive test of the interstellar nucleon spectra is provided by secondary antiprotons and positrons. Because they are secondary, they reflect the *large-scale* nucleon spectrum independent of local irregularities in the primaries.

We consider a case which matches the γ-ray data (Fig. 9) at the cost of a much harder proton spectrum than observed. The dashed lines in Fig. 7 (right) show the $\bar{p}/p$ ratio for the hard proton spectrum (with and without reacceleration); the ratio

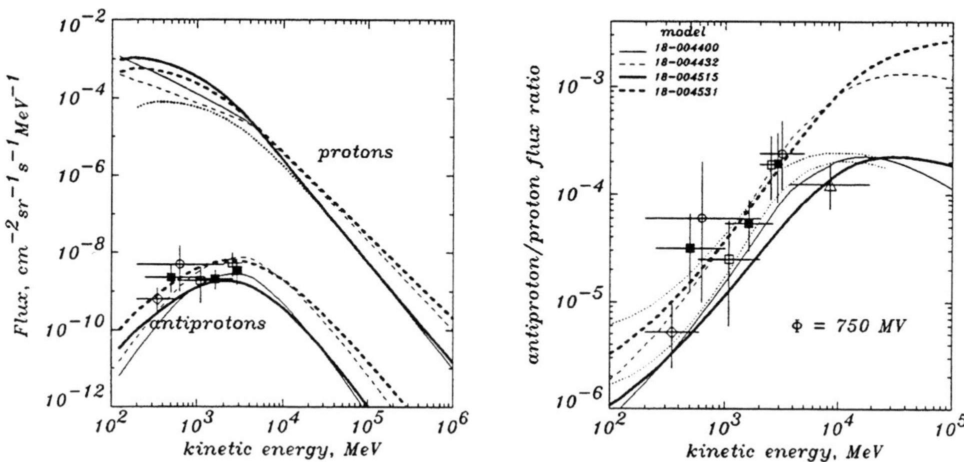

Fig. 7.— *Left panel:* Interstellar nucleon and antiproton spectra as calculated in nonreacceleration models (thin lines) and models with reacceleration (thick lines). Proton spectra consistent with the local one are shown by the solid lines, hard spectra are shown by the dashed lines. The local spectrum as measured by IMAX (Menn et al. 1997) is shown by dots. *Right panel:* $\bar{p}/p$ ratio for different ambient proton spectra. Lines are coded as on the left. The ratio is modulated with $\Phi = 750$ MV. Calculations of Simon et al. (1998) are shown by the dotted lines. Data: see references in Paper IV.

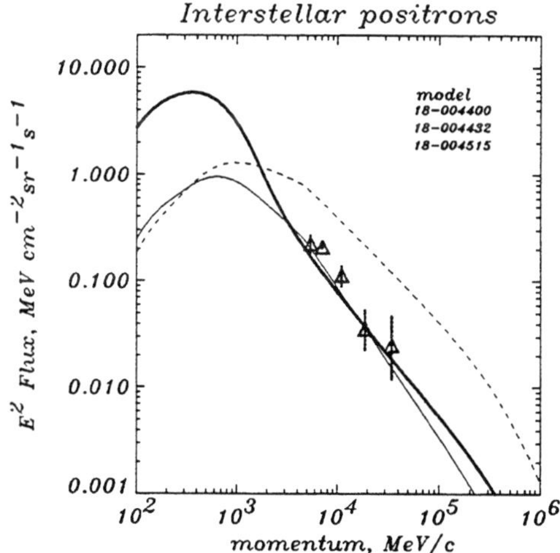

Fig. 8.— Spectra of secondary positrons for 'conventional' (thin line) and hard (dashes) nucleon spectra (no reacceleration). Thick line: 'conventional' case with reacceleration. Data: Barwick et al. (1998).

is still consistent with the data at low energies but rapidly increases toward higher

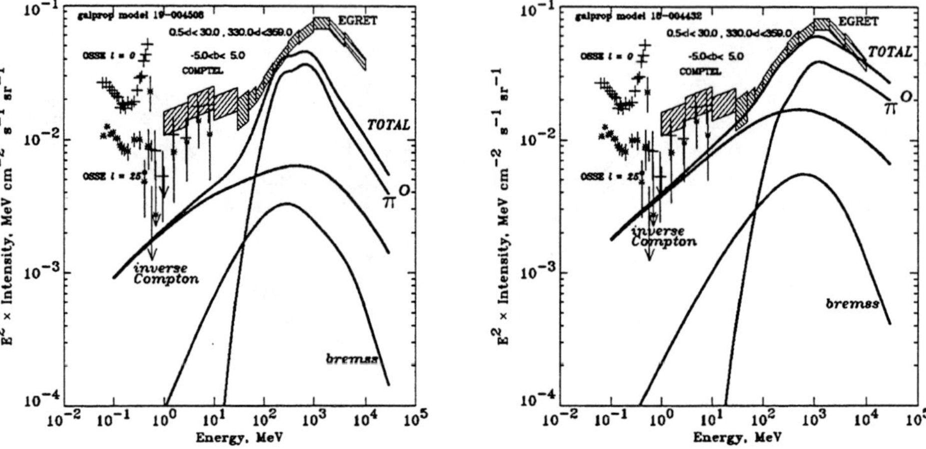

Fig. 9.— Gamma-ray energy spectrum of the inner Galaxy ($300° \leq l \leq 30°$, $|b| \leq 5°$) compared with our model calculations. Data: EGRET (Strong & Mattox 1996), COMPTEL (Strong et al. 1998), OSSE ($l = 0, 25°$: Kinzer et al. 1997). *Left panel:* Model with 'conventional' nucleon and electron spectra. Also shown are the contributions of individual components: bremsstrahlung, inverse Compton, and π^0-decay. *Right panel:* The same compared to the model with the *hard nucleon* spectrum (no reacceleration).

energies and becomes $\sim$4 times higher at 10 GeV. Up to 3 GeV it does not conflict with the data with their very large error bars. It is however larger than the point at 3.7–19 GeV (Hof et al. 1996) by about 5σ. Clearly we cannot conclude definitively on the basis of this one point[3], but it does indicate the sensitivity of this test. In view of the sharply rising ratio in the hard-spectrum scenario it seems unlikely that the data could be fitted in this case even with some re-scaling due to propagation uncertainties. More experiments are planned (see Paper IV for a summary) and these should allow us to set stricter limits on the nucleon spectra including less extreme cases than considered here, and to constrain better the interpretation of γ-rays.

Positrons also provide a good probe of the nucleon spectrum, but are more affected by energy losses and propagation uncertainties. Fig. 8 shows, in addition to the normal case, the positron flux resulting from a hard nucleon spectrum. The predicted flux is a factor 4 above the Barwick et al. (1998) measurements and hence provides further evidence against the 'hard nucleon spectrum' hypothesis.

---

[3]We do not consider here the older $\bar{p}$ measurement of Golden et al. (1984a) because the flight of the early instrument in 1979 was repeated in 1991 with significantly improved instrument and analysis techniques (see Hof et al. 1996 and a discussion therein).

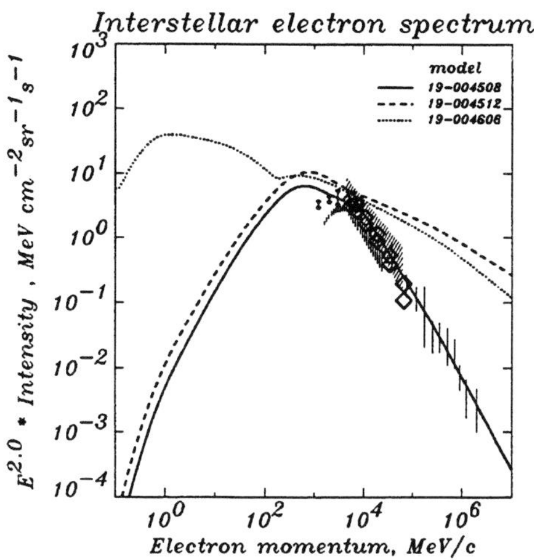

Fig. 10.— Electron spectra at $R_\odot = 8.5$ kpc in the plane, for 'conventional' (solid line), and hard electron spectrum models without (dashes), and with (dots) low-energy upturn. Data (direct measurements): Taira et al. (1993) (vertical lines), Golden et al. (1984b, 1994) (shaded areas), Ferrando et al. (1996) (small diamonds), Barwick et al. (1998) (large diamonds).

9. Diffuse Galactic continuum gamma rays

We can also use our model to study the diffuse γ-ray emission from the Galaxy. Recent results from both COMPTEL and EGRET indicate that inverse Compton (IC) scattering is a more important contributor to the diffuse emission that previously believed. COMPTEL results (Strong et al. 1997) for the 1–30 MeV range show a latitude distribution in the inner Galaxy which is broader than that of HI and H_2, so that bremsstrahlung of electrons on the gas does not appear adequate and a more extended component such as IC is required. The broad distribution is the result of the large z-extent of the interstellar radiation field[4] which can interact with cosmic-ray electrons up to several kpc from the plane. At much higher energies, the puzzling excess in the EGRET data above 1 GeV relative to that expected for π^0-decay has been suggested to originate in IC scattering from a hard interstellar electron spectrum (e.g., Pohl & Esposito 1998).

Fig. 9 (left) shows the γ-ray spectrum of the inner Galaxy for a 'conventional' case which matches the directly measured electron and nucleon spectra and is consistent with synchrotron spectral index data (Moskalenko & Strong 1998c, Paper V). Fig. 10 shows electron spectra at $R_\odot = 8.5$ kpc in the disk for this

---

[4]We have made a new calculation of the interstellar radiation field (Paper V) based on stellar population models and IRAS and COBE data.

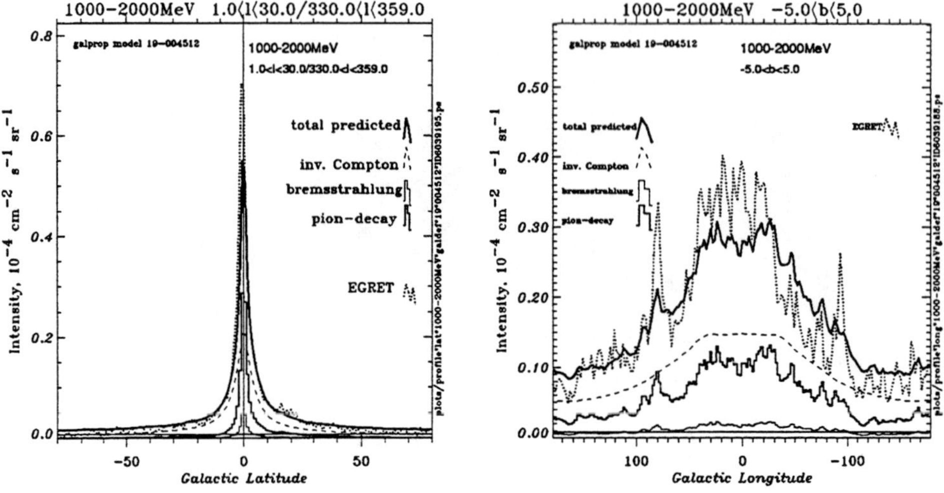

Fig. 11.— Distributions of 1–2 GeV γ-rays computed for a hard electron spectrum (reacceleration model) as compared to EGRET data (Cycles 1–4, point sources removed, see Paper V). Contribution of various components is shown as calculated in our model. *Left panel:* Latitude distribution ($330° < l < 30°$). *Right panel:* Longitude distribution for $|b| < 5°$.

model. It fits the observed γ-ray spectrum only in the range 30 MeV – 1 GeV. Fig. 9 (right) shows the case of π^0-decay γ-rays from a hard nucleon spectrum (but still the 'conventional' electron spectrum). This can improve the fit above 1 GeV but the high energy antiproton and positron data probably exclude the hypothesis that the local nucleon spectrum differs significantly from the Galactic average (see Section ??).

We thus consider the 'hard electron spectrum' alternative. The electron injection spectral index is taken as –1.7 (with reacceleration), which after propagation provides consistency with radio synchrotron data (a crucial constraint). Following Pohl & Esposito (1998), for this model we do *not* require consistency with the locally measured electron spectrum above 10 GeV since the rapid energy losses cause a clumpy distribution so that this is not necessarily representative of the interstellar average. For this case, the interstellar electron spectrum deviates strongly from that locally measured as illustrated in Fig. 10. Because of the increased IC contribution at high energies, the predicted γ-ray spectrum can reproduce the overall intensity from 30 MeV – 10 GeV but the detailed shape above 1 GeV is still problematic. Further refinement of this scenario is presented in Paper V.

Fig. 11 shows the model latitude and longitude γ-ray distributions for the inner Galaxy for 1–2 GeV, convolved with the EGRET point-spread function,

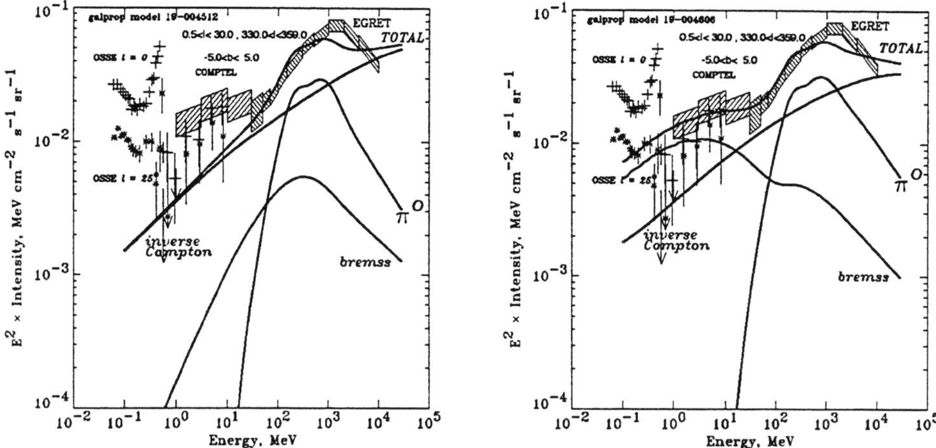

Fig. 12.— γ-ray spectrum of inner Galaxy compared to models with a hard electron spectrum without (left) and with low-energy upturn (right). Data as in Fig. 9.

compared to EGRET Phase 1–4 data (with known point sources subtracted). It shows that a model with large IC component can indeed reproduce the data. The latitude distribution here is not as wide as at low energies owing to the rapid energy losses of the electrons, so that an observational distinction between a gas-related π^0-component from a hard nucleon spectrum and the IC model does not seem possible on the basis of γ-rays alone, but requires also other tests such as consistency with antiproton and positron data (see Section **??**).

None of these models fits the γ-ray spectrum below $\sim$30 MeV as measured by the Compton Gamma-Ray Observatory (Fig. 12 left). In order to fit the low-energy part as diffuse emission (Fig. 12 right), without violating synchrotron constraints (Paper V), requires a rapid upturn in the cosmic-ray electron spectrum below 200 MeV (e.g., as in Fig. 10). However, in view of the energetics problems (Skibo et al. 1997), a population of unresolved sources seems more probable and would be the natural extension of the low energy plane emission seen by OSSE (Kinzer et al. 1997) and GINGA (Yamasaki et al. 1997).

10. Conclusions

Our propagation model has been used to study several areas of high energy astrophysics. We believe that combining information from classical cosmic-ray studies with γ-ray and other data leads to tighter constraints on cosmic-ray origin and propagation.

We have shown that simple diffusion/convection models have difficulty in accounting for the observed form of the B/C ratio without special assumptions chosen to fit the data, and do not obviate the need for an *ad hoc* form for the diffusion coefficient. On the other hand we confirm the conclusion of other authors that models with reacceleration account naturally for the energy dependence over the whole observed range. Combining these results points rather strongly in favour of the reacceleration picture.

We take advantage of the recent Ulysses Be measurements to obtain estimates of the halo size. Our limits on the halo height are $4 \text{ kpc} < z_h < 12 \text{ kpc}$. Our new limits should be an improvement on previous estimates because of the more accurate Be data, our treatment of energy losses, and the inclusion of more realistic astrophysical details (such as, e.g., the gas distribution) in our model. The gradient of protons derived from γ-rays is smaller than expected for SNR sources, and we therefore adopt a flatter source distribution in order to meet the γ-ray constraints. This may just reflect the uncertainty in the SNR distribution.

The positron and antiproton fluxes calculated are consistent with the most recent measurements. The $\bar{p}/p$ data point above 3 GeV and positron flux measurements seem to rule out the hypothesis that the local cosmic-ray nucleon spectrum differs significantly from the Galactic average (by implication adding support to the 'hard electron' alternative), but confirmation of this conclusion must await more accurate antiproton data at high energies.

Gamma-ray data suggest that the interstellar electron spectrum is harder then that locally measured, but this remains to be confirmed by detailed study of the angular distribution. The low-energy Galactic γ-ray emission is difficult to explain as truly diffuse and a point source population seems more probable.

REFERENCES

Barwick, S. W., et al. 1998, ApJ, 498, 779

Berezinskii, V. S., et al. 1990, Astrophysics of Cosmic Rays (Amsterdam: North Holland)

Blandford, R. D., & Ostriker, J. P. 1980, ApJ, 237, 793

Bloemen, H., Dogiel, V. A., Dorman, V. I., & Ptuskin, V. S. 1993, A&A, 267, 372

Bottino, A., et al. 1998, submitted (astro-ph/9804137)

Case, G., & Bhattacharya, D. 1996, A&AS, 120C, 437

Case, G., & Bhattacharya, D. 1998, ApJ, 504, 761

Connell, J. J. 1998, ApJ, 501, L59

Dermer, C. D. 1986, A&A, 157, 223

DuVernois, M. A., Simpson, J. A., & Thayer, M. R. 1996, A&A, 316, 555

Engelmann, J. J., et al. 1990, A&A, 233, 96

Ferrando, P., et al. 1996, A&A, 316, 528

Freedman, I., Kearsey, S., Osborne, J. L., & Giler, M. 1980, A&A, 82, 110

Ginzburg, V. L., Khazan, Ya. M., & Ptuskin, V. S. 1980, Ap&SS, 68, 295

Gleeson, L. J., & Axford, W. I. 1968, ApJ, 154, 1011

Golden, R. L., et al. 1984a, Astrophys. Lett., 24, 75

Golden, R. L., et al. 1984b, ApJ, 287, 622

Golden, R. L., et al. 1994, ApJ, 436, 769

Gralewicz, P., et al. 1997, A&A, 318, 925

Green, D. A. 1991, PASP, 103, 209

Guzik, T. G., et al. 1997, in Proc. 25th Int. Cosmic Ray Conference (Durban), 4, 317

Heinbach, U., & Simon, M. 1995, ApJ, 441, 209

Hof, M., et al. 1996, ApJ, 467, L33

Hunter, S. D., et al. 1997, ApJ, 481, 205

Johnston, S. 1994, MNRAS, 268, 595

Jokipii, J. R. 1976, ApJ, 208, 900

Jones, F. C. 1979, ApJ, 229, 747

Kinzer, R. L., et al. 1997, in AIP Conf. Proc. 410, Fourth Compton Symposium, ed. C. D. Dermer, M. S. Strickman, & J. D. Kurfess (New York: AIP), p.1577 (OSSE preprint #89)

Lerche, I., & Schlickeiser, R. 1982, A&A, 107, 148

Letaw, J. R., Silberberg, R., & Tsao, C. H. 1983, ApJS, 51, 271

Letaw, J. R., Silberberg, R., & Tsao, C. H. 1993, ApJ, 414, 601

Lukasiak, A., Ferrando, P., McDonald, F. B., & Webber, W. R. 1994a, ApJ, 423, 426

Lukasiak, A., Ferrando, P., McDonald, F. B., & Webber, W. R. 1994b, ApJ, 426, 366

Menn, W., et al. 1997, in Proc. 25th Int. Cosmic Ray Conference (Durban), 3, 409

Mori, M. 1997, ApJ, 478, 225

Moskalenko, I. V., & Strong, A. W. 1998a, ApJ, 493, 694 (Paper II)

Moskalenko, I. V., & Strong, A. W. 1998b, in Proc. 16th European Cosmic Ray Symp. (Alcala), GR-1.3, in press (astro-ph/9807288)

Moskalenko, I. V., & Strong, A. W. 1998c, in Proc. 3rd INTEGRAL Workshop "The Extreme Universe", paper #36, in press

Moskalenko, I. V., & Strong, A. W. 1998d, ApJ, submitted

Moskalenko, I. V., Strong, A. W., & Reimer, O. 1998, A&A, 338, L75 (Paper IV)

Press, W. H., et al. 1992, Numerical Recipes in FORTRAN, 2nd Edition (Cambridge: Cambridge University Press)

Pohl, M., & Esposito, J. A. 1998, ApJ, 507, 327

Porter, T. A., & Protheroe R. J. 1997, J. Phys. G.: Nucl. Part. Phys., 23, 1765

Ptuskin, V. S., & Soutoul, A. 1998, A&A, 337, 859

Ptuskin, V. S., Völk, H. J., Zirakashvili, V. N., & Breitschwerdt, D. 1997, A&A, 321, 434

Seo, E. S., & Ptuskin, V. S. 1994, ApJ, 431, 705

Simon, M., & Heinbach, U. 1996, ApJ, 456, 519

Simon, M., Molnar, A., & Roesler, S. 1998, ApJ, 499, 250

Simpson, J. A., & Garcia-Munoz, M. 1988, Spa. Sci. Rev., 46, 205

Skibo, J. G., et al. 1997, ApJ, 483, L95

Strong, A. W., & Mattox, J. R. 1996, A&A, 308, L21

Strong, A. W., & Moskalenko, I. V. 1997, in AIP Conf. Proc. 410, Fourth Compton Symposium, ed. C. D. Dermer, M. S. Strickman, & J. D. Kurfess (New York: AIP), p.1162 (Paper I)

Strong, A. W., & Moskalenko, I. V. 1998, ApJ, 509, in press (Paper III) (NB page numbering will be available in December)

Strong, A. W., Moskalenko, I. V., & Reimer, O. 1998, ApJ, submitted (Paper V)

Strong, A. W., et al. 1997, in AIP Conf. Proc. 410, Fourth Compton Symposium, ed. C. D. Dermer, M. S. Strickman, & J. D. Kurfess (New York: AIP), p.1198

Strong, A. W., et al. 1998, in Proc. 3rd INTEGRAL Workshop 'The Extreme Universe', paper #15, in press

Taira, T., et al. 1993, in Proc. 23rd Int. Cosmic Ray Conference (Calgary), 2, 128

Taylor, J. H., Manchester, R. N., & Lyne, A. G. 1993, ApJS, 88, 529

Webber, W. R. 1997, Spa. Sci. Rev., 81, 107

Webber, W. R., & Soutoul, A. 1998, ApJ, 506, 335

Webber, W. R., Kish, J. C., & Schrier, D. A. 1990, Phys. Rev. C, 41, 566

Webber, W. R., Lee, M. A., & Gupta, M. 1992, ApJ, 390, 96

Webber, W. R., Lukasiak, A., McDonald, F. B., & Ferrando, P. 1996, ApJ, 457, 435

Yamasaki, N.Y., et al. 1997, ApJ, 481, 821

Zirakashvili, V. N., Breitschwerdt, D., Ptuskin, V. S., & Völk, H. J. 1996, A&A, 311, 113

Propagation of Cosmic Rays in the Galaxy
Leaky Box Model, Diffusion Halo Model and Reacceleration

Manfred Simon

Universitit-GH Siegen, Department of Physics, D-57068 Siegen, Germany

1   Introduction

When the cosmic ray (CR) beam which resembles an accelerated sample of galactic matter propagates through interstellar medium of our galaxy it interacts in various ways depending of the type of cosmic ray particle and the constituents of this interstellar environment such as gas, magnetic fields or photons. These high energy interactions lead to the production of new particles, new photons of different wave lengths (e.g. radio, microwave, IR, optical, UV, X- and γ-rays) and also to spallation products (often called "secondary" nuclei) which result when the more abundant "primary" heavier nuclei break up while they interact with the interstellar gas. From a variety of different observations (radio, high energy photons, secondary/primary ratios (s/p ratio) and cosmic ray spectra) we have been gradually accumulating evidence that the existence of cosmic ray particles is a general feature everywhere in our galaxy and beyond.

In our own galaxy they spend more than 10^7 years before they escape into the intergalactic space. This suggests a diffusive process for their transport, since the confinement time along a line of sight through the disk of our galaxy would only be about 10^3 years. The high degree of isotropy which we observe in the cosmic radiation also suggests an effective diffusion since cosmic rays streaming freely out of the galaxy would be highly anisotropic with most of the flux coming from the direction of the central regions of the galaxy.

The distribution of synchrotron radio emission which occurs when high energy electrons spiral around magnetic field lines can be observed from edge-on spiral galaxies such as NGC 891 implying that the high energy particles do not strictly restrain to the thin galactic disk but propagate out into the halo. Thus the volume in which cosmic rays can be found is larger than that given by the thin galactic disk where most of the stars and energetic processes take place, and where probably also the cosmic ray sources are located.

Thus any model which aims to describe the transport of CR in the galaxy has to reproduce the observational data of many kinds which are related directly or indirectly to cosmic ray origin and propagation: directly via measurements of nuclei, electrons, positrons and antiprotons, indirectly via gamma rays and synchrotron radiation. All these various observational data provide many independent constraints and should be used to develop a realistic physical picture of CR propagation.

It has been pointed out many years ago (Ginzburg, Khazan & Ptuskin, 1980) that the sources of CR should be distributed within the thin galactic disk and that the escape from the disk into the halo and finally into the intergalactic space is determined by diffusion. In this Diffusion Halo Model (DHM) one expects a gradient of CR density away from the galactic disk implying a constant streaming of CR particles away from the galactic disk into the halo.

In the literature this DHM competes with the very popular Leaky Box Model (LBM). The LBM describes an equilibrium model, in which the cosmic ray sources, and the primary and secondary cosmic ray particles are homogeneously distributed in a confinement volume (box, galaxy) and constant in time with no gradient of CR density into any direction. Thus the transport of CR is not controlled by diffusion but by a hypothetic leakage process at the imaginary boundaries. The mechanism which particles use to escape from this confinement volume is not addressed as well as the physical size of this volume.

These two, the DHM and the LBM, are currently the basic competing models in CR propagation calculations. Although the DHM is more realistic in a physical sense, it is the LBM which is more popular and mostly used. The LBM gets its popularity by its simpler mathematical structure and by its ability to describe the data, at least those which are currently available. It is true that the precision and details of the data have not provided yet a tool to favour one model over the other. But as pointed out by Ginzburg, Khazan & Ptuskin (1980), the interpretation of secondary radioactive CR nuclei is model dependent and can be used to check on the physical reality of these models. I will explain how this works.

I will also address the aspect of reacceleration. While energy losses due to ionization processes in the interstellar gas are generally accepted, the idea of reacceleration does not find such level of acceptance since it introduces more complications to the mathematics and uses new parameters. But there are good physical reasons to take reacceleration into consideration since particles which scatter on turbulent magnetic field irregularities should naturally encounter acceleration via the 2nd order Fermi type of acceleration. I will address this aspect and will discuss and explain the consequences on the interpretation of the data.

In order to meet the objectives described above I will discuss the basic features of the LBM and the DHM and will illustrate in direct comparison the differences between them in terms of their physical framework and their results on the data. The reader is also refered to books and reviews (Gaisser, 1990; Longair, 1981; Cesarsky, 1987 and Webber, 1997). In this article I will first exclude all energy changing processes in order to make the mathematics simpler. In the last chapter I will then allow for reacceleration and energy losses.

2   Leaky-Box Model

In the LBM the particles propagate freely in the containment volume and the productions and losses of particles are balanced in time, thus the mathematical description of the LBM is given by a continuity equation. By ignoring energy changing processes and radioactive particles the equation becomes the following form:

$$N_i(E) \left\{ \frac{1}{\tau_{\mathrm{esc}}(E)} + \frac{1}{\tau_{\mathrm{int}}(E)} \right\} = {}_iQ_{\mathrm{prim}}(E) + \sum_{k>i} \frac{N_k(E)}{\tau_{\mathrm{int}}^{k \to i}(E)} \tag{1}$$

where $N_i(E)\,[\mathrm{cm}^{-3}\,\mathrm{GeV}^{-1}]$ and $N_k(E)\,[\mathrm{cm}^{-3}\,\mathrm{GeV}^{-1}]$ stands for the number densities of different types of nuclei of kinetic energy E. The left side of this equation (1) accounts for the losses of i-type nuclei and the right side for the sources. One can have primary sources, ${}_iQ_{\mathrm{prim}}\,[\mathrm{cm}^{-3}\,\mathrm{s}^{-1}\,\mathrm{GeV}^{-1}]$ and secondary sources. The secondary sources originate from spallation of k-type nuclei heavier than the i-type nuclei. The quantity $\tau_{\mathrm{int}}^{k \to i}(E)$ means the mean time which a k-type nuclei needs to produce an i-type secondary in the interstellar gas. This quantity depends on the production cross section and the interstellar gas in terms of density and composition. As one can see a huge number of nuclear cross sections are involved and not all are so well known. For more details on these cross sections I refer to the literature (Webber, Kish and Schrier, 1990; Isbert et al., 1993; Heinrich, W. et al. 1995). In the brackets on the left side one finds expressions for the losses and $\tau_{\mathrm{int}}(E)$ stands for the mean lifetime of i-type particles against interactions in the interstellar gas. Here again nuclear cross sections are involved. The quantity $\tau_{\mathrm{esc}}(E)$ stands for the mean escape time from the confinement volume, often called the age of cosmic rays. On a statistical basis, however, it is an exponential distribution which governs the escape of an individual particle and $\tau_{\mathrm{esc}}(E)$ is the mean of it. The exponential distribution results since the probability for a particle to escape from the box in time dt is given by $dt/\tau_{\mathrm{esc}}(E)$. The escape time $\tau_{\mathrm{esc}}(E)$ is an important astrophysical parameter since within this time the CR population in the containment volume has to be replenished by the cosmic ray sources, thus it sets the power of the cosmic ray sources.

For practical purpose one has to rearrange eq. (1) since one measures fluxes $I\,[\mathrm{cm}^{-2}\,\mathrm{s}^{-1}\,\mathrm{GeV}^{-1}\,\mathrm{ster}^{-1}]$ and not densities $N\,[\mathrm{cm}^{-3}\,\mathrm{GeV}^{-1}]$. This relation can be easily found since we assume isotropy in the cosmic radiation:

$$N(E) = \frac{4\pi}{v} I(E) \tag{2}$$

where v is the velocity of the particle.

In order to be able to calculate s/p ratios one has further to replace the characteristic times $\tau_{\mathrm{esc}}(E)$, $\tau_{\mathrm{int}}(E)$, $\tau_{\mathrm{int}}^{k \to i}(E)$ by lambdas which characterize the matter traversed in $[\mathrm{g/cm}^2]$. These relations are given by:

$$\lambda_{\mathrm{esc}}(E) = \langle m \rangle \cdot n\,(\mathrm{cm}^{-3}) \cdot c \cdot \beta \cdot \tau_{\mathrm{esc}}(E) \tag{3}$$

$$\lambda_{\mathrm{int}}(E) = \langle m \rangle \cdot n\,(\mathrm{cm}^{-3}) \cdot c \cdot \beta \cdot \tau_{\mathrm{int}}(E) \tag{4}$$

$$\lambda_{\mathrm{int}}^{k \to i}(E) = \langle m \rangle \cdot n\,(\mathrm{cm}^{-3}) \cdot c \cdot \beta \cdot \tau_{\mathrm{int}}^{k \to i}(E) \tag{5}$$

where $n\,(\mathrm{cm}^{-3})$ refers to the mean interstellar gas density which the particles penetrate, $\langle m \rangle$ means the mean mass of the gas and $c \cdot \beta$ is the velocity of the particle.

The above equilibrium eq. (1) can be solved by different mathematical techniques and I refer to the literature. I like to note that care has to be taken when energy

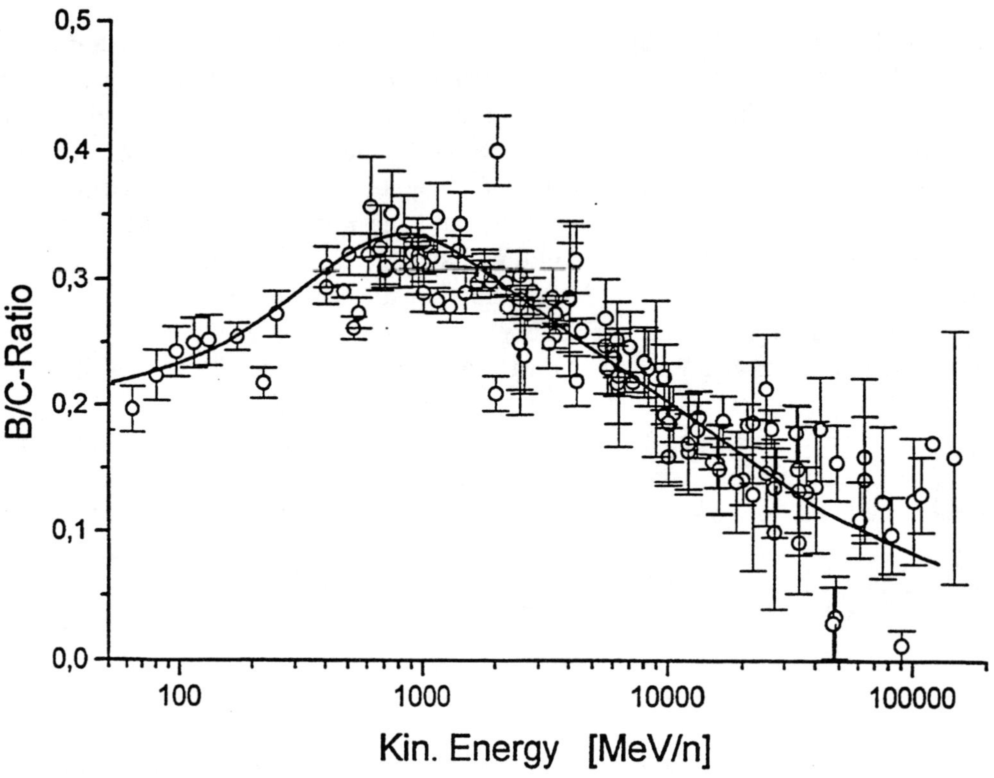

Figure 1: Collection of measured B/C ratios at different energies (Garcia-Munoz et al.,1987). The curve represents a calculated fit to the data.

changing processes are involved (Heinbach & Simon, 1995; Stephens & Streitmatter, 1998; Garcia-Munoz et al, 1987; Gaisser & Schaefer, 1992). By comparing calculated s/p ratios, such as B/C, with the measured ratio one can adjust $\lambda_{esc}(E)$ as the remaining free parameter by a fitting process to the data. Fig. 1 shows such a fit to a compilation of measured B/C ratios. Such calculations have been done by many authors and in Fig. 2 I show some of the recently published curves on the rigidity dependence of $\lambda_{esc}(E)$. As can be seen from this Fig. 2 the particles around some GeV/nucleon traverse roughly $10\,\mathrm{g/cm^2}$ interstellar matter before they escape. Although there are quite remarkable differences between these curves, particular at low energies, which may be partly due to cross section uncertainties, they all agree in their general shape. They peak around some GeV/nucleon and fall off to higher energies as well as to lower energies. Note that $\lambda_{esc}(E)$ is used as a free parameter and this characteristic shape is not based on physics. In the LBM it results ad hoc by the fitting procedure.

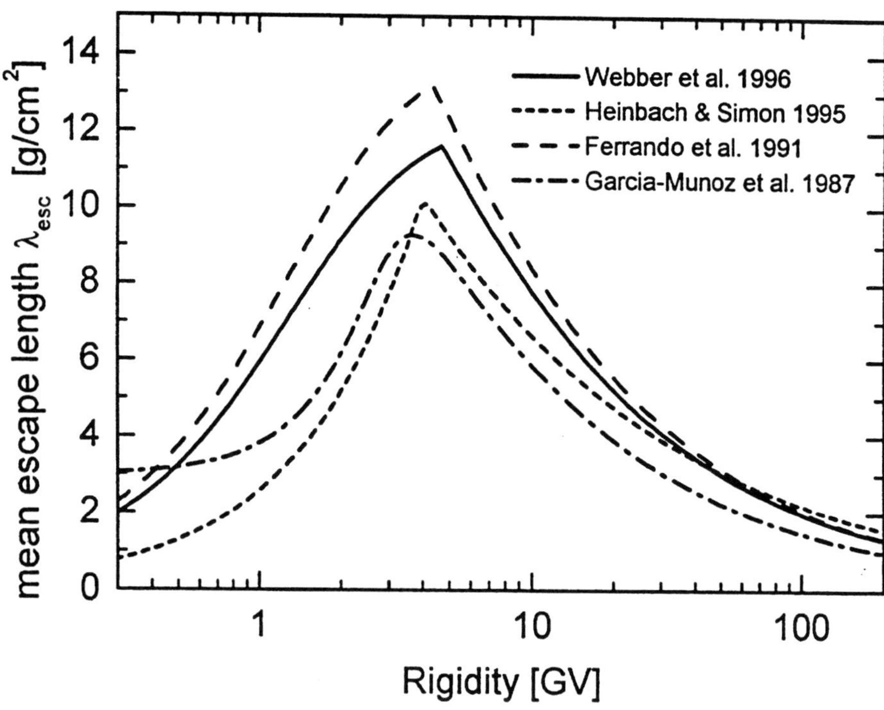

Figure 2: Energy dependence of the mean escape length $\lambda_{esc}(E)$ as published by different authors. This quantity is a free parameter in the LBM calculations. These characteristic shapes result by fitting the calculated s/p ratio to the actual data (see Fig. 1).

In the LBM one uses these derived $\lambda_{esc}(E)$ values to obtain the mean age or escape time $\tau_{esc}(E)$ of cosmic rays by making use of eq. (3). One then readily runs into the problem of not knowing what mean gas density $n\,(\text{cm}^{-3})$ one should use. The interstellar gas density of 1H-atom/cm^3, which we find in the thin galactic disk is probably too high since we know that the particles occupy a larger space beyond the disk. A test particle actually is needed which is not only sensitive to the total matter traversed in units of g/cm^2 but also to the gas density through which it traverses. Secondary radioactive isotopes are such test particles since secondary radioactive isotopes are not only produced along the mean path of $\lambda_{esc}(E)$, they can also decay on their journey, and the number of actually survived isotopes depends on the gas density around them. In a high density gas more isotopes survive than in a thin density gas. Good candidates are those nuclei with decay times τ_d comparable to the expected escape time $\tau_{esc}(E)$ such as:

$$\begin{array}{lll}
\text{Be-10} & \tau_d = 2.3 \cdot 10^6 & \text{years} \\
\text{Al-26} & \tau_d = 1.0 \cdot 10^6 & \text{years} \\
\text{Cl-36} & \tau_d = 4.5 \cdot 10^5 & \text{years} \\
\text{Mn-53} & \tau_d = 5.4 \cdot 10^5 & \text{years}
\end{array}$$

Thus, by comparing the measured surviving fraction of Be-10 or the Be-10/Be-9 ratio (Be-9 is a stable isotope) with the calculated one the mean gas density $n\,(\text{cm}^{-3})$ can be derived. For such a calculation one has to add a further loss term $1/\gamma\,_i\tau_d$ to the bracket of the left side of eq. (1). The quantity γ means the Lorentz factor and $_i\tau_d$ is the decay time seen in the rest frame of the radioactive nuclei. In Fig. 3 I show the result of such a calculation and compare it with the measured Be-10/Be-9 ratio, see also Lukasiak et al. (1994). The fit to the data, shown here, allows a mean gas density of $n = 0.2\,(\text{cm}^{-3})$. A higher density would lift this calculated curve up and a lower density would lower it down. This small gas density is interpreted such that the particles indeed reside in a volume which is beyond the size of the galactic disk. But remember in the LBM one assumes that radioactive Be-10 is capable to probe the full size of the volume in which all particles reside including the stable nuclei. This issue becomes important when we discuss this situation in the DHM. The rise of the calculated Be-10/Be-9 ratio with energy as shown in Fig. 3 reflects the relativistic time dilatation. The faster Be-10 isotopes are the longer they life and the more they survive.

Taking this gas density of $n = 0.2\,(\text{cm}^{-3})$ with a mean mass of $\langle m \rangle = 2 \cdot 10^{-24}\,\text{g}$ and assuming the escape length to be $\lambda_{esc}(E) = 10\,\text{g/cm}^2$ then the cosmic rays around some GeV/nucleon will have an escape time of $\tau_{esc} = 2 \cdot 10^7$ years.

The net result is that we learn something about the mean time, within which all cosmic rays have to be replenished in the Leaky Box but we do not learn a great deal about the actual size of the hypothetic box neither something about the mechanism which controls the escape from the box. This will become different in the DHM which will be the subject in the next chapter. But it is worthwhile mentioning that the LBM despite its unrealistic physical picture does a very good job in explaining the observation of stable cosmic ray nuclei. This applies to the various s/p ratios of heavy nuclei but also to the $\bar{p}/p$ ratio (Gaisser & Schaefer, 1992; Webber & Potgieter,

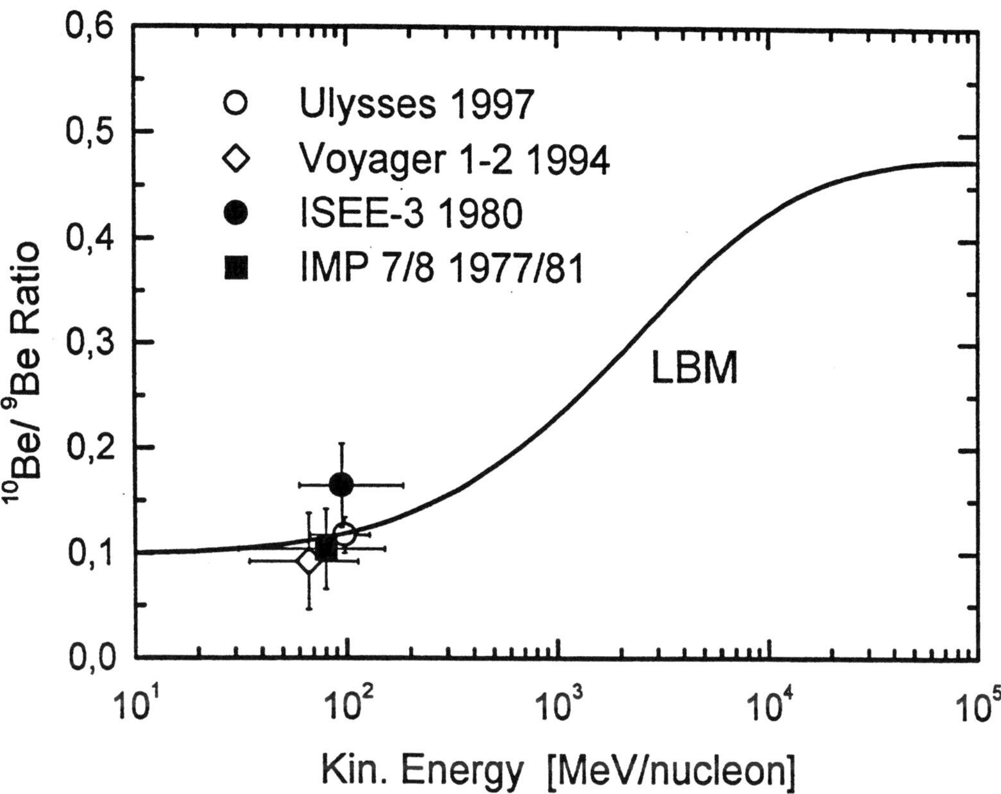

Figure 3: Comparison between the calculated Be-10/Be9 ratio and data. This calculation was performed in the framework of the LBM. The fit to the data provide information on the mean gas density through which the CR particles propagate. This curve implies a mean gas density of 0.2 H-atoms/cm^3. The data refer to the spacecrafts with which these measurements were made, IMP 7/8 (Garcia-Munoz et al., 1977, 1981), ISEE-3 (Wiedenbeck & Greiner, 1980), Voyager 1-2 (Lukasiak et al., 1994) and Ulysses (Connell, 1997).

1989; Simon, Molnar & Roesler, 1998), to the He-3/He-4 ratio (Reimer et al., 1997; Beatty et al., 1993; Wefel et al., 1995) to the $e^+/(e^+ + e^-)$ ratio (Moskalenko and Strong, 1998; Protheroe, 1982) and also to the spectra (Heinbach & Simon, 1995; Engelmann et al., 1990).

In the next chapter I will show that this does not come as a surprise since the DHM, which has the more physical concept, will provide the same results for stable nuclei although the physical interpretation is very different.

3   The Diffusion Halo Model

Fig. 4 sketches the physical picture of the DHM. The shaded area illustrates the thin galactic disk of height h and the quantity H stands for the height of the halo. It is assumed that the cosmic ray sources are placed in the thin galactic disk, where most of the interstellar gas is located but the cosmic rays themselves diffuse out and may spend appreciable portions of their lifetime in the halo. A three dimensional diffusion equation would be the proper approach to the problem. It could cover physical details in our galaxy such as the spatial gas distribution of atomic and molecular hydrogen, could link the cosmic ray sources to the distribution of supernova remnants (SNR), and could also add aspects such as galactic winds and convection (Strong & Moskalenko, 1998; Ptuskin et al., 1997; Jones, 1979). This readily illustrates that the DHM accounts for much more physical detail than the LBM.

For our discussion here in this chapter it is sufficient to make the picture somehow simpler. I will allow that the cosmic ray sources and the interstellar gas is homogenously distributed throughout the thin galactic disk and will ignore convection and energy changing processes. This provides symmetry and one can describe this scenario with a one dimensional diffusion equation:

$$\frac{\partial N_i(z,t)}{\partial t} = \frac{\partial}{\partial z}\left\{ D(z)\frac{\partial}{\partial z}N_i(z,t)\right\} - N_i(z,t)\left\{\frac{1}{_i\tau_{\mathrm{int}}(E)} + \frac{1}{\gamma_i\tau_d}\right\} +$$

$$+ _iQ_{\mathrm{prim}}(z) + \sum_{k>i}\frac{N_k(z,t)}{\tau_{\mathrm{int}}^{k\to i}} \tag{6}$$

Various quantities and terms can be found. $N_i(z,t)$ and $N_k(z,t)$ describe the density of i-type and k-type particles at position z at time t with k heavier than i. The first term of the right side describes the diffusion and $D(z)$ means the diffusion coefficient at position z. For simplicity I allow D to be independent of position. The second bracket on the right side of eq. (6) accounts for the losses of i-type particles similar to those quantities described in the last chapter. The first term in the bracket stands for the losses against interactions and the second term for losses due to radioactive decay. In the last two terms one finds the sources for the i-type particles. One can have primary sources $_iQ_{\mathrm{prim}}(z)$ as well as secondary sources by spallation of k-type nuclei expressed by the last term on the right side of eq. (6). Various interesting

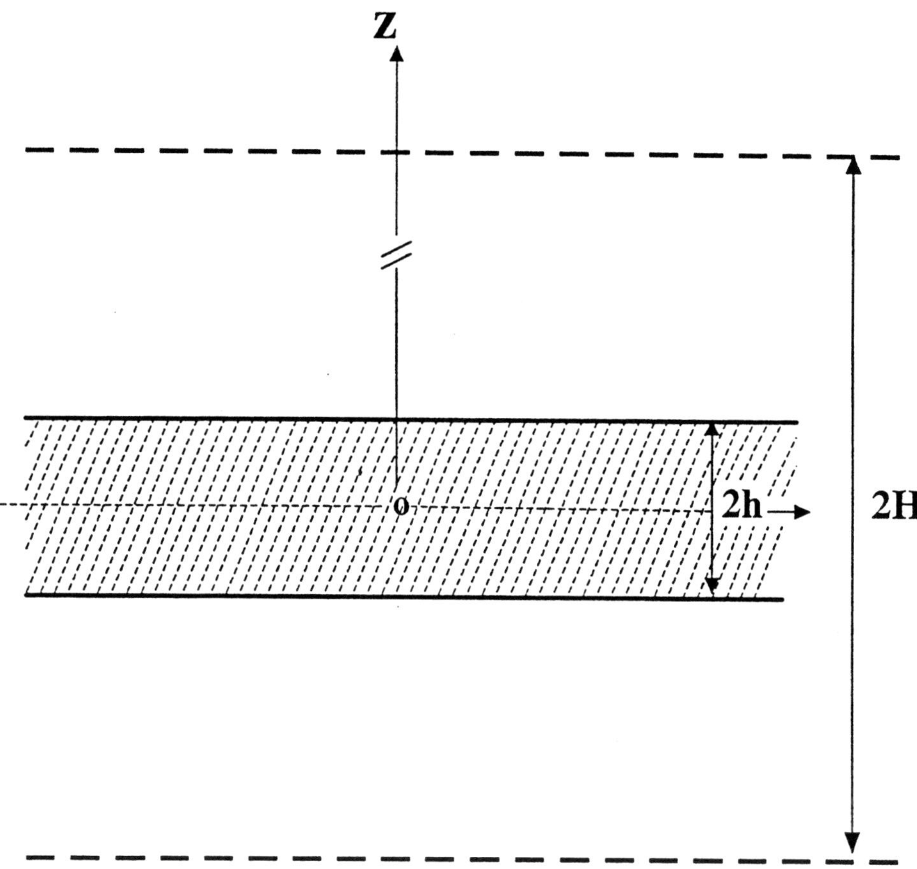

Figure 4: Schematic description of the physical concept of the DHM.

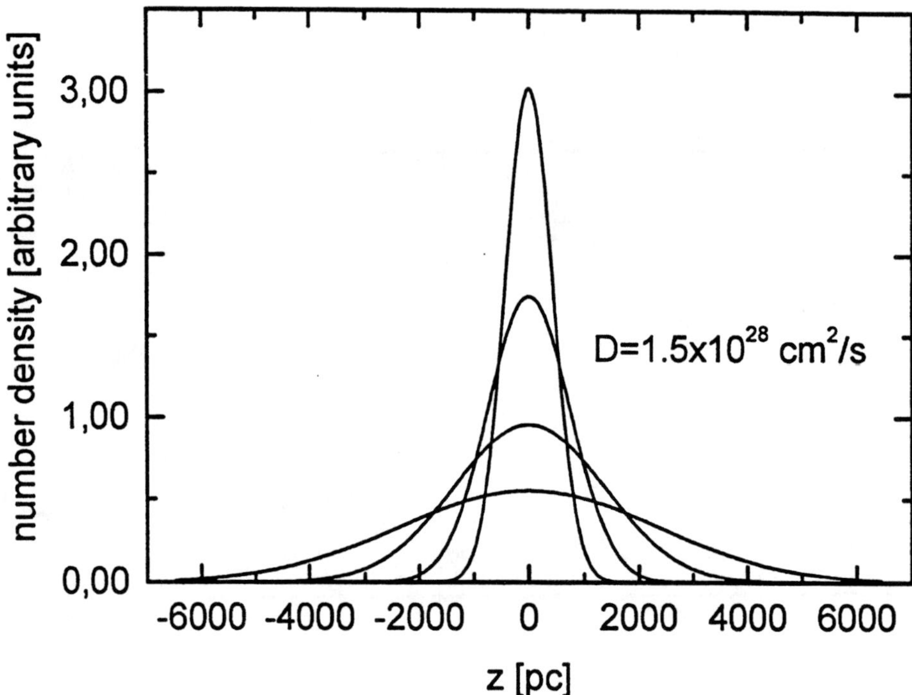

Figure 5: Particles which are injected at the center line of the galactic disk ($z = 0$) diffuse out into the halo according to the diffusion coefficient D. Each curve represents for different elapsed times the probability of finding the particle at position z. From top to bottom the elapsed times are: 10^6 years, $3 \cdot 10^6$ years, 10^7 years and $3 \cdot 10^7$ years.

results can be obtained by solving eq. (6) under different sets of parameters and boundary conditions.

I will begin with Fig. 5. These curves solely illustrate the diffusion of particles out of the galactic disk into the halo as a function of time after being injected at $t = 0$ and $z = 0$ in a single shot. Losses and secondary productions as expressed on the right side of eq. (6) were ignored. The diffusion coefficient was chosen to $D = 1.5 \cdot 10^{28}$ cm^2/s.

These curves are represented by the following analytic formula:

$$N(z, t) = \frac{1}{2(\pi \cdot D \cdot t)^{1/2}} \cdot \exp\left\{-\frac{z^2}{4Dt}\right\} \quad . \tag{7}$$

They allow to derive the mean square of the displacement which the particles encounter on a statistical basis by diffusing away from the center line of the galactic plane into the halo as a function of time. This is given by

$$\langle z^2 \rangle = \int_{-\infty}^{\infty} z^2 N(z, t) \mathrm{d}z = 2Dt \quad . \tag{8}$$

If one interprets $\sqrt{\langle z^2 \rangle}$ with the halo size H one obtains the mean time $t_{\rm esc}$ which a particle needs to escape freely into the intergalactic space:

$$t_{\rm esc} = \frac{H^2}{2D} \ .$$
(9)

A three dimensional calculation would make it to $t_{\rm esc} = 2H^2/3D$.

As can be seen in order to calculate such an escape time in the framework of the DHM one needs information on the halo size H and the diffusion coefficient. This information can be obtained directly from the data and I will show how. Fig. 6 illustrates the equilibrium state for two stable particles for the primary carbon and the secondary boron. These curves were obtained by solving eq. (6) numerically in an iterative procedure until equilibrium is reached. The solution comes in two steps. One first fixes the equilibrium state for the primary carbon such that it agrees with the observation and then calculates the boron which follows as the secondaries from carbon. In this calculation we allowed a mean gas density n_d of 1 H-atom/cm^3 in the thin galactic disk and in the halo we set it to zero. In the insert of Fig. 6 we give the numbers for H and D which were used in this calculation. As can be seen in Fig. 6 the B/C ratio is position dependent. It is smaller in the halo than in the disk. This is very different to the LB situation where everything in the box is the same everywhere.

By inspecting calculated B/C ratios under different combinations of H and D one realizes that these ratios only depend on the D/H ratio. A smaller diffusion coefficient D can be compensated by a smaller halo size H, only the ratio counts. One can understand this in the following way. The calculated carbon spectrum in the disk is fixed to the measured one, therefore, independent of D and H. D and H only effects the gradient of carbon into the halo but not the flux within the disk. The secondary boron is only produced in the disk where the interstellar gas is and proportional to the fixed carbon flux in that region. The amount of boron which, however, remains in the galactic disk depends on the velocity under which they stream away into the halo and this streaming velocity depends on the ratio of D/H.

Thus a measurement of B/C or any other stable s/p ratio can only settle a ratio of these two parameters. In order to disentangle this correlation one needs another kind of measurement which also responses to these two parameters but in different way. The solution comes from the secondary radioactive nuclei such as Be-10. This works as follows.

Secondary radioactive nuclei are produced in the galactic disk where the gas is concentrated. These particles diffuse into the halo according to their diffusion coefficient $D(E)$. On their way out, however, they decay and the shapes of their equilibrium states reach a form which is illustrated in Fig. 7. The three inner curves refer to Be-10 at different energies and the outer curve to the stable Be-9 isotope. In this calculation we kept the halo size fixed, i.e. $H = 7\,{\rm kpc}$, and the diffusion coefficients which correspond to the different Be-10 energies are given in the insert of this Fig. 7. We finally normalized these curves to one at the center line of the galactic disk. Various things can be seen.

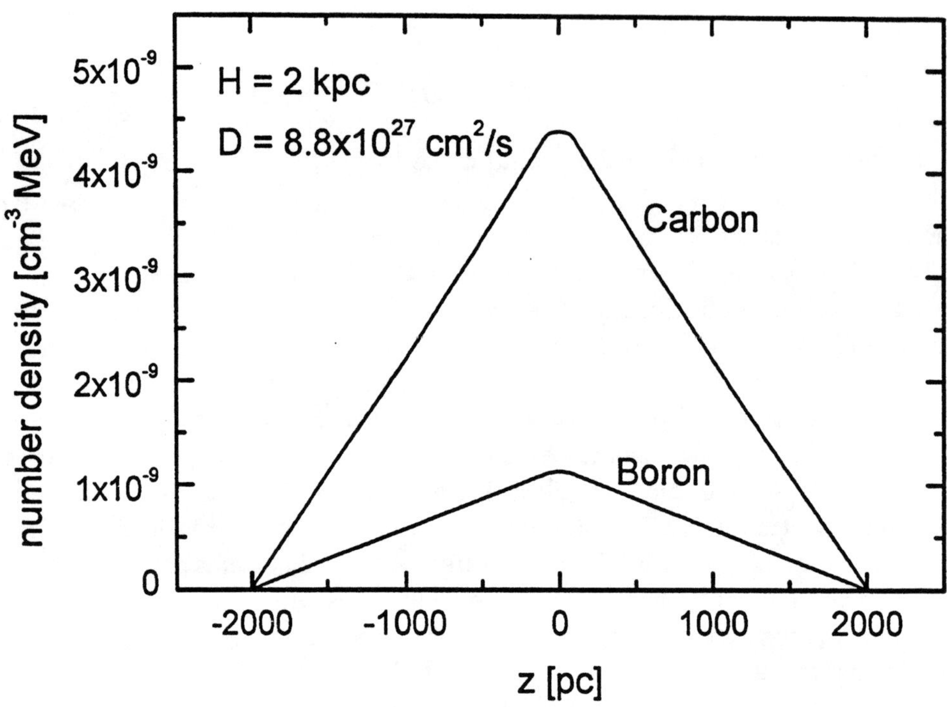

Figure 6: Calculated equilibrium states in the framework of the DHM for the primary particle carbon and the secondary boron. For a given halo size H (in this example H was fixed to 2 kpc) the diffusion coefficient D remains as a free parameter which can be fixed by adjusting the calculated B/C ratio at $z = 0$ (where the measurements are made) to the data. Note that s/p ratios can only settle the ratio of D/H.

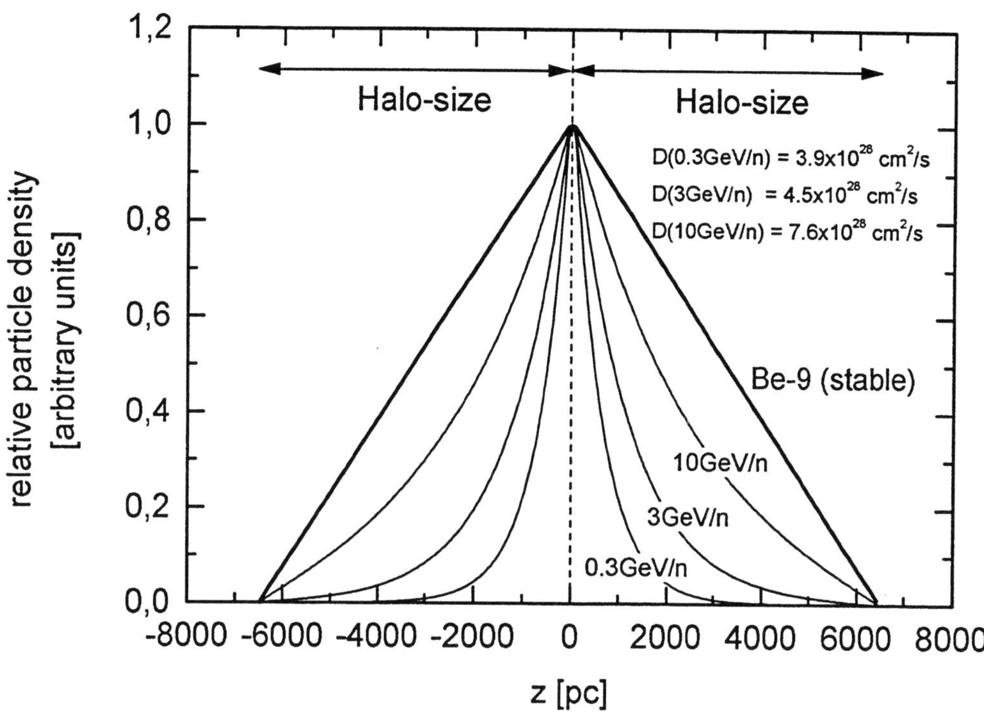

Figure 7: Equilibrium states for the radioactive Be-10 isotope at different energies (different diffusion coefficients) and for the stable Be-9. All curves are normalized to one at $z = 0$.

- The equilibrium states for radioactive nuclei are different from those of stable nuclei.

- In this normalized presentation the curve which represents the stable Be-9 isotope remains unchanged for different diffusion coefficients.

- Due to the small diffusion coefficient and the short lifetime the low energy Be-10 isotopes preferentially decay and only a small portion may reach the outer edge of the halo.

- Due to the last point it becomes clear that the inner curves which represent the radioactive Be-10 will change with energy, i.e., with the diffusion coefficient.

As a consequence of these points one realizes that the radioactive Be-10 isotope indeed responds differently to combinations of H and D than the stable Be-9. Thus a fit to a s/p ratio of stable nuclei and simultaneously to the measured Be-10/Be-9 ratio will allow to determine H and D.

This comparison is shown in Fig. 8. There are actually two curves shown. The upper one refers to the Be-10/Be-9 ratio as calculated in the LBM (see Fig. 3) and the lower one shows the results as obtained from the DHM. A number of interesting and important conclusions can be drawn from this Fig. 8.

1. The LBM and the DHM provide different dependences on the Be-10/Be-9 ratio or on the surviving fraction of Be-10. These quantities are model dependent and a measurement of these quantities over a larger energy range can actually tell what model describes the cosmic ray propagation. Such a measurement is currently under way with the new ISOMAX experiment which will extend these measurements up to about $5\,\mathrm{GeV/nucleon}$ (Streitmatter, 1993).

2. From Fig. 8 one also derives a diffusion coefficient of $D = 1.5 \cdot 10^{28}\,\mathrm{cm^2/s}$. With this value one fits the data. Combining this result on D with the D/H ratio as obtained by fitting the B/C ratio one fixes $H = 3.5\,\mathrm{kpc}$. With these two numbers one calculates the mean time which the particles spend in the volume which is bounded by the full size of the halo, see eq. (9)

$$t_{\mathrm{esc}} = \frac{H^2}{2D} = 1.2 \cdot 10^8 \text{ years} \quad . \tag{10}$$

Thus this time is 10 times longer than that deduced in the LBM and one can give the following explanation for this difference. On the basis of the physical picture of the DHM the Be-10 isotopes cannot probe the full halo volume which the stable particles pervade because they decay on their way out from the galactic disk where they are physically produced. Fig. 7 illustrates this explicitly. Thus any attempt to deduce the mean gas density from the Be-10 isotopes, as one does in the LB approach, will overestimate this number. The true mean gas density which the stable particles encounter is much smaller since they fill a larger volume. That is the reason that one underestimates the escape time in the LB approximation.

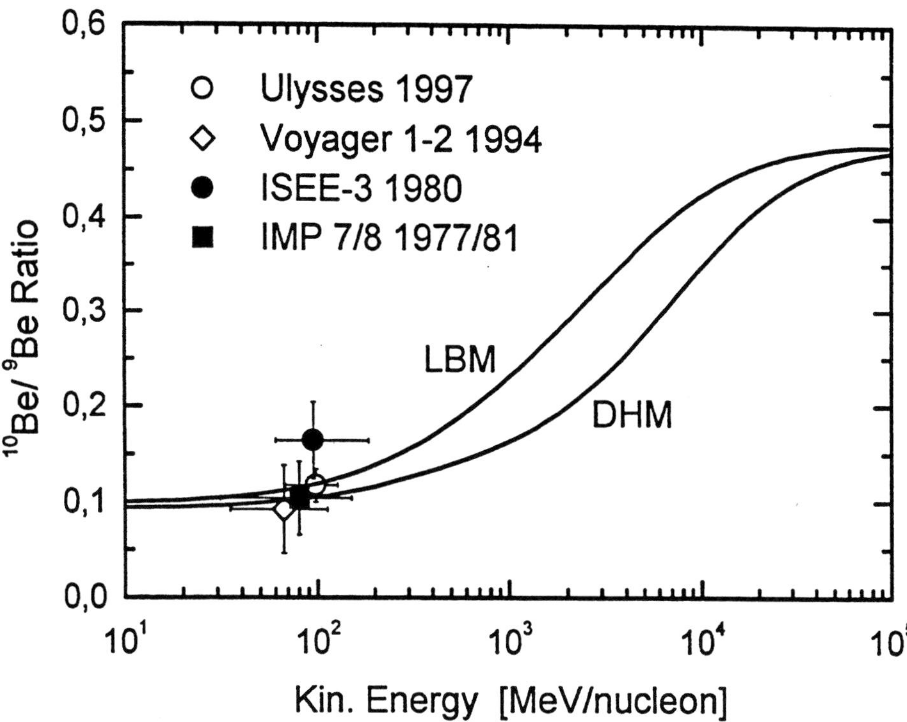

Figure 8: The two curves show calculated Be-10/Be-9 ratios in the framework of the two propagation models. Both calculations fit the data at low energies. One realizes that the calculated curves are model dependent.

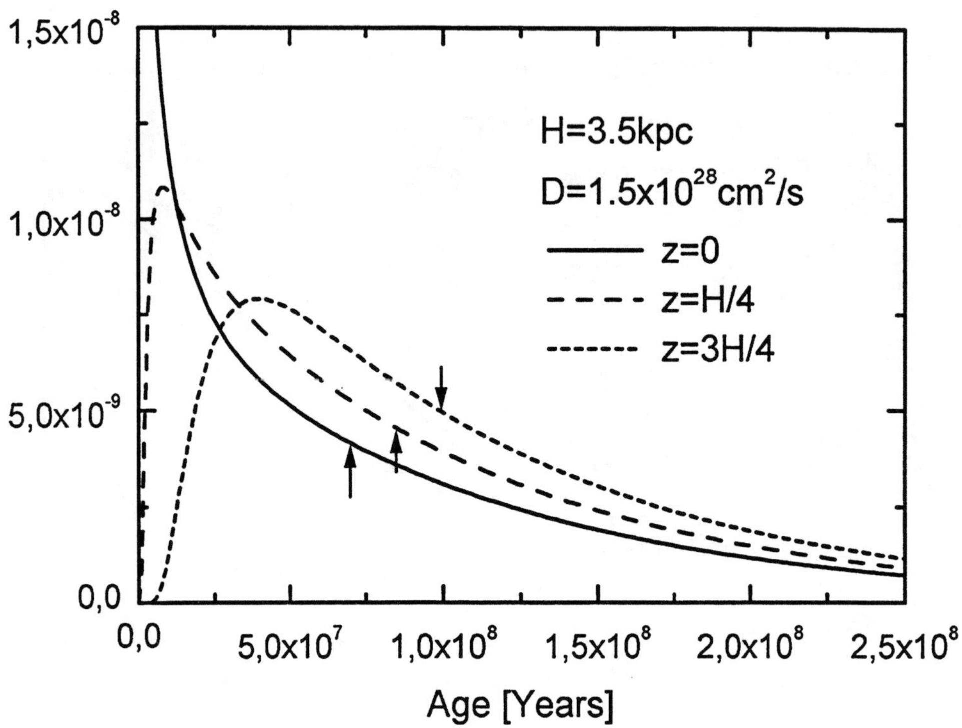

Figure 9: Age distributions of cosmic ray particles at different z-positions. In the galactic plan ($z = 0$) the particles are younger than further out in the halo. These results are very different to the situation in the LBM which deals with one exponential distribution everywhere. The arrows mark the mean age of the individual distribution.

I like to address another important difference between the LBM and the DHM and that is the age distribution. In the LBM all particles everywhere within the containment volume encounter the same exponential age distribution. In the DHM there does not exist such a single distribution. The age distributions are neither exponential nor everywhere the same. Close to the galactic disk the particles are younger and further out in the halo they become older. This is illustrated in Fig. 9. In this figure one finds for a given set of parameters ($H = 3.5\,\mathrm{kpc}$, $D = 1.5 \cdot 10^{28}\,\mathrm{cm}^2/\mathrm{s}$) calculated normalized age-distributions at different z-positions. It becomes obvious that the young particles are located closer to the galactic center than further out in the halo. The arrows mark the mean age of these distributions.

Despite these various differences in the physical concepts between the LBM and the DHM it is true, if a good fit to the data is obtained in the DHM with parameters h, H, n_d and $D(E)$, the calculation in the LBM will also fit the data with an escape length $\lambda_{\mathrm{esc}}(E)$ which satisfies the following relation

$$\lambda_{\mathrm{esc}}(E) = \frac{\langle m \rangle \cdot n_d \cdot h \cdot c \cdot \beta \cdot H}{2D(E)} \tag{11}$$

where n_d stands for the gas density in the disk. This relation can be verified in the following way. The stable nuclei pervade the full halo size with a mean escape time $t_{\rm esc}(E) = \frac{H^2}{2D(E)}$, see eq. (9), and they encounter on their journey a mean gas density of $n = n_d \cdot \frac{h}{H}$ since $H \gg h$. Using these inputs in eq. (3) one obtains eq. (11).

For a much broader discussion on the DHM I like to refer to the book of Berezinskii et al. (1990).

4   Reacceleration

The concept of acceleration of cosmic rays during propagation has its origin dating back to the idea of Fermi (1949). In this picture cosmic ray acceleration occurs gradually by repeated interactions on large scale magnetic turbulences. In later years these ideas were largely supplanted by shock acceleration theories that were developed in the late 1970s (Blanford & Eichler, 1987; Cesarsky, 1980). But models that allow cosmic ray acceleration solely to occur gradually during propagation suffer from the well known problem that they predict an increasing s/p ratio with energy which contradicts the observation.

For this reason a one shot acceleration, probably by shock acceleration mechanism at SNR (Drury, 1983; Axford, 1981), was favoured over the idea of a gradual continuous acceleration. But there are good physical reasons to believe that propagating particles somehow get reenergized during propagation. This can occur, as Fermi suggested, on large scale magnetic turbulences or by the cosmic ray diffusion process itself when the particles scatter on turbulent magnetic field irregularities. Thus an energy changing process is almost inevitable under this physical scenario, the question is just to what amount.

In the last 15 years various authors have attempted to incorporate such an idea in their propagation calculations (Simon et al., 1986; Osborne & Ptuskin, 1987; Seo & Ptuskin, 1994; Heinbach & Simon, 1995; Letaw et al., 1984). In all these attempts one allowed that the cosmic rays are preaccelerated at their outsets in a strong single shot but one allows that they get moderately reenergized during propagation and it could be shown that reacceleration is compatible with the data. Even more, it does not only describe the data, it leads to very interesting conclusions. In the following I will present and discuss those and will do the calculation in the framework of the LBM.

If one allows for reacceleration one has to add further terms to eq. (1) (Ginzburg & Syrovatskii, 1964):

$$Q_i(E) - \frac{N_i(E)}{\lambda_{\rm esc}^{\rm reac}(E)} - \frac{N_i(E)}{\lambda_{\rm int}^i(E)} + \sum_{k>i} \frac{N_k(E)}{\lambda_{\rm int}^{ki}(E)} \tag{12}$$

$$-\frac{\partial}{\partial E}\left[\left(\left\langle\frac{\partial E}{\partial x}\right\rangle_i^{\rm ion} + \left\langle\frac{\partial E}{\partial x}\right\rangle^{\rm reac}\right) N_i(E)\right] + \frac{1}{2}\frac{\partial^2}{\partial E^2}\left[\left\langle\frac{(\Delta E)^2}{\Delta x}\right\rangle^{\rm reac} N_i(E)\right] = 0 \;,$$

where $\lambda_{\rm esc}^{\rm reac}(E)$ means the mean escape length in the presence of reacceleration and $\langle\partial E/\partial x\rangle_i^{\rm ion}$ is the average energy loss (Salamon, 1990) of particle species i

per unit path length due to the ionization and excitation of interstellar gas atoms. The term $\langle \partial E/\partial x \rangle^{\text{reac}}$ represents the average energy gain per unit path length due to reacceleration and $\langle (\Delta E)^2/\Delta x \rangle^{\text{reac}}$ represents the diffusion coefficient in energy space and accounts for the fluctuation in the reacceleration process. The remaining terms are explained above.

In a paper by Heinbach & Simon (1995) it is shown that a solution of eq. (12) can be found which fits the s/p ratios and the spectra under the following conditions:

$$\left\langle \frac{\mathrm{d}E}{\mathrm{d}x} \right\rangle^{\text{reac}} = 0.6 \cdot E_{\text{tot}} \cdot R^{-1/3}(MV) \left(\frac{\mathrm{MeV/n}}{\mathrm{g/cm}^2} \right) \tag{13}$$

$$\left\langle \frac{\Delta E^2}{\mathrm{d}x} \right\rangle^{\text{reac}} = 0.6 \cdot E_{\text{tot}}^2 \cdot \beta^2 \cdot R^{-1/3}(MV) \left(\frac{(\mathrm{MeV/n})^2}{\mathrm{g/cm}^2} \right) \tag{14}$$

$$\lambda_{\text{esc}}^{\text{reac}}(E) = 86^* \cdot R^{-1/3}(MV)(\mathrm{g/cm}^2) \tag{15}$$

*Remark: in the original paper a curve for $\lambda_{\text{esc}}^{\text{reac}}(E)$ and a formula, $\lambda_{\text{esc}}^{\text{reac}}(E) = 103 \cdot R^{-1/3}$ was shown. The curve is correct but the formula carries a computation error. The 103 should be changed to 86.)

These relations on E_{tot}, R and β do not come just arbitrarily. They follow directly from the physical processes which are involved in the diffusive reacceleration process. It is the 2nd order Fermi acceleration process, the rigidity dependence of the diffusion coefficient and the relation between $\lambda_{\text{esc}}^{\text{reac}}(E)$ and $D(R)$, see eq. (11). The value of 0.6 is a free parameter and expresses the strength of reacceleration and was obtained by the fit to the data. For more details I refer to the original paper. But the net result is that one finds a fit to the data with an escape length dependence, $\lambda_{\text{esc}}^{\text{reac}}(E)$, see eq. (15), which stands as a single power law in rigidity over the whole energy range with an exponent of 1/3. Thus very different to those which one obtains without reacceleration, which peak around some GeV/nucleon and then fall off to higher but also to lower energies, see Fig. 2. This decrease to lower energies was always a puzzle. Why should the escape probability increase to lower energies while the diffusion coefficient $D(E)$ is not expected to increase with decreasing energy. There is another problem with these curves in Fig. 2. The decrease to higher energies goes in form of a power law with an exponent close to 0.6, much too high to understand the high degree of anisotropy as found in the cosmic radiation. Reacceleration overcomes both problems. It provides a single power law for $\lambda_{\text{esc}}^{\text{reac}}(R)$ and with an exponent of 1/3 which is much weaker than 0.6. The exponent of 1/3 points to a Kolmogorov spectrum of hydromagnetic waves.

In Fig. 10 I display the amount of energy change which the cosmic ray particle encounter under these reacceleration conditions. The curves in Fig. 10 show the energy distribution of cosmic ray carbon which started with an energy of 1 GeV/nucleon and which penetrated different layers of interstellar material. One clearly sees the diffusive character of this reacceleration process. The more material the particle traverse the more the mean energy increases indicated by the arrows, but the wider this distribution also becomes. As one can see carbon which started with 1 GeV/nucleon

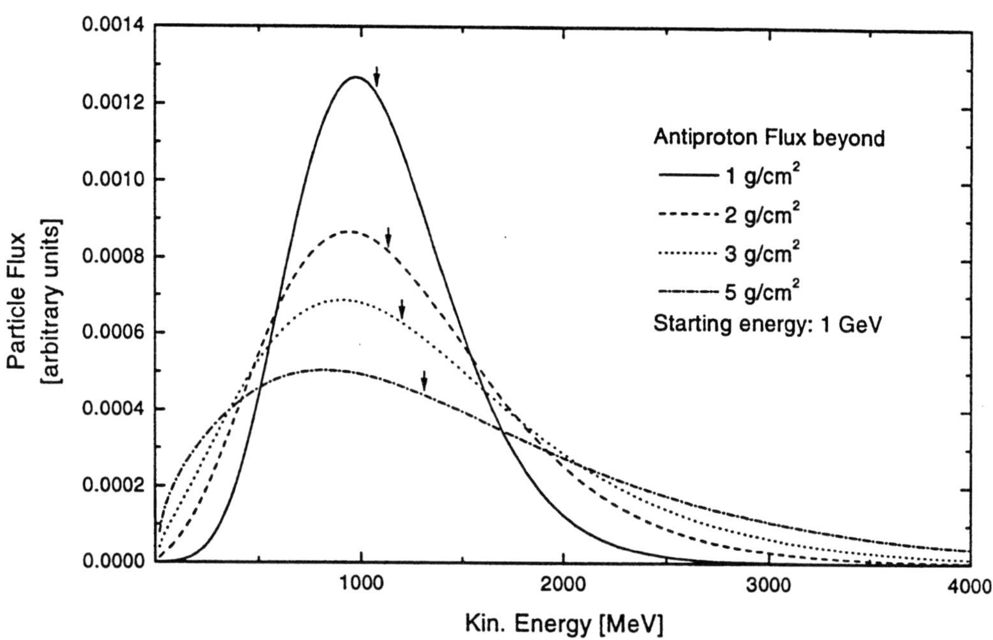

Figure 10: The energy change of cosmic ray carbon particles which start with an initial energy of $1\,\mathrm{GeV}$/nucleon and which propagate through different layers of interstellar material under conditions of diffusive reacceleration (see eq. (13), (14) and (15)). The arrows mark the mean energy of the distribution. One clearly sees the stochastic character of the diffusive reacceleration.

gains energy by 40% after traversing $5\,\mathrm{g/cm^2}$ interstellar matter. This is a moderate energy gain and since the diffusion coefficient increases with energy the relative energy gain at higher energies is even smaller.

In the following I will explain how reacceleration actually effects the calculation so that the data can be fitted with an escape length $\lambda_{\mathrm{esc}}^{\mathrm{reac}}(E)$ so different from that without reacceleration. For that purpose let us rearrange eq. (12) by introducing three new lambdas.

$$\frac{N_s(E)}{\lambda_{\mathrm{ion}}} = \frac{\partial}{\partial E}\left\{\left\langle\frac{\mathrm{d}E}{\mathrm{d}x}\right\rangle_{\mathrm{ion}} \cdot N_s(E)\right\} \tag{16}$$

$$\frac{N_s(E)}{{}_1\lambda_{\mathrm{reac}}(E)} = -\frac{\partial}{\partial E}\left\{\left\langle\frac{\mathrm{d}E}{\mathrm{d}x}\right\rangle^{\mathrm{reac}} \cdot N_s(E)\right\} \tag{17}$$

$$\frac{N_s(E)}{{}_2\lambda_{\mathrm{reac}}(E)} = \frac{1}{2}\frac{\partial^2}{\partial E^2}\left\{\left\langle\frac{\Delta E^2}{\Delta x}\right\rangle^{\mathrm{reac}} \cdot N_s(E)\right\} \tag{18}$$

Let us then consider only one type of primary particle which produces one type of secondary particle. Since the secondary particles have no direct primary source, i.e.

$Q_i^{\text{sec}}(E) = 0$, eq. (12) becomes the following form:

$$\frac{N_s(E)}{\lambda_{\text{eff}}^{\text{reac}}(E)} = \frac{N_p(E)}{\lambda_{\text{int}}^{p \to s}(E)} \tag{19}$$

where $\lambda_{\text{int}}^{p \to s}(E)$ stands for the mean interaction length for the primary particle to produce a secondary nuclei in the interstellar gas. This quantity is given by nuclear physics and has a fixed value almost independent of energy. Thus the energy dependence of the s/p ratio is determined by the energy dependence of $\lambda_{\text{eff}}^{\text{reac}}(E)$. This effective lambda in the presence of reacceleration carries all the other lambdas introduced above

$$\frac{1}{\lambda_{\text{eff}}^{\text{reac}}(E)} = \frac{1}{\lambda_{\text{esc}}^{\text{reac}}(E)} - \frac{1}{{}_1\lambda_{\text{reac}}(E)} - \frac{1}{{}_2\lambda_{\text{reac}}(E)} + \frac{1}{\lambda_{\text{int}}(E)} + \frac{1}{\lambda_{\text{ion}}(E)} \quad . \tag{20}$$

Without reacceleration the two reacceleration terms will be zero. Thus if a good fit to the data is obtained without reacceleration with a mean escape length dependence $\lambda_{\text{esc}}^{NR}(E)$, the model with reacceleration will also fit these data if the new escape length $\lambda_{\text{esc}}^{\text{reac}}(E)$ satisfies

$$\frac{1}{\lambda_{\text{esc}}^{NR}(E)} = \frac{1}{\lambda_{\text{esc}}^{\text{reac}}(E)} - \left\{ \frac{1}{{}_1\lambda_{\text{reac}}(E)} + \frac{1}{{}_2\lambda_{\text{reac}}(E)} \right\} \quad . \tag{21}$$

This relation readily illustrates the difference between these two situations. In the case of no reacceleration $\lambda_{\text{esc}}^{NR}(E)$ alone determines the s/p ratio while in the presence of reacceleration the s/p ratio is determined by the interplay of all three lambdas on the right side of eq. (21). Simon & Molnar (1996) went a step further and calculated these lambdas according to eq. 16, 17 and 18. They used boron as the secondary particle and carbon as the primary. Fig. 11 and Fig. 12 shows the energy dependences of these various lambdas individually and in Fig. 12 one sees the full shape of the energy dependence of $\lambda_{\text{eff}}^{\text{reac}}(E)$, i.e., that quantity which determines the s/p ratio. Various interesting things can be seen:

- At high energies $\lambda_{\text{eff}}^{\text{reac}}(E)$ is solely dominated by the escape length $\lambda_{\text{esc}}^{\text{reac}}(E)$ because all other contributions are small. Thus at high energies the s/p ratio should follow a pure $R^{-1/3}$ dependence as can be seen in Fig. 13 where $\lambda_{\text{eff}}^{\text{reac}}(E)$ indeed flattens to a 1/3 dependence at higher energies.

- In the medium energy range, between 2 GeV/nucleon and about 50 GeV/nucleon, one has to take the reacceleration lambdas into account since their contributions increase to lower energies. Since their contribution enters with a negative sign the effective lambda rises faster with energy than with the power of 1/3 as it should to fit the data. This can also be seen in Fig. 13. This is actually the reason why the data can be fitted while the escape length $\lambda_{\text{esc}}^{\text{reac}}(E)$ still keeps its exponent of 1/3.

- Fig. 13 demonstrates very nicely that reacceleration is able to reproduce the full shape of the s/p ratio which is needed to fit the data, see Fig. 1.

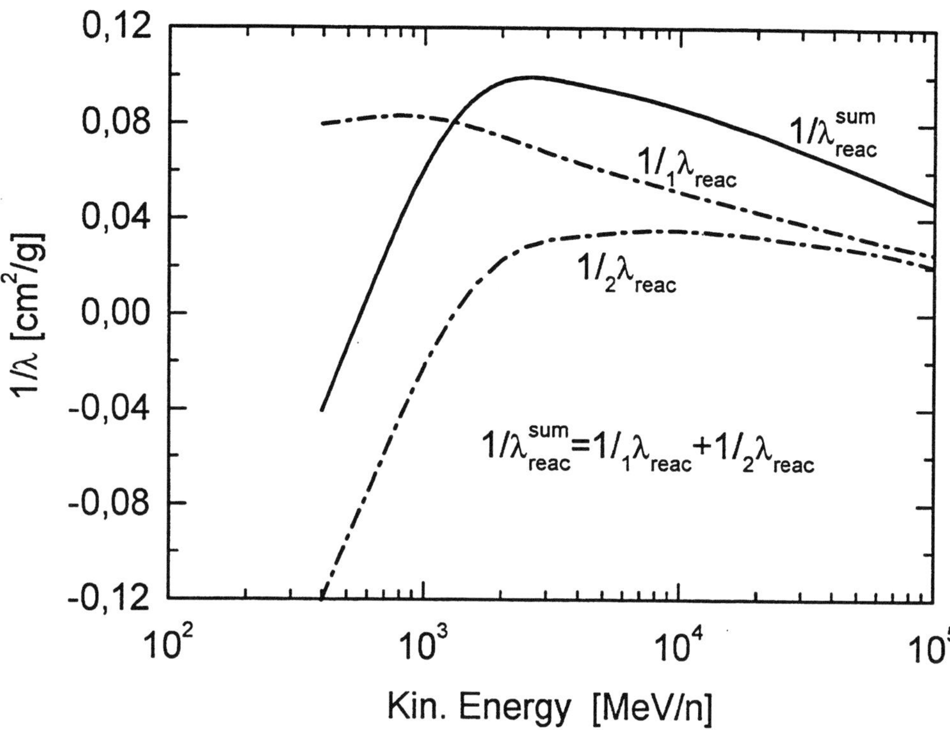

Figure 11: Calculated lambdas which represent the reacceleration process for the secondary boron according to eq. (17) and (18).

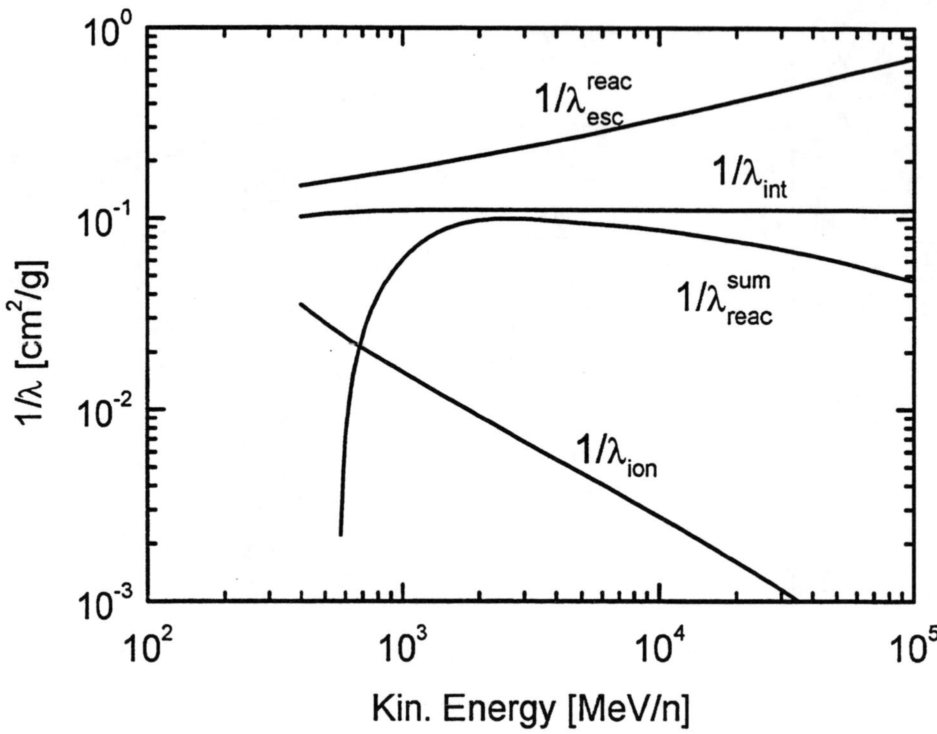

Figure 12: All the relevant lambdas which determine in the presence of reacceleratin the B/C ratio according to eq. (20).

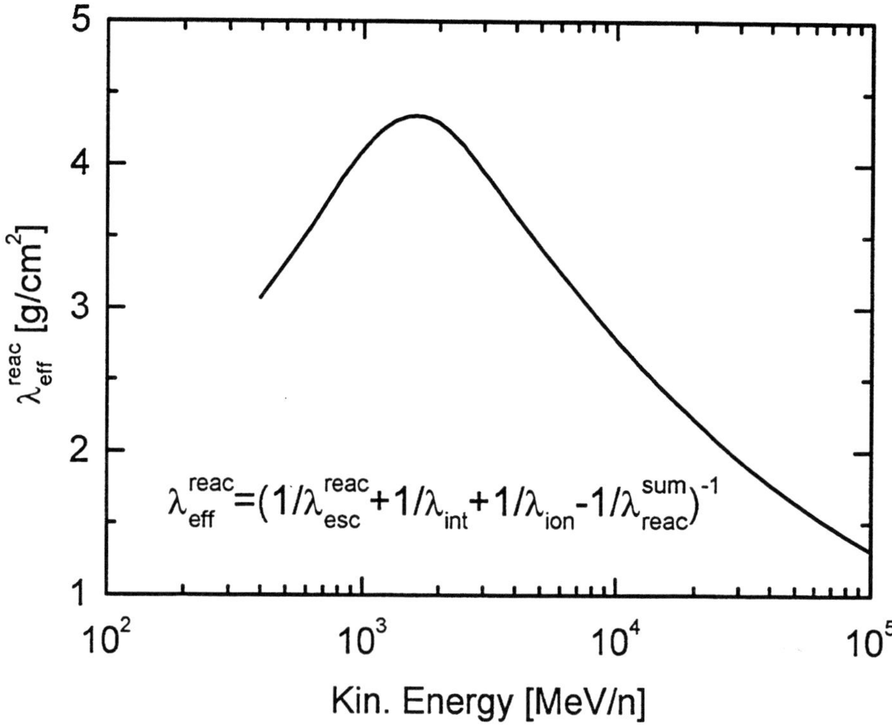

Figure 13: The effective lambda which determines the B/C ratio.

- A physical aspect can also be seen. At high energies one reaches $\lambda_{\text{esc}}^{\text{reac}}(E) \cdot \lambda_{\text{reac}}(E) = 55\,(\text{g/cm}^2)^2$ which reflects the physical interrelation between the escape length (or diffusion coefficient) and the characteristic reacceleration length. These two processes are correlated. When the diffusion coefficient increases at higher energies (or the escape length decreases) the mean reacceleration length increases.

These results show that reacceleration is compatible with the data of stable nuclei, but if reacceleration is at work some things have to be reconsidered. Reacceleration will effect the surviving fraction of secondary radioactive nuclei since the Lorentz factor γ becomes a function of travel time. On the average the Lorentz factor is smaller under condition of reacceleration than without it. This should lead to more decay and thus to a smaller surviving fraction and a smaller Be-10/Be-9 ratio which is actually shown in Fig. 14. The upper curve in Fig. 14 refers to a LB-calculation in which no reacceleration is allowed. But we included ionization losses which explains the difference between this curve and that shown in Fig. 3. The lower curve in Fig. 14 shows the result of a LB-calculation which included the ionization losses as well as reacceleration. For both calculations an interstellar gas density of $n = 0.2$ H-atoms/cm^3 was used. As can be seen, the absolute value and the energy dependence of the Be-10/Be-9 ratio changes in the presence of reacceleration. Reacceleration will also effect the source spectrum which one concludes from the data. According to eq. (19) the following relation combines the primary source spectrum $Q_p(E)$ with the measured equilibrium spectrum $N_p(E) = \lambda_{\text{eff}}(E) \cdot Q_p(E)$. Measurements show that these equilibrium spectra of cosmic ray nuclei follow a single power law in energy with an exponent of about -2.75 at least for energies above a few GeV/nucleon (Engelmann et al., 1990). Thus the source spectrum $Q_p(E)$ can then be derived depending on the energy dependence of $\lambda_{\text{eff}}(E)$. In the case of no reacceleration, ($\lambda_{\text{eff}} = \lambda_{\text{eff}}^{NR}(E) \sim E^{-0.6}$, see Fig. 2) the source spectrum is described by a single power law, i.e., $Q_p(E) \sim E^{-2.15}$. In the presence of reacceleration it is different. The source spectrum has not a single power law shape. At high energies, see Fig. 13 and 12, it is a power law with an exponent of about $-2.4 (\lambda_{\text{eff}} = \lambda_{\text{eff}}^{\text{reac}}(E) \sim E^{-1/3})$, but to lower energies the source spectrum should somehow flatten according to the faster rise of $\lambda_{\text{eff}}^{\text{reac}}(E)$ with decreasing energy, see Fig. 13.

In a recent paper Strong & Moskalenko, 1998, carried this idea of reacceleration further and confirmed that calculations in the framwork of the DHM also lead to the same conclusions as presented in this article. The reader is also referred to a review by Cesarsky (1987) in which further aspects of reacceleration are addressed.

5 Conclusion

In this article I tried to address some of the main issues which researchers are today confronted with on questions concerning the conditions of CR propagation. The currently two competing models, LBM and DHM, are presented and the main characteristic features are explained. It is not yet settled what model describes better the physical reality. By theoretical arguments the DHM should be superior. A direct experimental prove is, however, still open. In the near future we might get better data on the surviving fraction of secondary radioactive isotopes. As I mentioned above the new ISOMAX experiment is aiming for it and is capable to do this kind of measurement up to about 5 GeV/nucleon. At low energies the data should also become greatly improved by the CRIS experiment on the ACE satellite.

As far as the reacceleration is concerned there is also so far no direct convincing experimental evidence for its existence, although it is much more appealing in its conclusions. Even when reacceleration allows to interpret the existing data we have no information how far down in energy and how far up in energy reacceleration might work. There are theoretical arguments that particles can only gain energy when their drift velocity is lower than roughly the Alfven speed v_A in the interstellar medium (Jokipii, 1976; Wentzel, 1974). Otherwise they will generate their own turbulences and will stumble over it.

Thus more experimental observations are needed to check these questions. In my judgement I would see the following measurements as the most promising ones which could help to check on the idea of reacceleration:

- Measurements of s/p ratios to high energies at least to energies above 100 GeV/nucleon with good statistics so that the predicted flattening to a dependence of 1/3 could be seen. There are measurements of the sub-iron/iron ratio

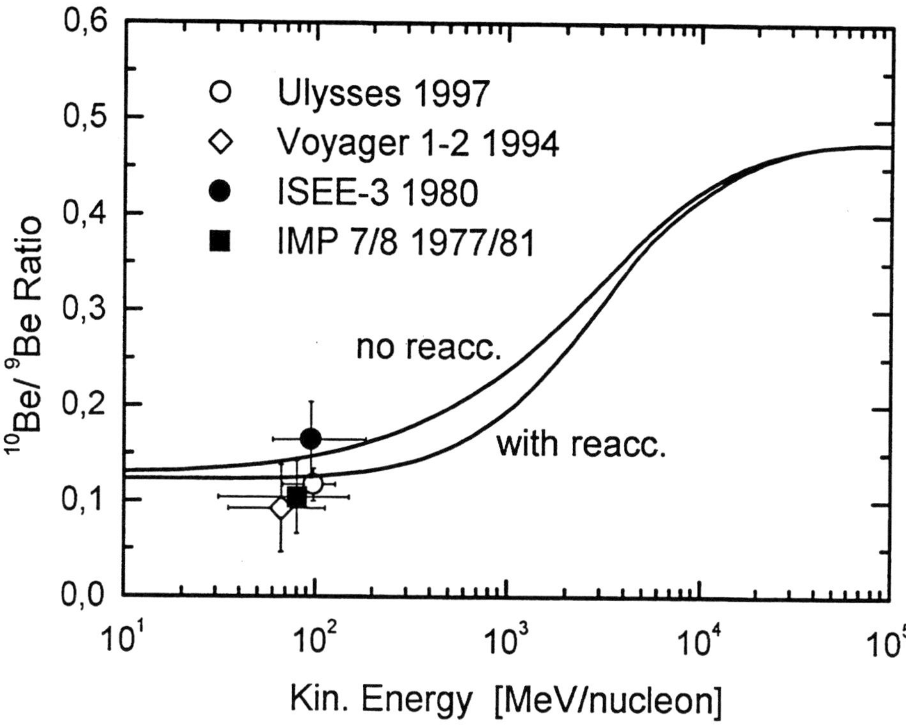

Figure 14: Calculated Be-10/Be-9 ratios in the framework of the LBM. The lower curve includes reacceleration and ionization losses and the upper curve only ionization losses. The data and the references are identical with those shown in Fig. 3.

currently available (Ichimura et al., 1993) but they are not precise enough to give an answer.

- Measuring certain k-capture isotopes, such as V-49, Cr-51 and Fe-55 which are produced entirely as secondaries. The electron attachment probability is strongly energy dependent and can serve as a probe for the energy at which the attachment occurs (Webber, 1997).

- Possible, but probably less sensitive due to solar modulation effects, is also a precise measurement on the low energy antiprotons (Simon & Heinbach, 1996).

The next step of improvements on the understanding of CR propagation should come when more attempts are made to develop models which incorporate more physical details, such as the interstellar gas distribution or the distribution of SNR and which simultaneously reproduces observational data of many kinds, not only particles but also gamma rays and synchrotron radiation. There are attempts along that line (Strong & Moskalenko, 1998).

Acknowledgement

I wish to thank A. Molnar for many useful discussions on this topic and for his help in preparing this article.

References

Axford, W.J. 1981, Proc. 17th ICRC (Paris), 12, 155

Beatty, J.J., et al. 1993, ApJ, 413, 268

Berezinskii, V.S., et al. 1990, Astrophysics of Cosmic Rays (Amsterdam: North Holland)

Blandford, R., & Eichler, D. 1987, Physics Reports, 154, 1

Cesarsky, C.J. 1980, Ann.Rev.Astron.Astrophys., 18, 289

Cesarsky, C.J. 1987, Rapporteur talk, 20th ICRC, 8, 87

Connell, J.J. 1997, 25th ICRC, Vol. 3, 385

Connell, J.J. 1998, ApJ, 501, L59

Dury, L.Oc. 1983, Rep.Prog.Phys. 46, 973

Engelmann, J.J., et al. 1990, A & A, 233, 96

Fermi, E. 1949, Phys.Rev., 75, 1169

Ferrando, P., et al. 1991, A & A, 247, 163

Gaisser, T.K. 1990, Cosmic rays and particle physics, Cambridge University Press

Gaisser, T.K., & Schaeder, R.K. 1992, ApJ, 394, 174

Garcia-Munoz, M., Mason, G.M. & Simpson, J.A. 1977, Ap.J. 217, 859

Garcia-Munoz, M., Simpson, J.A. & Wefel, J.P. 1981, 17th ICRC, 2, 72

Garcia-Munoz, M., et al. 1987, Ap.JS, 64, 269

Ginzburg, V.L., Khazan, Ya.M., & Ptuskin, V.S. 1980, Ap&SS, 68, 295

Heinbach, U., & Simon, M. 1995, ApJ, 441, 209

Heinrich, W., et al. 1995, Radiation Measurement, 25, 203

Ichimura, M. et al. 1949, Phys.Rev. D48

Isbert, J., et al. 1993, 23th ICRC, 2, 167

Jokipii, J.R. 1996, Ap.J., 208, 900

Jones, F.C. 1979, ApJ, 229, 747

Letaw, J.R., Silberberg, R., & Tsao, C.H. 1984, Astrophys.J.Suppl., 56, 369

Longair, M.S. 1981, High energy astrophysics, Cambridge University Press

Lukasiak, A., Ferrando, P., McDonald, F.B., & Webber, W.R. 1994a, ApJ, 423, 426

Moskalenko, J.V., and Strong, A.W. 1998, ApJ, 493, 694

Osborne, Y.L. & Ptuskin, V.S. 1987, 20th ICRC, Moscow, 2, 218

Protheroe, R.J. 1982, ApJ, 254, 391

Ptuskin, V.S., Volk, H.J., Zirakashvili, V.N., & Breitschwerdt, D. 1997, A&A, 321, 434

Reimer, O., et al. 1998, ApJ, 496, 490

Salamon, M.H., et al. 1990, ApJ, 349, 78

Seo, E.S. & Ptuskin, V.S. 1994, Ap.J. 431, 705

Simon, M., & Heinbach, U. 1996, ApJ, 456, 519

Simon, M., Heinrich, W. & Mathis, K.D. 1986, Ap.J., 300, 32

Simon, M. & Molnar, A. 1997, Advances in Space Research, Vol. 19, No. 5, 781

Simon, M., Molnar, A., and Roesler, S. 1998, ApJ, 499, 250

Stephens & Streitmatter 1998, 20. Sept., ApJ, 505

Streitmatter et al. 1993, 23th ICRC, 2, 623

Strong, A.W., & Moskalenko, J.V. 1998, 509

Webber, W.R., & Potgieter, M.S. 1989, ApJ, 344, 779

Webber, W.R., Kish, J.C. & Schrier, D.A. et al. 1990, Phys.Rev.C 41, 520

Webber, W.R., et al. 1990, Phys.Rev.C 41, 553

Webber, W.R., et al. 1996, ApJ, 457, 435

Webber, W.R. 1997, Space Science Reviews, 81, 107

Wefel, J.P., et al. 1995, 24th ICRC, 2 630

Wentzel, D.G. 1974, Ann.Rev.Astron.Astrophys., 12, 71

Wiedenbeck, M.E. & Greiner, D.E. 1980, Ap.J.Lett., 239, L139

Observations of Energetic Particles Upstream of the Earth's Bow Shock

J. R. Dwyer

Department of Physics, University of Maryland, College Park, MD 20742 USA;
dwyer@umstep.umd.edu

ABSTRACT

The Earth's Bow shock provides a unique laboratory for studying the acceleration of ions up to energies of about 0.5 MeV/nucleon. Upstream of the bow shock, energetic particles are typically seen as intense bursts lasting a few tens of minutes. Although these particles have been observed for about 30 years, how and where they are produced, e.g. by first-order Fermi acceleration at the bow shock or by magnetospheric leakage, is still not well understood. In recent years, new instruments have become available for studying upstream events, allowing for the first time detailed measurements of the composition, spectra and anisotropies. Instruments onboard the WIND spacecraft have observed thousands of these upstream events since its November 1994 launch, and now with the launch of ACE in August 1997, simultaneous observations from separate locations upstream are possible.

1. Introduction

Energetic particles are produced in a wide range of astrophysical settings, and are detectable at Earth either directly as in the case of galactic cosmic-rays or indirectly as in the case of radio synchrotron and diffuse gamma-ray emissions. In order to understand how particles are injected from a thermal population and accelerated to high energies, it is desirable to have measurements of the source particle population as well as the energy spectra and composition of the energized component. In fact, it would be nice to also have measurements of the intermediate stages when the particles are in the process of gaining energy. Of course, such measurements are difficult, if not impossible for many of the most interesting astrophysical sites, such as within supernova remnants where galactic cosmic-rays are thought to be produced.

For energies below about 1 GeV/nucleon, because of their rapid absorption in the Earth's atmosphere, cosmic-rays are only measurable using space based instrumentation. These measurements may be statistically very precise due to the high cosmic-ray intensities at these energies but are difficult to compare with models since cosmic-rays experience large changes in their energy spectra as they travel inward through the expanding solar wind to reach the Earth. This energy loss, called solar modulation, is solar cycle dependent and can result in changes in the particle intensities by several orders of magnitude at the lowest energies. In addition, cosmic-ray spectra and composition change dramatically as the particles propagate from their place of origin, through the galaxy to our solar system by the processes of ionization energy losses, nuclear spallation and from an energy dependent escape mechanism out of our galaxy (e.g. Berezinskii *et al.* 1990).

Once we apply propagation models to the observations and derive a source spectrum and composition of galactic cosmic-rays, we can then compare this result with the predictions of models such as diffusive shock acceleration at the supernova shocks. Another problem, however, is that the plasma conditions at remote acceleration sites are not well known. As a result, model predictions are for the large part qualitative and are not strongly constrained by the observations. In fact, diffusive shock acceleration, or first-order Fermi acceleration, which is a widely used model for cosmic-ray acceleration, is supported largely on theoretical grounds, and there are few observational facts that pinpoint this particular model as being the dominant mechanism for cosmic-ray production.

As a nearby test-bed for shock acceleration models, consider the Earth's bow shock, which is produced as the supersonic solar wind (300 km/sec $\leq V_{sw} \leq$ 800 km/sec) flows around the Earth's magnetosphere (Sagdeev & Kennel 1991). The result is a roughly paraboloid shaped standing shock, with a nose located about 15 R_E (Earth radii) upstream (toward the Sun) of the Earth. Spacecraft have flown through the bow shock for decades and many of the important plasma parameters and the magnetic field configurations are known at and near the shock, including measurements of the ambient particle populations that are injected at the shock. If we hope to understand how particles are accelerated at exotic sites many light years away, perhaps a good warm up problem would be to understand particle acceleration at the Earth's bow shock. For example, will first-order Fermi acceleration produce energetic particles at this shock? The answer, as I shall attempt to show, is by no means straight forward. In section 2, I shall give an overview of energetic particle observations upstream of the bow shock. In sections 3, I shall cover recent observations from the WIND spacecraft. Finally, in section 4, I will briefly discuss future work, made possible with the launch of the ACE spacecraft.

2. Overview of Upstream Events

Spacecraft such as AMPTE, IMP-7 & 8, ISEE-1, 2 & 3, and later WIND and ACE, have observed energetic particle populations just upstream of the bow shock since the early 1970's. These particles, which can extend up to a few MeV in total energy, are called upstream ion events. The reason that the name contains the word "event" is that, depending upon where the spacecraft is located, the upstream ions are observed as short spikes in the particle intensity, usually lasting anywhere from a few minutes up to a few hours.

A complicating factor in the study of upstream events is that the Earth's magnetosphere sits directly behind the bow shock. The magnetosphere is a well known reservoir of energetic particles, and it is possible that the particles that are observed upstream of the bow shock were not actually energized at or near the shock but instead leaked out of the magnetosphere. Certainly there are times when the conditions appear ripe for Fermi acceleration at the bow shock and no energetic particles are observed. For example, Sarris & Krimigis (1988) reported time periods with no observed upstream ions even though there seemed to be a good magnetic connection between the spacecraft and the shock. Of course, isolated examples such as this cannot be used alone to discredit shock acceleration for all circumstances. On the other hand, models of shock acceleration need to account not only for the particles that we observe, but also for cases when no acceleration takes place. In addition, during some upstream events, energetic electrons (>220 keV) are also observed, presenting a further challenge to Fermi acceleration models (Krimigis *et al.* 1978).

In discussing upstream events, it is useful to distinguish between observations made near the bow shock (within ~10 R_E) and those made farther away (up to several hundred R_E). Close to the bow shock, upstream events can be loosely divided in three categories: the reflected, intermediate and diffuse components. The reflected component (Asbridge *et al.* 1968) is seen as a low energy (<10 keV/charge) beam of particles which almost certainly originates from a single reflection of the solar wind off the quasi-perpendicular part of the shock, where the interplanetary magnetic field is nearly perpendicular to the shock normal angle. This is located on the dusk side of the bow shock, near the nose, because the average interplanetary magnetic field is oriented at a 45° angle at 1 AU. The diffuse component (Gosling *et al.* 1978), on the other hand, is a nearly isotropic population, observed up to about 150 keV/charge, and is usually seen near the quasi-parallel (dawn) side of the shock. The diffuse component sometimes shows an inverse velocity dispersion in the arrival of the particles, meaning the lower energy particle intensities increase more rapidly than the higher energy particles. A natural interpretation of this is that the process of particle acceleration is being observed *in situ* since higher energy particles will take longer to be energized than lower energy particles. Another feature of diffuse ions is the association of low-frequency fluctuations in the solar wind density and field strength (Paschmann *et al.* 1979). The intermediate component possesses some of the properties of both the reflected and diffuse components in terms of energy, anisotropies and spatial location and may represent particles that are evolving with time from the reflected into the diffuse components by some acceleration process (Paschmann *et al.* 1979). A fourth component, called magnetospheric bursts, is sometimes added to distinguish events that are magnetospheric in origins from those, i.e. the diffuse component, that appears to be produced at the bow shock. However, some researchers believe that nearly all ions, including the diffuse component are magnetospheric and so such a distinction is not necessary (e.g. Sarris *et al.* 1987).

Mitchel & Roelof (1983) performed a statistical study of 4 years of IMP 7 and 8 observations and concluded that θ_{Bn}, the angle between the interplanetary magnetic field and the shock normal angle, is an important parameter for determining the occurrence of upstream events and that the local structure of the bow shock dominates the production of upstream events. They also found a correlation of the probability of upstream events exceeding a given flux and the 3-hour K_p-index, which gives a broad measure of magnetospheric activity. Ipavich *et al.* (1984), using ISEE-1 observations, reported a correlation of the H/He ratio in the upstream events with the ratio in the solar wind. They concluded that their results were consistent with Fermi acceleration of a seed population extracted from the solar wind. Scholer *et al.* (1981) analyzed 35 proton events observed by ISEE-1 & 3 and found two classes of events: The first group was associated with ≥ 75 keV electrons and proton energies extending up to ≥ 300 keV. The second group had exponential spectra with a mean e-folding energy of ~15 keV and is not accompanied by energetic electrons. They attributed the former group to magnetospheric bursts and the latter to diffuse ions of bow shock origin.

Farther away from the bow shock (>10 R_E), upstream events with energies up to several hundred keV/nucleon are observed as short duration spikes in the ion time-intensity profile with the intensity often increasing by a factor of a thousand. These events show very large anisotropies away from the Earth (Scholer *et al* 1980) with only slight decreases in the intensity with distance, indicating that they are free-streaming (scatter free). This is in contrast to near the shock; for instance, Ipavich *et al.* (1981), using ISEE-1 data measured within ~8 R_E of the bow shock, found that the intensity of ~30 keV upstream protons decreased with distance from the shock with an e-folding length of 7 R_E, implying that the free-escape boundary is somewhere beyond this distance. Presumably upstream events, seen far from the bow shock, correspond to the diffuse (and/or magnetospheric burst) components after passing through a free escape boundary upstream. Often times no velocity dispersion is observed during these events, which suggests that the time structure, and to some extent the intensity, of upstream events is probably governed by the magnetic connection of the spacecraft to the bow shock and not the acceleration process. One consequence of this is that the role of other parameters, such as the distance upstream, can be difficult to interpret.

3. Recent Observations

The WIND spacecraft, launched in November 1994, carries a large assortment of instruments designed to measure magnetic field (MFI, Lepping *et al.* 1995), plasma (SWE, Ogilvie *et al.* 1995), wave (WAVES, Bougeret *et al.* 1995) and energetic particle data (EPACT, von Rosenvinge *et al.* 1995; 3DP, Lin *et al.* 1995; STICS, Gloeckler *et al.* 1995). For the first month after launch, WIND remained relatively close to the Earth in a highly elliptical orbit. Then, after a lunar swing-by, it moved farther upstream for its initial pass around the first Lagrangian (L1) point. The large amount of time that WIND spends in the Earth's foreshock region, upstream, allow extended observations of energetic ions originating from and/or passing through the Earth's bow shock.

The Supra-Thermal through Energetic Particle (STEP) sensor is part of the EPACT experiment onboard WIND and consists of two identical time of flight (TOF) mass spectrometers, with a combined geometrical factor of 0.8 cm^2 sr. The mass and kinetic energy per nucleon of individual ions are determined by the TOF through the telescopes and the total kinetic energy deposited in a solid state detector. STEP was designed to measure the ion species hydrogen through iron, over the energy range of ~20 keV/nucleon up to several MeV/nucleon, making it ideal for measuring upstream events.

Figure 1 shows the time-intensity profile of 24-40 keV/nucleon C+N+O measured by STEP for the first six days of 1995. During this time period, WIND was between 100 and 150 R$_E$ from the Earth. In the figure, upstream events appear as sharp spikes in the intensity. The slowly varying component of the intensity, clearly seen between days 1 and 3.5, is from energetic particles associated with a corotating interaction region (CIR). CIRs are large interplanetary structures produced by the interaction of fast and slow solar wind streams. In addition to their unique time structure, upstream events (at these geocentric distances) are also characterized by soft energy spectra and large anisotropies directed away from the Earth.

Figure 2 shows the C+N+O time-intensity profile of 24-40 keV/nucleon measured by STEP, as well as the solar wind speed and the distance of the WIND spacecraft from the Earth, for the entire year of 1995. As was reported by Mason *et al.* (1996), upstream events, which appear as dense clusters of spikes in figure 2, are often associated with high speed solar wind streams. This observation is consistent with the results of Mitchel & Roelof (1983) since the K$_p$-index and the solar wind speed are also correlated. This again underlines the difficulty in determining the source of upstream events. At higher solar wind speeds, acceleration at the Earth's bow shock should be more efficient since the shock strength increases and higher levels of magnetic turbulence occur in high speed wind streams. On the other hand, the K$_p$-index and hence magnetospheric activity also tend to be higher during these time periods which might result in more leakage of energetic particles upstream. Mason *et al.* (1996) suggested another possibility for this correlation of upstream events with high speed solar wind streams: Because corotating interaction regions, which often produce large energetic particle enhancements, are also associated with high speed solar wind streams, these CIR suprathermal ions may serve as a seed population for particle acceleration at the bow shock. The question of whether CIR particles or solar wind particles provide the seed population for upstream events directly addresses the fundamental problem of particle injection during shock

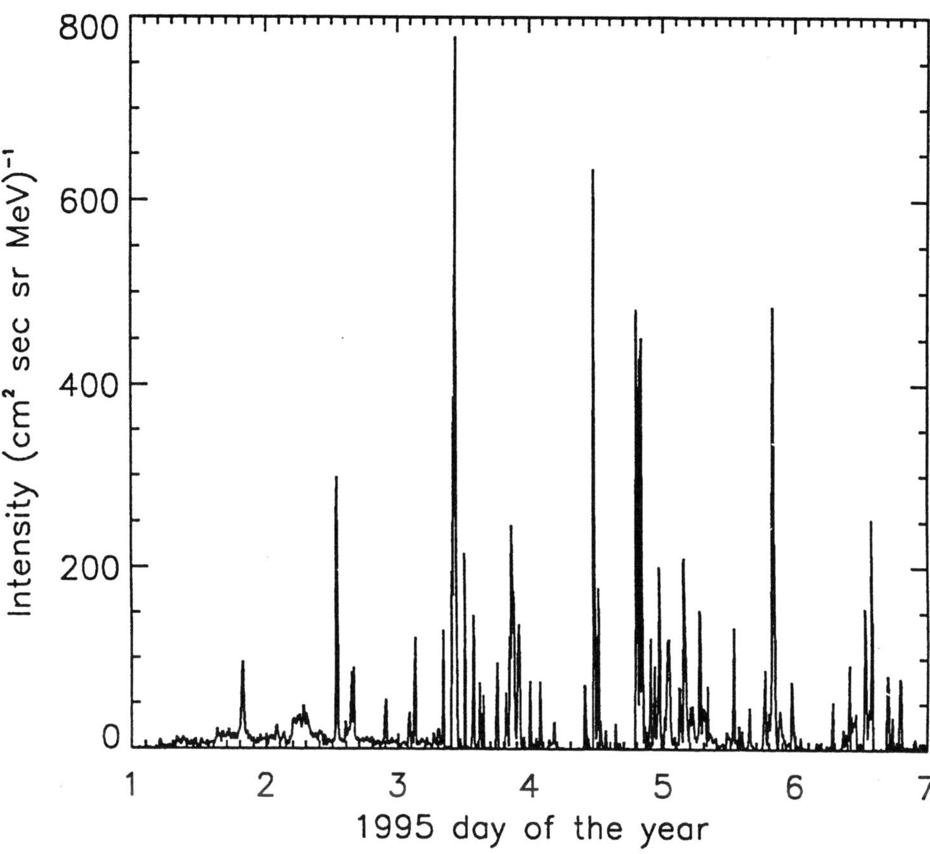

Figure 1. Time-intensity profile of 24-40 keV/nucleon C+N+O measured by STEP for the first six days in 1995. The upstream events appear as short duration spikes in the particle intensity, usually lasting no more than 1 hour. Note that the upstream events are very intense compared to the slowly varying background (at the level of $\sim 10/cm^2$ sec sr MeV) between days 1 and 3.5, which originates from a corotating interaction region (CIR).

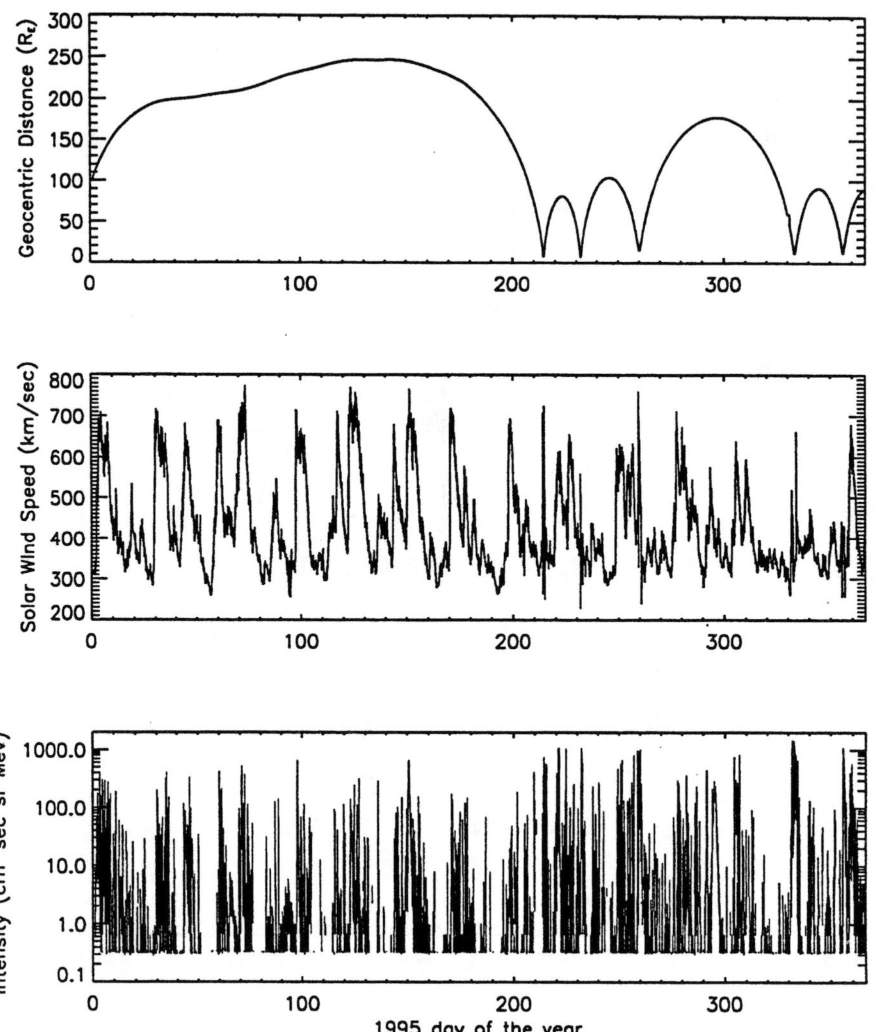

Figure 2. *Top Panel:* Geocentric distance of the WIND spacecraft for the year 1995. For most of the year WIND remained upstream of the bow shock in interplanetary space. *Middle Panel:* Solar wind speed from the SWE instrument onboard WIND (Ogilvie *et al.* 1995). The large fluctuations in the solar wind speed are due to the passage of high speed solar wind streams at 1 AU. *Bottom Panel:* Time-intensity profile of 24-40 keV/nucleon C+N+O measured by STEP. The STEP data consists of hundreds of short duration spikes, each an upstream event. As reported by *Mason et al.* (1996), the upstream events tend to cluster around time periods with elevated solar wind speeds.

acceleration. That is, do wave-particle interactions at the Earth's bow shock accelerate particles directly out of the thermal population or must an additional high energy component be present in order to overcome an injection threshold? This injection issue has significance for a wide range of astrophysical phenomena, e.g. the acceleration of galactic cosmic-rays by supernova remnants (Axford 1981).

Composition observations of upstream events also provide information for identifying the seed population, and hence give clues to the origin and acceleration mechanism of these particles. Composition measurements for several upstream events that occurred in February 1995, during a high speed wind stream, were reported by Mason *et al.* (1996). They found C:O abundances distinctly different from magnetospheric ring current abundances measured on AMPTE; rather, the abundance ratios they found were consistent with both the solar wind and energetic particles in CIRs. The ambiguity in this result is due to the general similarity between solar wind and CIR particle compositions. CIR energetic particles are believed to be accelerated directly out of the high speed solar wind (Fisk & Lee 1980) with a small (~15% for He at 1 AU) contribution from interstellar pickup ions (Chotoo *et al.* 1996).

In order to study the acceleration of ions near the Earth, it is desirable to have an energetic component present with a composition, unlike CIRs, that is very different from that of the solar wind. Fortunately, the Sun provides just such a population in the form of solar energetic particles from impulsive flares. On December 26 (day 360), 1994 a ^{3}He and Fe-rich solar energetic particle (SEP) event was observed by STEP lasting approximately 2 days. During this time, STEP was 50 R_E from the Earth and observed several intense upstream events with significantly elevated intensities above the level of the flare. Because this SEP event had a very distinct composition, with a ^{3}He/^{4}He enhancement of ~10^5, a factor of 16 increase in Fe/O, and a factor of 2 decrease in C/O with respect to the solar wind, the presence or absence of the SEP population within the upstream events can be clearly identified.

Table 1 lists the abundance ratios, reported by Dwyer *et al.* (1997), measured by STEP for the solar flare and for the upstream events that occurred in coincidence with the solar energetic particles. Due to the near absence of ^{3}He in the solar wind, this sizable ^{3}He abundance detected in the upstream events of 12/26/94-12/27/94 must have been accelerated from the flare population. The other important feature of the upstream event

Table 1. Relative Abundances of Ion Species

Species	keV/nucleon	^{3}He-rich SEP event 12/26/94 - 12/27/94	Upstream events 12/26/94 - 12/27/94	In-ecliptic solar wind[*]
^{3}He /^{4}He	80-320	0.19 ± 0.02	0.15 ± 0.05	$(4.9 \pm 0.5) \times 10^{-4}$
He/CNO	40-80	24.66 ± 3.77	49.70 ± 2.73	40.56 ± 11.04
	80-320	15.31 ± 1.59	38.34 ± 7.07	
NeMgSi/CNO	40-80	1.05 ± 0.17	0.19 ± 0.03	0.28 ± 0.03
	80-320	1.19 ± 0.13	0.24 ± 0.15	
Fe/CNO	40-80	1.09 ± 0.17	0.04 ± 0.01	0.10 ± 0.05
	80-320	1.24 ± 0.13	0.02 ± 0.13	

[*] ^{3}He /^{4}He: Coplan *et al.* (1984); others: Geiss *et al.* (1994)

composition is the solar wind-like abundances along with the notable lack of an intensity increase for Fe, whose intensity barely increases above the SEP background, in stark contrast to He and CNO. Taken at face value, this implies the puzzling property that only the He and CNO from the SEP are accelerated at the bow shock, while the Fe is not. These Fe results, however, are not necessarily inconsistent since solar wind and SEP Fe is expected to be less ionized than the lighter elements. The resulting higher rigidity of Fe (at the same energy/nucleon) means that Fe would be less efficiently accelerated due to its larger gyroradius and the limited size of the Earth's bow shock ($\sim$30 R_E). A hint of the importance of the bow shock size comes from the very soft spectra of upstream events, which have either exponential spectra or power law spectra with indices typically in the range $\sim$3 - 8. These values are much softer than the $\sim$1 spectral index predicted by non-relativistic Fermi acceleration for strong shocks (like the bow shock) of infinite extent (Jones & Ellison 1991).

Another instrument onboard WIND is the Three-dimensional Plasma and Energetic Particle Experiment (3DP). This instrument makes measurements of the three-dimensional distribution of electrons and ions in the suprathermal energy range from the solar wind plasma up to low energy cosmic-rays. At low energies (~3 eV to 30 keV), electrons and ions are measured with top-hat symmetrical spherical section electrostatic analyzers with microchannel plate detectors. At higher energies (>20 keV), they are measured by three arrays of pairs of double-ended semi-conductor telescopes. Using 3DP observations, Sanderson *et al.* (1996) presented for the first time the complete energy spectrum of upstream event ions from a few eV up to a few MeV. They found that the bulk of the energy density of the ions was at 1 keV and concluded that it was most likely these particles that are responsible for exciting the low-frequency waves that usually accompany upstream events. This wave particle interaction is important for models of energetic particle production at the bow shock since it is these waves that scatter the particles and allowing diffusive shock acceleration to occur (Lee 1982).

4. Future Work

In August 1997 the ACE spacecraft was launched and is now in a halo orbit about the first L1 point between the Earth and the Sun. Like WIND, ACE has a large assembly of instruments designed to measure plasma, magnetic field and, particularly, energetic particle data. One such instrument is the Energetic Proton and Alpha Monitor (EPAM). EPAM uses solid state detectors and consists of five telescope apertures of three different types designed to measure the intensity and anisotropies of >30 keV electrons, >50 keV ions, and the elemental composition of ions (Gold *et al.* 1998). Recently, the EPAM/ACE instrument has been used in collaboration with the STEP/WIND instrument to study upstream events. Figure 3 shows the position of the ACE and WIND spacecrafts during times when simultaneous upstream events were observed by both instruments (Haggerty *et al.* 1998). The events were observed during an extended time period when the interplanetary magnetic field remained nearly radial. The large separation of the spacecraft during these events illustrates the large region of space filled by upstream event ions and suggests that particles are emitted over a very large area of the bow shock and not just from isolated "hot spots".

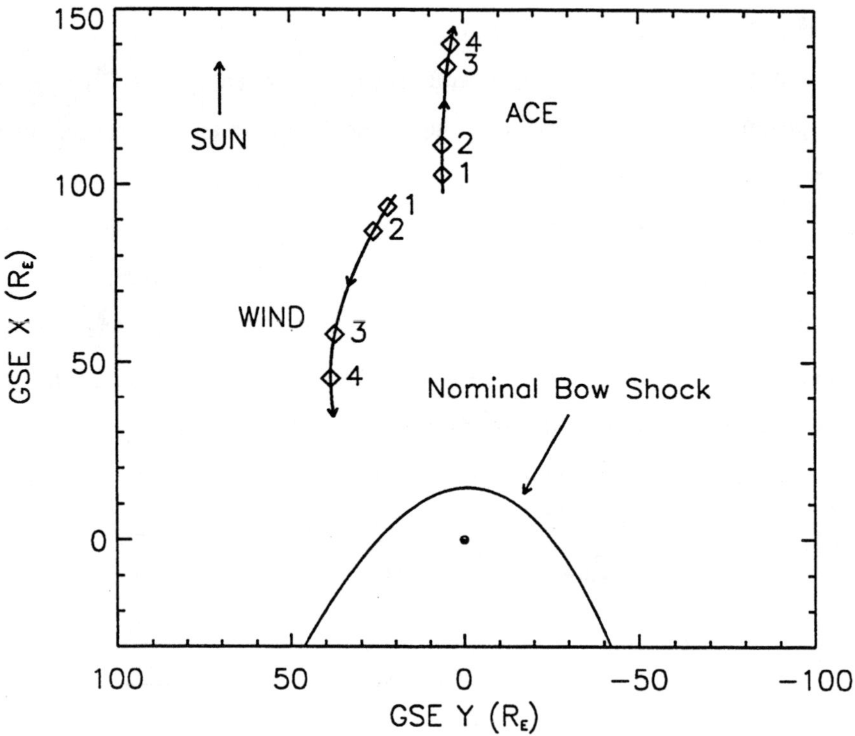

Figure 3. The position of the ACE and WIND spacecrafts projected onto the X-Y plane in GSE coordinates from day 243 to day 249 in 1997. Also shown is the location of the Earth and the nominal location of the Earth's bow shock. The diamonds on each spacecraft trajectory, labeled 1-4, denote the position of each spacecraft at the time that simultaneous upstream events were observed (Haggerty *et al.* 1998).

Instruments onboard ACE and WIND should continue to observe upstream events for a number of years, providing detailed measurements of composition, spectra and anisotropies as well as event topology in the space upstream of the shock. These measurements will occur over a wide range of plasma and field conditions as the solar cycle progresses and should give us a much better understanding of the processes that accelerate particles near the Earth and, more generally, the processes that produce energetic particles at remote astrophysics sites.

Acknowledgments

I wish to thank the members of the Space Physics Group, Department of Physics at the University of Maryland, and the EPACT instrument team, Goddard Space Flight Center, for the construction of the instrumentation, and, in particular, I thank G. M. Mason, J. E. Mazur and T. T. von Rosenvinge for the STEP data presented here. I also wish to thank K. W. Ogilvie, R. B. Torbert, and A. J. Lazarus for the SWE solar wind data cited here. This work was supported in part by NASA contract NAS5-30927, and grant NAG 5-7228.

References

Asbridge, J. R., Bame, S. J., & Strong, I. B., 1968, *J. Geophys. Res.*, 73, 5777

Axford, W. I., 1981, *Proc. 17th. Int'l Cosmic Ray Conf.* (Paris), 12, 155

Berezinskii, V. S., Bulanov, S. V., Dogiel, V. A., Ginzburg, V. L., & Ptuskin, V. S., 1990, *Astrophysics of Cosmic Rays*, ed. V. L. Ginzburg, North-Holland, The Netherlands

Bougeret, J.-L., *et al.*, 1995, *Space Sci. Rev.*, 71, 231

Chotoo, K., Galvin, A. B. & Gloeckler, G., 1996, *Trans. Am. Geophys. U.*, 77, S212

Coplan, M. A., Ogilvie, K. W., Boschler, P., & Geiss, J., 1984, *Solar Phys.*, 93, 415

Dwyer, J. R., Mason, G. M., Mazur, J. E., & von Rosenvinge, T. T., 1997, *Geophys. Res. Lett.*, 24, 1, 61

Fisk, L. A., & Lee, M. A., 1980, *ApJ*, 237, 620

Geiss, J., Gloeckler, G., & von Steiger, R., 1994, *Phil. Trans. R. Soc. Lond.*, 349, 213

Gloeckler, G., *et al.*, 1995, *Space Sci. Rev.*, 71, 79

Gold, R. E. et al., 1998, *Space Sci. Rev.* (In Press)

Gosling, J. T., Asbridge, J. R., Bame, S. J., Paschmann, G., & Sckopke, N., 1978, *Geophys.Res. Lett.*, 5, 11, 957

Haggerty, D. K., Desai, M. I, Mason, G. M., Dwyer, J. R., Gold, R. E., Krimigis, S. M.,Mazur, J. E., & von Rosenvinge, T. T., 1998, *Geophys. Res. Lett.* (submitted)

Ipavich, F. M., Galvin, A. B., Gloeckler, G., Scholer, M., & Hovestadt, D., 1981, *J. Geophys.Res.*, 86, A6, 4337

Ipavich, F. M., Gosling, J. T., & Scholer, M., 1984, *J. Geophys. Res.*, 89, A3, 1501

Jones, F. C., & Ellison, D. C., 1991, *Space Sci. Rev.*, 58,259

Krimigis, S. M., Venkatesan, D., Barichello, J. C., & Sarris, E.T, 1978, Geophys. Res. Lett., 5,961

Lee, M. A., 1982, J. *Geophys. Res.*, 87, A7, 5063

Lepping, R. P., *et al.*, 1995, *Space Sci. Rev.*, 71, 207

Lin, R. P., *et al.* 1995, *Space Sci. Rev.*, 71, 125

Mason, G. M., Mazur, J. E., & von Rosenvinge, T. T., 1996, *Geophys. Res. Lett.*, 23, 10, 1231

Mitchell, D. G., Roelof, E. C., 1983, *J. Geophys. Res.*, 88, A7, 5623

Oglivie, K. W., *et al.*, 1995, *Space. Sci. Rev.*, 71, 55

Paschmann, G., Sckopke, N., Bame, S. J., Asbridge, J. R., Gosling, J. T., Russell, C. T., &Greenstadt, E. W., 1979, *Geophys. Res. Lett.*, 6, 209

Sagdeev, R. Z., & Kennel, C. F., 1991, *Scientific American*, April, p106

Sanderson, T. R., *et al.*, 1996, *Geophys. Res. Lett.*, 23, 10, 1215

Sarris, E.T, Anagnostopoulos, G. C., & Krimigis, S. M., 1987, *J. Geophys. Res.*, 92, A11,12083

Sarris, E.T, & Krimigis, S. M., 1988, *Geophys. Res. Lett.*, 15, 233

Scholer, M., Ipavich, F. M., Gloeckler, G., Hovestadt, D., & Klecker, B., 1980, *Geophys. Res.Lett.*, 7, 73

Scholer, M., Hovestadt, D., Ipavich, F. M., & Gloeckler, G., 1981, *J. Geophys. Res.*, 86, A11,9040

von Rosenvinge, T. T., *et al.*, 1995, *Space Sci. Rev.*, 71, 155

Nuclear Cross Sections and the Composition, Transport, and Origin of Galactic Cosmic Rays

C. H. Tsao, A. F. Barghouty[1] and R. Silberberg

Physics Department, Roanoke College, Salem, VA 24153

ABSTRACT

Recent cross section measurements and the new semiempirical calculations for proton-nucleus and nucleus-nucleus reactions are reviewed. Both the measured and calculated cross sections provide critical information on isotopic products - covering nuclei of charge $Z = 3 - 82$. The relation of the generated radioactive nuclei in cosmic rays to their galactic confinement time is discussed. A robust cross-section database and propagation calculations are needed to critically evaluate the cosmic-ray source composition and the competing models for the origin of cosmic rays. One of these models, with acceleration of *freshly* nucleosynthesized material, for example, would have trans-uranic nuclei and a high U/Th ratio (due to shorter lived isotopes of U), contrary to LDEF measurements, and can probably be eliminated. The importance and implications to include reacceleration in the propagation calculations are also discussed.

Subject headings: cosmic rays − nuclear reactions, nucleosynthesis, abundances

1. Introduction

Nuclear cross sections, both partial and total, are essential for understanding the composition, transport, and origin of galactic cosmic rays. This can readily be appreciated in Fig. 1 which shows the cosmic-ray abundances from He to Ni, and compares these to the solar system abundances. To a first approximation, the solar system abundances, especially those of the solar corona and the solar wind, resemble the cosmic ray abundances at the source regions. However, the elements Li, Be, B, F, Cl, K, Sc, Ti, V, Cr, Mn and much of N, Na, Al, P in cosmic rays are

---

[1]Present address: Space Radiation Laboratory, MC 220-47, Caltech, Pasadena, CA 91125

produced by nuclear spallation of the more abundant elements of the cosmic-ray source component, i.e., C, O, Ne, Mg, Si, S, Ca and Fe.

With the help of nuclear propagation (species transformation) terms in the galactic transport equations for cosmic rays, in which the nuclear cross sections are essential ingredients, we can determine the mean path length traversed by cosmic rays before escape from the galaxy, the energy (or rigidity) dependence of this path length, and the source composition of cosmic rays. These topics are discussed briefly here and in greater detail in other contributions to this volume. In this contribution, however, emphasis is on nuclear cross sections and their permeating role in the studies of the composition, propagation, and origin of galactic cosmic rays. The studies on the origin of cosmic rays are still to a large degree model dependent. With more precise estimates of the relevant cross sections, however, we are able to critically evaluate the merits of one model against the others, as we demonstrate below.

The partial inelastic cross sections σ_{ij} (for the production of species j from spallation of species i) turn out to have systematic regularities that permit the formulation of predictive semiempirical equations. (Species tranformation is essentially determined by σ_{ij}.) Rudstam (1966) first noted that these systematic regularities depend on the mass difference of the target and product nuclides and on the neutron-proton ratio of the product nuclides. On the basis of such regularities he was able to formulate a set of tractable, semiempirical equations to estimate the various σ_{ij}. Silberberg and Tsao (1973a and b) followed suit and constructed semiempirical equations resembling Rudstams's (1966) but with additional parameters, and were able to extend the applicability regions for both target and product mass intervals. Of particular importance to cosmic-ray investigations was the extension to lighter targets like C, N, O, Ne, Na, Mg and to light products Li, Be and B, as well as to peripheral reactions with $\Delta Z < 3$, with ΔZ being the charge change in the reactions. As experimental data from heavy ion beams became available, the equations and parameters were further refined as described in Silberberg et al. (1985) and Silberberg and Tsao (1990). For nuclei with $Z \leq 28$, Webber et al. (1990) carried out detailed measurements (with a hydrocarbon target) of the production of both stable and radioactive isotopes, and derived parametric semiempirical equations. Recently the Transport Collaboration of Chen et al. (1995,1997a,1997b), Knott et al. (1993, 1996, 1997) and Tull et al. (1993) carried out highly precise measurements with heavy ion beams incident on a hydrogen target. Spallation of ultra-heavy nuclei like Kr, Ag, La, Ho and Au has been explored by Binns et al. (1987), Cummings et al. (1990), Garrard et al. (1995), Geer et al. (1995), Nilsen et al. (1993) and Waddington et al. (1997) from the University of Minnesota, Washington University and Caltech Collaboration, with elemental (i.e. no isotopic) resolution. The new semiempirical proton-nucleus relations we briefly describe below have been updated in light of these recent measurements.

The calculations have also been extended to nucleus-nucleus reactions. These are important in astrophysics, planetary and atmospheric sciences, reactions in particle and gamma-ray detectors, reactions in astronaut tissue, in evaluating radiation risks, and in interactions in computer components, with the related single-event upsets. About 10% by number of the interstellar gas is helium, hence, about 20% of cosmic-ray generated nuclear spallation products are formed in nucleus-helium interactions. In analysis of cosmic-ray interactions with atmospheric nuclei (mainly nitrogen and oxygen on earth and CO_2 on Mars, for example), an accurate description of nucleus-nucleus interactions is essential. In investigations of cosmic-ray and solar-flare particle interactions and in radiation transport calculations in lunar material (where SiO_2 is an important component) nucleus-nucleus cross sections are needed. On Mars, under its thin atmosphere, and on asteroids, knowledge of nucleus-nucleus partial cross sections is also required. Particle and gamma-ray detectors contain NaI, CsI, Si and Ge, and occasionally W collimators as well. Nucleus-nucleus reactions and the radiation transport in these materials have to be accurately evaluated. Similarly, nuclear interactions and radiation transport in spacecraft materials, e.g., Al, need to be known with high precision as well.

In addition to the partial cross sections, semiempirical expressions for the total inelastic cross sections, which enter as source and sink terms in the transport equations, are discussed in Silberberg et al. (1998) and Tsao et al. (1998). Non-nuclear contributions to the interactions' cross sections, i.e., electromagnetic dissociation, are briefly addressed here and in these recent updates as well.

2. Updated Proton-Nucleus Reactions' Cross Sections

A new set of modifications has been added to the semiempirical equations[2] (Silberberg et al. 1998). A few illustrative samples of comparing the calculated results with experimental data are reproduced below.

Figs. 2a and 2b show such a comparison with data on elemental products from the Transport Collaboration, Knott et al. (1993) for the spallation of Si at energies 500 and 1300 MeV/n and Ca from 400 to 800 MeV/n. In Figs. 3a and 3b similar comparisons near 400 MeV/n are shown for the spallation of Cr and Fe.

The spallation of the so-called ultra-heavy nuclei is illustrated for Au at 0.56,

---

[2]The corresponding code is available from http://spdsch.phys.lsu.edu.

0.92 and 10.6 GeV/n in Fig. 4, comparing the calculated cross sections with the measured ones from Geer et al. (1995) and Garrard et al. (1995).

Extensive cross section measurements of isotopes from the spallation of beams ranging from ^{11}B to ^{58}Ni were carried out by Webber et al. (1990) at energies near 600 MeV/n. Figs 5a,b,c and d display the breakup of ^{20}Ne into isotopes of F, O, N and C.

Webber et al. (1990) have developed parametric equations[3] to estimate partial cross sections. While the above semiempirical calculations have a standard deviation of about 20%, those of Webber et al. (1990) have about 10% for the set in which their data were used to derive the parametric equations. However, for those cross sections not measured by Webber et al. (1990), the parametric equations yield a considerably poorer fit. The cross section of ^{22}Ne into ^{18}O, for example, is underestimated by a factor of 4, whereas the new semiempirical calculations yield a much better fit (Fig. 7 of Silberberg et al. 1998). Also, the equations of Webber et al. (1990) are not applicable to heavy ($Z > 28$) nuclei.

3. Partial Cross Sections of Nucleus-Nucleus Reactions

The earlier semiempirical calculations (Silberberg and Tsao, 1990 and Sihver et al. 1993) have been updated by Tsao et al. (1998), using features of a theoretical simulation model (Barghouty et al. 1991), and in light of recent heavy-ion cross section measurements. The idea is essentially to scale the proton-nucleus projectile-fragment cross sections, discussed above, to the nucleus-nucleus case using the so-called weak factorization property of projectile fragments (Tsao et al. 1993).

Figs. 6a,b,c, and d compare the calculated results with the experimental cross sections of Fe (at 1.05 GeV/n) with C. The measured values are from Zeitlin et al. (1997), Westfall et al. (1979), Cummings et al. (1990), and Webber et al. (1990).

The isotopic yields are displayed in Fig. 7, which shows the data of Webber et al. (1990) for the spallation of Fe on a carbon target into Mn, Cr, V and Ti.

The charge-pickup cross sections in nucleus-nucleus collisions are shown in Fig. 8 for La and Au on targets of C, Al and Cu. The calculated results are compared with the data of Nilsen et al. (1994) and Geer et al. (1995).

The electromagnetic dissociation cross sections, due to virtual photon exchange via the dipole resonance, is large in interactions between heavy nuclei. The cross

---

[3]The code for the parametric calculations is also available from http://spdsch.phys.lsu.edu.

section is roughly proportional to Z^2. At beam energy of 1.26 GeV/n, Hill et al. (1988) measured the electromagnetic dissociation $\sigma(^{138}\mathrm{La} + {}^{197}\mathrm{Au} \rightarrow {}^{196}\mathrm{Au})$ to be 2000 mb. Brechtmann and Heinrich (1988) measured $\sigma(^{32}\mathrm{S} + \mathrm{Pb})$ to be 4600 mb for the dissociation of Pb by a $^{32}\mathrm{S}$ beam at 200 GeV/n. At ultra-high beam energies, about 100 GeV/n, the virtual pion is sufficiently energetic for photo-pion reactions, giving rise to spallation-like reactions with multiple emission of nucleons (Brechtmann and Heinrich, 1988). Fig. 9 shows the data of Brechtmann and Heinrich (1988) for $^{32}\mathrm{S}$ at 200 GeV/n incident on Al, Cu, Ag and Pb. The calculated spallation cross sections with and without electromagnetic dissociation are also shown. Norbury and Townsend (1993) and Norbury and Muller (1994) collected data on electromagnetic dissociation for single and double nucleon removal and developed relations to estimate the corresponding cross sections.

4. Galactic Propagation of Cosmic Rays

About half of the cosmic rays heavier than carbon suffer nuclear breakup before they escape from the galaxy. Hence, knowledge of nuclear cross sections becomes essential before determining the conditions of cosmic ray propagation. Prior to detection, cosmic-ray energy spectra as well as relative abundances are further affected by ionization losses, degree of reacceleration, escape from the galaxy, abundance of helium (relative to hydrogen) in the interstellar gas, and solar modulation.

The path length of material traversed by cosmic rays can be estimated from observed secondary-to-primary ratios, e.g. B/C and (Sc-Mn)/Fe. From the energy or rigidity dependence of these ratios one obtains the escape path length as well as its energy dependence. At energies near 10 GeV/n, the escape path length is about 4 g/cm^2, while near 1 GeV/n, the estimated values range from 9 to 12 g/cm^2. These estimates have been arrived at neglecting the effects of reacceleration. Fig. 10 shows the corresponding data for the escape path length vs. energy. The curves shown are those of Engelmann et al. (1990) and Heinbach and Simon (1995).

We consider next the case of reacceleration. The principal acceleration mechanism of cosmic rays occurs on a relatively rapid time scale; otherwise the relative abundance of secondary nuclei would increase with energy (Hayakawa 1969). It was noted (Silberberg et al. 1983) that some cosmic-ray relative abundances at lower energies could be fitted well if some weak reacceleration or distributed acceleration takes place after the principal acceleration. This was further developed into a quantitative theory of reacceleration (Wandel et al. 1988). At the same time Simon et al. (1986) developed a theory of weak stochastic reacceleration. Giler et al. (1989) explored further tests for reacceleration. Letaw,

Silberberg and Tsao (1993) and Silberberg et al. (1993) noted that the problem of the original model of Wandel et al. (1988) could be overcome by using (1) a spread of weak reacceleration shock strengths, based on Axford (1982) and (2) introducing an energy dependence of mean free path for reacceleration. They obtained excellent fits to about 500 data points of cosmic rays from Li to Ni, in energy intervals 10 to 10^6 MeV/nucleon.

Stephens and Golden (1990) noted that for lightly ionizing particles like H and He, weak reacceleration competes too strongly with ionization losses if the original reacceleration parameters of Wandel et al. (1988) are used. The same problem affects also the calculated electron spectra (Stephens, 1990). Seo and Ptuskin (1994) found, however, that the inevitable stochastic reacceleration of cosmic rays by random hydromagnetic waves in interstellar space yields calculated spectra of H, He and heavy nuclei that fit well the observational data. Silberberg et al (1995,1998) have developed weak reacceleration parameters and a novel spread of shock strengths that fit simultaneously both electrons and heavy ions.

The principal advantages of the reacceleration model are: (1) It allows the use of a single galactic escape path length relation of the form: $\lambda_e = \lambda_{e,o} (R/10 \text{ GV})^\epsilon$, in g/cm^2, rather than two separate expressions for $R > R_o$ and $R < R_o$, where R is rigidity in units of GV and R_o is some arbitrary rigidity cutoff parameter. Seo and Ptuskin (1994) and Heinbach and Simon (1995) also find that the exponent ϵ for leakage from the galaxy fits the value $\epsilon = -1/3$, which is consistent with a Kolmogorov type spectrum resulting from scattering by hydromagnetic waves. Fig. 11, based on Heinbach and Simon (1995) compares the escape path length distribution of the standard and distributed (or diffusive reacceleration) models. (2) The model admits a smaller value of $|\epsilon|$ (< 0.6 of the standard model) which seems to be required to explain the observed small, near-zero anisotropy of cosmic rays, even at air shower energies.

A test of the distributed reacceleration models is shown in Figs. 12 and 13 taken from Letaw, Silberberg and Tsao (1993). Fig. 12 compares the ratio of $(B/C)/(B/C)_s$ as a function of energy, where $(B/C)s$ refers to that obtained from the standard model. As can be seen in Fig. 13, at high energies, the reacceleration models predict a flatter B/C ratio than what the standard leaky box model would suggest. A further test for the distributed reacceleration models is provided by the decay of the electron capture nuclides ^{49}V into ^{49}Ti and ^{51}Cr into ^{51}V. According to Lukasiak, McDonald and Webber (1997) and Soutoul et al. (1995), about one third of ^{49}V and ^{51}Cr has decayed at energies of a few hundred MeV/n. Based on these isotopes' production cross sections, the original abundances of ^{49}Ti and ^{51}V are low. It appears that their abundances have been boosted by a factor of about 2.5. It is interesting to note that the energy dependence of ^{49}Ti/^{49}V and ^{51}V/^{51}Cr is sensitive to distributed reacceleration.

Based on the surviving fraction of the radioactive isotope ^{10}Be, the galactic confinement time of cosmic rays has been estimated from the ratio ^{10}Be/^{9}Be. Recently, the isotopes ^{26}Al, ^{36}Cl and ^{54}Mn have also been used to infer the confinement time. With only the measurements with the smallest standard deviations are given, Table 1 below lists some of the inferred confinement times of cosmic rays using these isotopes as chronometers. The average confinement time is about $14 \pm 3 \times 10^6$ years. Based on that, the mean density of the confinement volume is about 0.3 atoms cm^{-3}. The half-life of ^{10}Be is 1.6×10^6 years. Taking time dilation into account, at 10 GeV/n most of the ^{10}Be nuclei should survive. Now, if the confinement time scales with the mean pathlength traversed, a confinement time of $\leq 10^7$ years can then be inferred from the decay of ^{10}Be at 10 GeV/n. The half-life of ^{54}Mn, when stripped of electrons, is 1×10^6 years.

5. Source Composition and the Origin of Cosmic Rays

The elemental and isotopic abundances of cosmic rays (CR), when corrected for spallation in the interstellar gas back to the source(s) tend to resemble the general abundances, e.g. Grevesse and Anders (1989), but also display some significant differences. The general abundances (GA) are based on solar spectra and carbonaceous chondrite meteorites. The elements H, He and N are underabundant in CR by a factor of 20-30 relative to GA. The nuclide ^{20}Ne is underabundant in CR by a factor of 6, the elements O, S, Ar and Kr by a factor of about 4, and C, Zn, Se, Xe and the nuclide ^{22}Ne by a factor of 2. Also, the r-process part of lanthanides and Pt is overabundant in CR by a factor of 2, and the r-process actinides (Th with U) are overabundant in CR by a factor of 4.

Three main models have been proposed to explain the source composition. One model (Model 1) assumes the initial injection at normal stars like the sun, with photosphere-to-corona particle escape dependent on the first ionization potential (FIP). If the FIP is < 10 eV, which corresponds to a temperature of 10^4 K, these low-FIP elements are charged, and tend to escape more easily along the magnetic field lines; hence the higher abundances. This stellar-injected component is supplemented by the Wolf-Rayet star component which contributes significantly to the abundance of ^{12}C, ^{16}O and ^{22}Ne, as well as a fraction of ^{4}He. The FIP-dependent suppression of high-FIP elements was proposed by Havnes (1971), Casse, Goret, Cesarsky (1973) and Meyer (1985). Flare stars as injectors of cosmic-ray nuclei was proposed by Shapiro (1997). The Wolf-Rayet contribution was proposed by Meyer (1981). Fig. 14 illustrates the main feature of this model showing the ratio (CR source)/GA vs. FIP. An additional process is, clearly, needed to account for features in the elements H, He, N and the nuclide ^{20}Ne. Silberberg and Tsao (1990) proposed such a procedure, wherein a rigidity-dependent suppression of ions near 1 MeV/n, i.e. roughly at flare-particle energy is assumed to take place. Heavier nuclei, due to multiple electron pickup have $Z_{eff}/Z < 1$, hence they have a relatively higher rigidity, and, in turn, undergo easier escape from the stellospheres. Equation (1) of Silberberg and Tsao (1990) relates CR source to GA, and their Table 1 compares the calculated and observed source abundances. The fit between the 18 elemental and isotopic abundances is about 20%.

The data of Binns et al. (1989) on the heaviest nuclei imply that the r-process nuclei beyond the r-process Te-peak are enhanced by a factor of 2 (i.e. lanthanides and Pt), and those beyond the r-process Pt peak (i.e. the actinides) are enhanced by a factor of 4. This could either be due to galactic evolution after the formation of the solar system, or due to the concentration of the sun in a region where the heaviest r-process elements are produced less abundantly.

In another model (Model 2), preferential acceleration of non-volatile or refractory elements, whose condensation temperature is about 1000 K, is assumed. In this model the grains are accelerated by supernova remnant shock waves, broken up, and then accelerated by supernova shock waves together with the relatively less abundant volatile nuclei. This model was proposed by Bibring and Cesarsky (1981), Sakurai (1990), and discussed in detail by Meyer, Drury and Ellison (1997). Fig. 15, taken from the latter paper, shows the CR-source/solar-abundances vs. atomic mass number A for refractories, semivolatiles, volatiles and highly volatiles elements. The ratios can be represented by the simple relation CR-source/GA $= a_i A^{b_i}$, with $i = 1, 2, 3, 4$, but with special cases for the Wolf-Rayet components of ^{12}C, ^{16}O, ^{22}Ne. The exponent b_i implies a mass or charge dependent acceleration process. Volatility is related to FIP, but not without exceptions, which should permit a critical evaluation of the merits of Model 1 vs. Model 2. The elements Ge and Pb are low-FIP but volatile. The elements Na, Ga and Cu are low-FIP but semivolatile, and P is high-FIP and semivolatile. The abundances of Ge and Pb, when based on meteoritic GA values, favor Model 2; but when based on solar spectrum GA values, however, they tend to favor Model 1. Ga and Cu abundances are consistent with Model 1. The abundance of Na, on the other hand, is consistent with both models if one adopts the Silberberg and Tsao (1990) equation for light-ion suppression. The abundance of P favors Model 2, alas by less than two standard deviations. A critical test between Models 1 and 2 can be provided by measuring the relative abundance of Ge in solar flares or solar wind to see whether it agrees with the solar spectral estimate. Also, the unpublished cross section measurement of S into P of the Transport Collaboration should be as critical.

The third model (Model 3) was recently proposed by Lingenfelter, Ramaty and Koslovsky (1998), where cosmic rays originate in freshly formed material from supernovae, particularly grains in young supernova remnants. In this model, the "age" of cosmic rays after nucleosynthesis should be similar to the confinement time deduced from, e.g. ^{10}Be. With such a relatively "short" age, 10^7 to 2×10^7 years, trans-uranic nuclei like Pu and Cm should survive. From Blake and Schramm (1974) one can deduce that for an "age" of 10^7 to 2×10^7 years, the ratio (U+Pu+Cm)/Th is ≈ 7. The observed ratio from LDEF (Long Duration Exposure Facility) from Thompson et al. (1993), Domingo et al. (1995) and Keane et al. (1997) is (U ≈ 6.5)/(Th ≈ 8.5) = 0.7 ± 0.3 with Cm $\approx 0 \approx$ Pu. Fig. 16 shows the observed LDEF value along with the predicted value from Model 3, which is, again, ≈ 7 for 10^7 to 2×10^7 years confinement time. The observed value implies an origin in stars or interstellar medium, derived from supernovae over 10^9 years ago.

The model of Lingenfelter et al. (1998) was proposed on the basis of the linear increase of Be relative to heavier nuclei in stars formed at various stages of galactic nucleosynthesis. (Be is formed by spallation of, or induced by, cosmic rays.) If cosmic rays are derived from the interstellar medium (either directly or via flare particles), the early-generation cosmic rays would have such a preponderance of H and He, so much so that not enough Be (or B) would be formed. This is especially so because the early interstellar medium also has few C,N,O nuclei. On that basis, one may conclude that cosmic rays are likely to have formed out of a concentrated (i.e. relatively undiluted) and relatively freshly nucleosynthesized source.

How is this to be reconciled with the data on the composition of actinides of Fig. 16? The model of pre-supernova stellar wind particles with enhanced H- and He- burning products, i.e., He, C, N, O, of Silberberg et al. (1990) provides an adequate concentration of newly nucleosynthesized C, N, O, that when accelerated by the shocks of the supernova remnant later on, does yield the observed abundances of Be and B nuclei from spallation. This aspect of the model can be further combined with some subsequent contributions, as described by Models 1 and/or 2, to form a consistent, with observed composition, model for the origin of galactic cosmic rays.

6. Summary

We have attempted to briefly underscore the importance and relevance of nuclear cross sections to various aspects of cosmic rays studies, from species transformation in their transit, to their very source and origin. ¿From this particular exposé, we can list the following main points:

- Based on recent (after $\sim$1990) experimental cross sections, the semiempirical cross section predictions for proton-nucleus reactions have been improved. The calculations have also been extended to include nucleus-nucleus interactions, in addition to further refinements, e.g., contribution from electromagnatic dissociation.

- Based on the work of Blake and Schramm (1974), as well as recent observational data on cosmic-ray actinides, the nucleosynthesis of most actinide nuclei seem to have occurred over 10^9 years ago, including the nuclei which subsequently became cosmic rays about 10^7 years ago.

- The observed small anisotropy of ultra-high energy cosmic rays favors a small ($\approx 1/3$) exponent for the rigidity-dependent leakage of cosmic rays from the galaxy, consistent with weak distributed (or diffusive reacceleration) models.

- A test for FIP dependent vs. volatility dependent injection of cosmic rays is suggested; from the relative abundance of Ge in the solar wind or solar flares. Also, the source abundance of phosphorus is important and needs to be determined. The publication of $\sigma(S \to P)$, from the hydrogen-target experiments, in particular, is critical to further refine our (often competing) models for the origin of galactic cosmic rays.

The authors wish to acknowledge the support of NASA (NAG5-5053,5165) and an enlightening preprint on cosmic-ray origin from Dr. Ramaty. A.F.B. acknowledges the support of NASA-JOVE grant no. NAG8-1208.

REFERENCES

Axford, W.I. (1982), 17th ICRC (Paris) 12, 155.

Barghouty, A.F., Fai, G. and Keane D. (1991), Nucl. Phys. A, 535, 715.

Bibring, J.P. and Cesarsky, C.S. (1981), Proc. 17th ICRC (Paris) 2, 289.

Binns, W.R. Garrard, T.L., Israel, M.H., Klarmann, J., Stone, E.D. and Waddington, C.J. (1987), Phys. Rev. C36, 1820.

Binns, W.R. et al. (1989), Ap. J. 346, 997.

Blake, J.B. and Schramm, D.N. (1974), Ap. Space Sci. 30, 275.

Brechtmann, C. and Heinrich (1988a), Z. Phys. 331, 463.

— — —.(1988b), Z. Phys. 331, 407.

Casse, M., Goret, P. and Cesarsky, C.S. 1975, Proc. 14th ICRC (Munich) 2, 646.

Chen, X. et al. (1995), Proc. 24th ICRC (Rome) 3, 184.

− − −.(1997a), Phys. Rev. C56, 1536.

− − −.(1997b), Ap. J. 179, 504.

Connell, J.J. 1997, 25th ICRC (Durban) 3, 385.

Connell, J.J. and Simpson, J.A. (1977) 25th ICRC (Durban) 3, 393.

Connell, J.J., DuVernois, M.A.. and Simpson J.A. 1997, 25th ICRC (Durban) 3, 797

Cumming, J. B., Stoenner, R.W. and Haustein, P.E. (1976), Phys. Rev. C14, 1554.

Cummings, J.R. et al. (1980), Phys. Rev. C42, 2530.

Domingo, C. Font, S., Buixeras, C. and Fernandez, F. (1995), 24th ICRC (Rome) 2, 572.

DuVernois, M.A. (1997) Ap. J. 481, 241.

Engelmann, et al. (1990) Astron. and Ap. 233, 98.

Garrard, T.L., et al. (1995) Proc. 24th ICRC (Rome), 3, 192.

Geer, L.Y., Klarmann, J., Nilsen, B.S., Waddington, C.J., Binns, W.R. Cummings, J.R. and Garrard, T.L. (1995) Phys. Rev. C52, 334.

Giler, M., Osborne, J.L., Ptuskin, V.S., Szabelska, B., Wdowczyk, S. and Wolfendale, A.W. (1988) Astron. Ap. 217, 311.

Grevesse, N. and Anders, E. (1989), in Cosmic Abundances of Matter, ed. J.C. Waddington, AIP Conf. Proc. 183, New York, P. 1.

Havnes, O. (1971), Nature 229, 548.

Hayakawa, S. (1969), in Cosmic Ray Physics, New York, Wiley.

Heinbach, N. and Simon, M. (1993), 23rd ICRC (Calgary) 2, 263.

− − −.(1995) Ap. J. 441, 209.

Hill, J.C., Wohn, F.K., Winger, J.A., Khayat, M., Leininger, K. and Smith, A.R. (1988), Phys. Rev. C38, 1322.

Keane, A.J., Thompson, A., O'Sullivan, D., O'C.Drury, L. and Wentzel, K.P. (1997) 25th ICRC, (Durban) 3, 361.

Knott, C.N. et al. (1993) Proc. 23rd ICRC (Calgary) 2, 187.

− − −.(1996), Phys. Rev. C53, 347.

− − −.(1997), Phys. Rev. C56, 398.

Letaw, J.R., Silberberg, R. and Tsao, C.H. (1993), Ap. J. 414, 601.

Lingenfelter, R. E., Ramaty, R. and Kozlovsky, B. (1998), Ap. J. Letters 500, 153.

Lukasiak, A., McDonald, F.B. and Webber, W.R. (1997) 25th ICRC (Durban), 3, 357.

Meyer, J.P. (1981), 17th ICRC (Paris) 2, 265.

— — —.(1985), Ap. J. Suppl. 57, 173.

Meyer, J.P., O'C.Drury, L., and Ellison, C. (1997), Ap. J. 487, 182.

Nilsen, B.S., Waddington, D. J.,, Cummings, J.B., Garrard, T.L. and Klarmann, J. (1993), Phys. Rev. C52, 3277.

Norbury, J.W. and Townsend, L.W. (1993) Ap. J. Suppl. 86, 302.

Norbury, J.W. and Muller, C.M. (1994), Ap. J. Suppl. 90, 115

Rudstam, G. (1966) Z. Naturforsch. A21, 1027.

Sakurai, K. (1990), in Cosmic Rays, Supernovae and the Interstellar Medium, ed. M.M. Shapiro, R. Silberberg and J.P. Wefel, Kluwer Acad. Publ., Dordrecht.

Shapiro, M.M. (1994) 25th ICRC (Durban) 4, 353.

Seo, E.S. and Ptuskin, V.S. (1994), Ap. J. 431, 705.

Sihver, L., Tsao, C.H., Silberberg, R. Kanai, T., and Barghouty, A.F. (1993), Phys. Rev. C47, 1225.

Silberberg, R. and Tsao, C.H. (1973a), Ap. J. Suppl. 25, 315.

— — —.(1973b), Ap. J. Suppl. 25, 335.

— — —.(1990a), Physics Reports 191, 351.

— — —.(1990b), Ap. J. Letters 352, L49.

Silberberg, R., Tsao, C.H. and Letaw, J.R. (1985), Ap. J. Suppl. 58, 873.

Silberberg, R., Tsao, C.H., Letaw, J.R. and Shapiro, M.M. (1983), Phys. Rev. Letters 51, 1217.

Silberberg, R., Tsao, C.H. Letaw, J.R. (1993), in Particle Astrophysics and Cosmology, ed. M.M. Shapiro, R. Silberberg and J.P. Wefel, Kluwer Acad. Publ., Dordrecht, P. 1.

Silberberg, R., Tsao, C.H., Shapiro, M.M. and Biermann, P.L. (1990) Ap. J. 363, 265.

Silberberg, R., Tsao, C.H. and Shapiro, M.M. (1995), 24th ICRC (Rome) 3, 41.

— — —.(1998) in *Toward the New Millenium in Astrophysics*, eds. M.M. Shapiro, R. Silberberg and J.P. Wefel, World Scientific Publisher, Singapore, to be published.

Silberberg, R. Tsao, C.H. and Barghouty, A.F. (1998), Ap. J. 501, 911.

Simon, M., Heinrich, W. and Mathis, K.D. (1986), Ap. J. 300, 32.

Soutoul, A. Legrain, R. and Webber, W.R. (1995), 24th ICRC (Rome) 3, 200.

Stephens, S.A. (1990) 21st ICRC (Adelaide), 3, 241.

Stephens, S.A. and Golden, R.L. (1990), 21st ICRC (Adelaide) 3, 353.

Thompson, A. O'Sullivan, P., Wentzel, K.P., Bosch, J., Keegan, R., Domingo, C. and Jansen, F. (1993) 23rd ICRC (Calgary) 1, 603.

Tsao, C.H., Silberberg, R. and Barghouty, A.F. (1998) Ap. J. 501, 920.

Tsao, C.H., Silberberg, R., Barghouty, A.F., Sihver L., and Kanai, T. (1993), Phys. Rev. C47, 1257.

Tull, C.E. et al. (1993), 23rd ICRC (Calgary), 2, 163.

Waddington, C. J., Cummings, J.R., Garrard, T., Hink, P. and Nilsen B.C. (1997) 25th ICRC (Durban) 3, 345.

Wandel, A. Eichler, D.S., Letaw, J.R., Silberberg, R. and Tsao, C.H. (1988) Ap. J. 316, 676.

Webber, W.R., Kish, J.C. and Schrier, D.A. (1990), Phys. Rev. C41, 547.

Westfall, G.D., Wilson, L.W. Lindstrom, P.J., Crawford, H.J., Greiner, D.E. and Heckman, H.H. (1979), Phys. Rev. C14, 1309.

Wiedenbeck, M.E. and Greiner, D.E. (1980), Ap. J. 239, L139.

Zeitlin, C. et al. (1997), Phys. Rev. C56, 388.

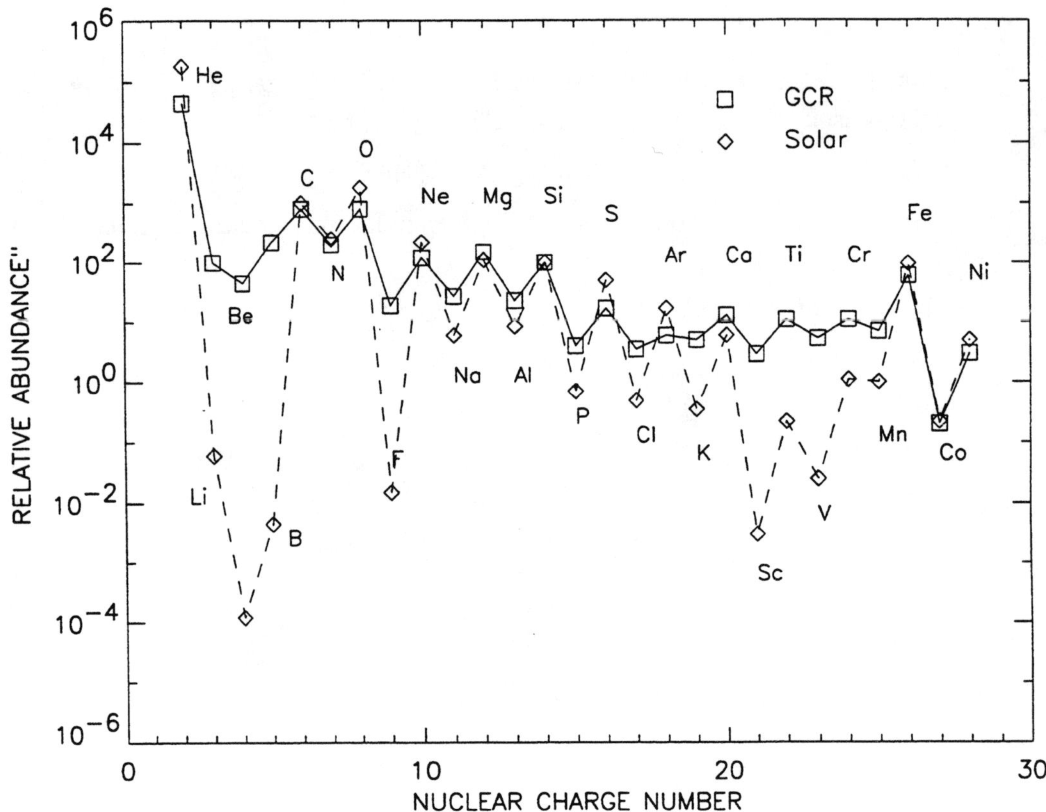

Fig. 1.— The cosmic-ray elemental abundances (He to Ni) measured at earth, at energies 0.1 to 2 GeV/n, compared to solar system abundances, normalized at Si at 100. Squares, connected by the solid line, depict the cosmic ray abundances, while diamonds, connected by the dashed line, depict the solar system abundances.

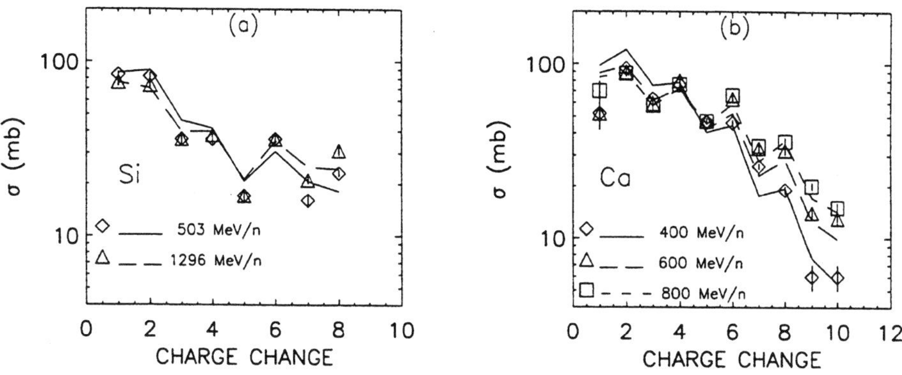

Fig. 2.— The charge-change cross sections of Si and Ca. The measurements are from Knott et al. (1997). The calculated values are connected by lines to help visualize the trend in the data.

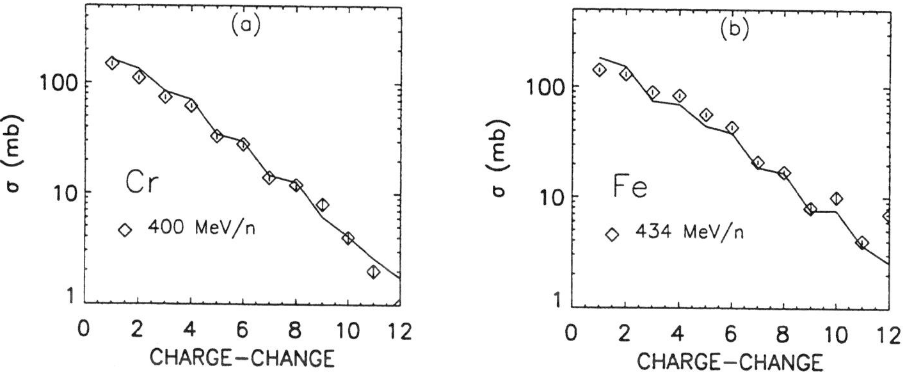

Fig. 3.— The charge-change cross sections of Cr and Fe. The data are from Knott et al. (1997). The calculated values are connected by lines.

Table 1. Inferred confinement time of cosmic rays

Isotope	Half-life	Confinement time (10^6 yr)	Reference
^{10}Be	1.6×10^6 yr	8.4 (+4.0, −2.4)	Wiedenbeck and Greiner (1980)
^{10}Be	1.6×10^6 yr	18 (± 3)	Connell (1997)
^{26}Al	7.4×10^5 yr	15.6 (+2.5, −2.6)	Connell and Simpson (1997)
^{36}Cl	3.1×10^5 yr	11 (± 4)	Connell et al. (1997)
^{54}Mn	1.0×10^6 yr	14 (+6, −4)	Du Vernois (1997)
			Connell et al. (1997)

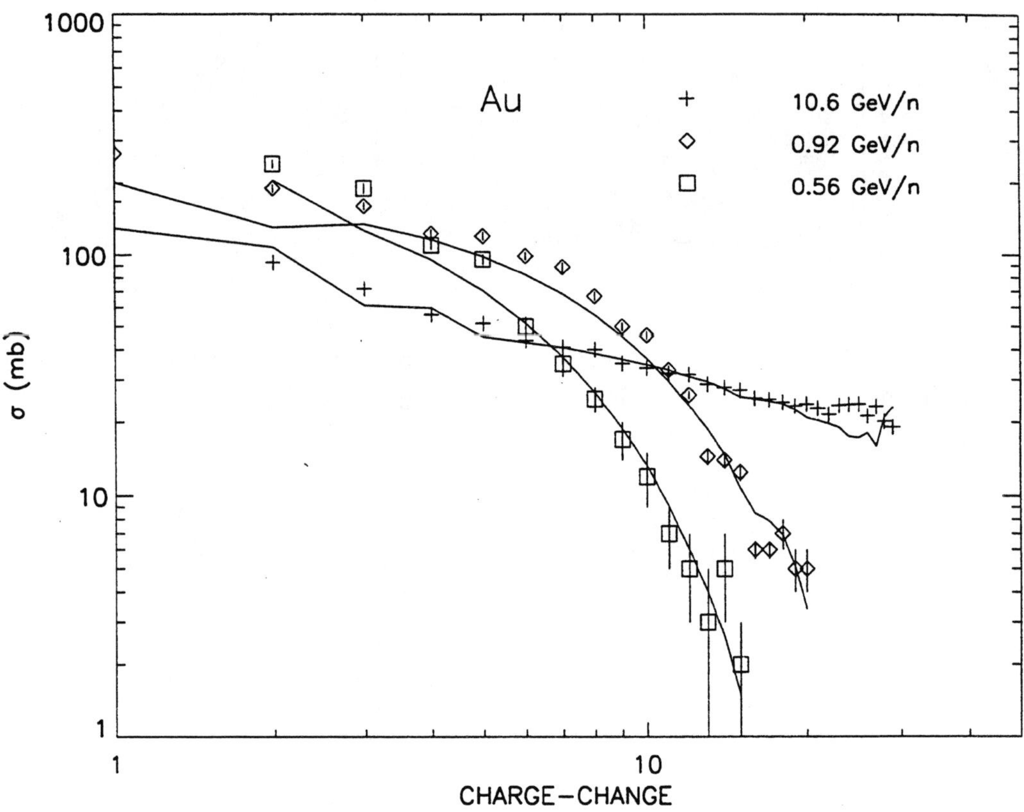

Fig. 4.— The charge-change cross sections of Au. The data are from Geer et al. (1995) and Garrard et al. (1995). Solid lines connect the calculated values.

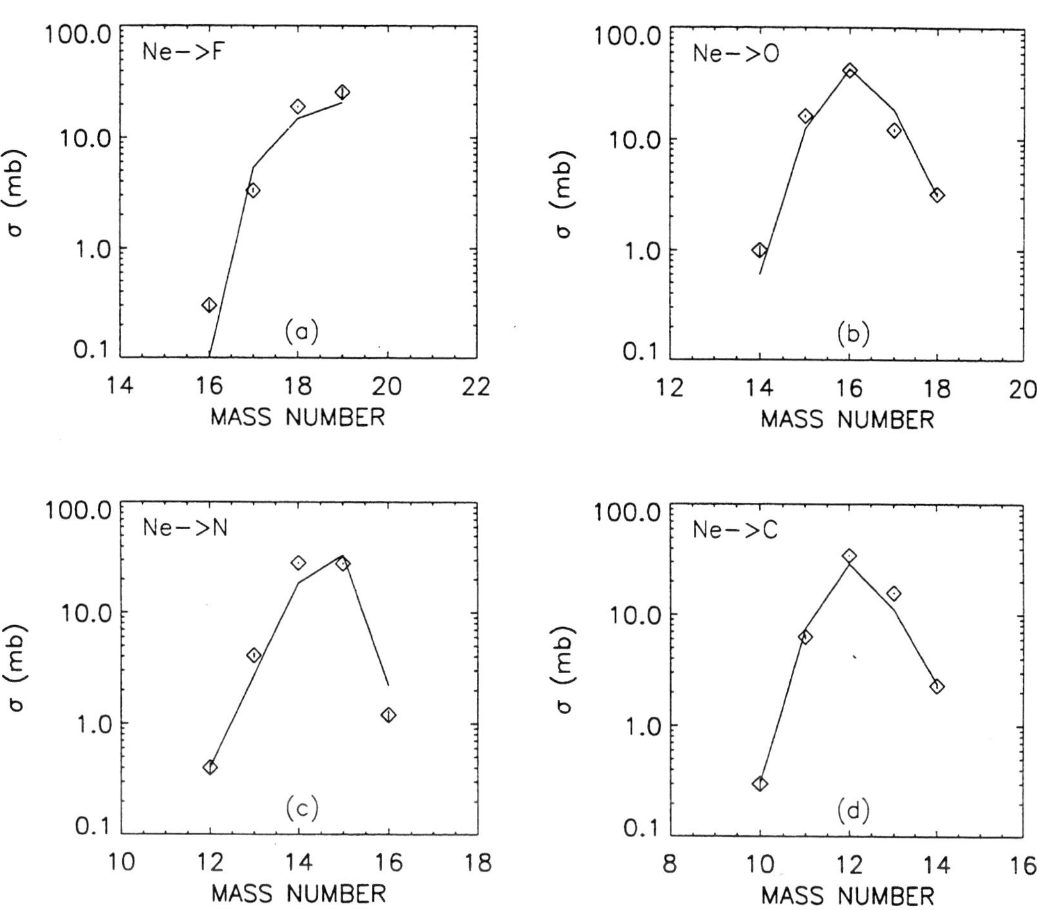

Fig. 5.— The isotopic yields of F, O, N and C from ^{20}Ne at 0.6 GeV/n. Solid lines connect the calculated values. The measurements are those of Webber et al. (1995).

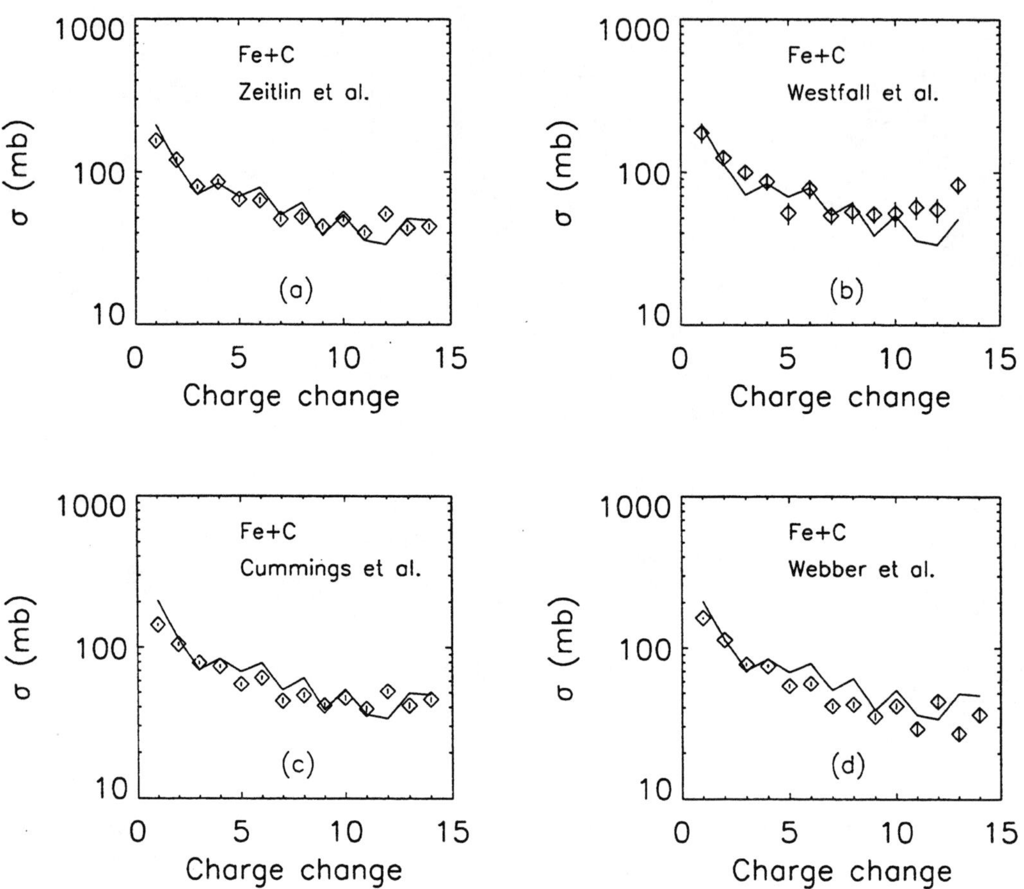

Fig. 6.— Charge-change cross sections of Fe (1.06 GeV/n) on C, measured respectively by Zeitlin et al. (1997), Westfall et al. (1979), Cummings et al. (1990) and Webber et al. (1990). Solid lines connect the calculated values.

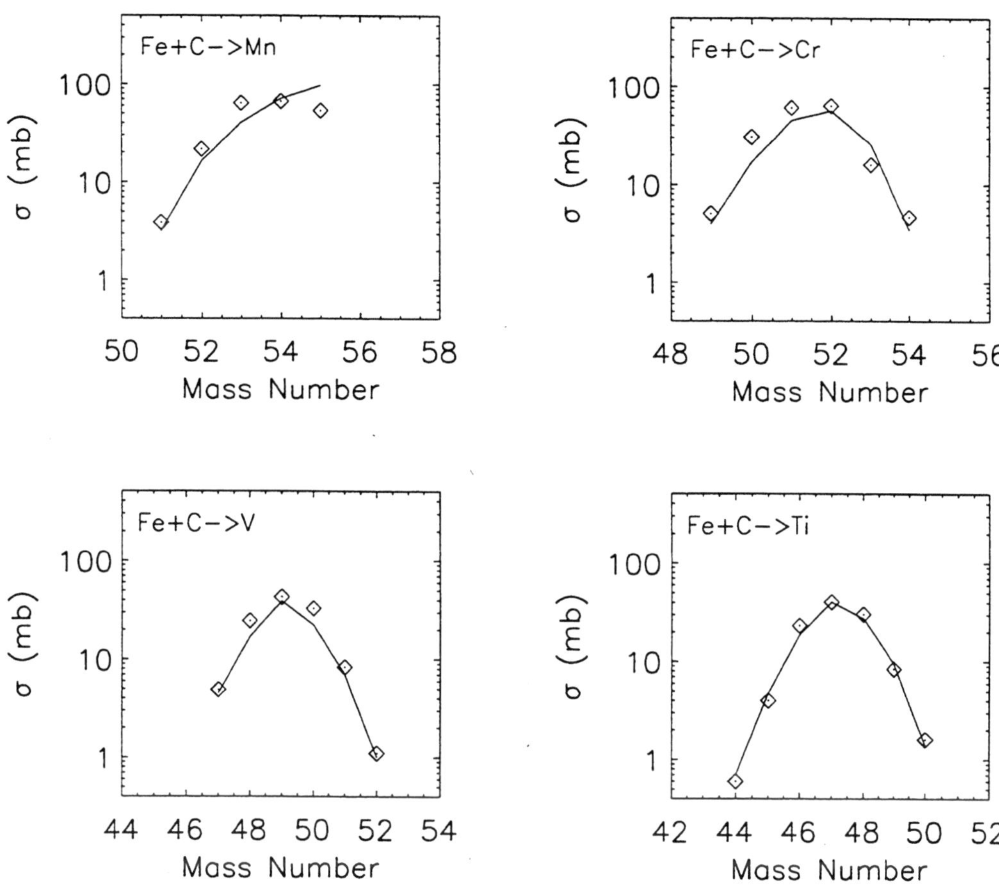

Fig. 7.— Isotopic yields of Fe at 0.6 GeV/n, incident on C, breaking into Mn, Cr, V and Ti. The measured values are from Webber et al. (1990). The lines connect the calculated values.

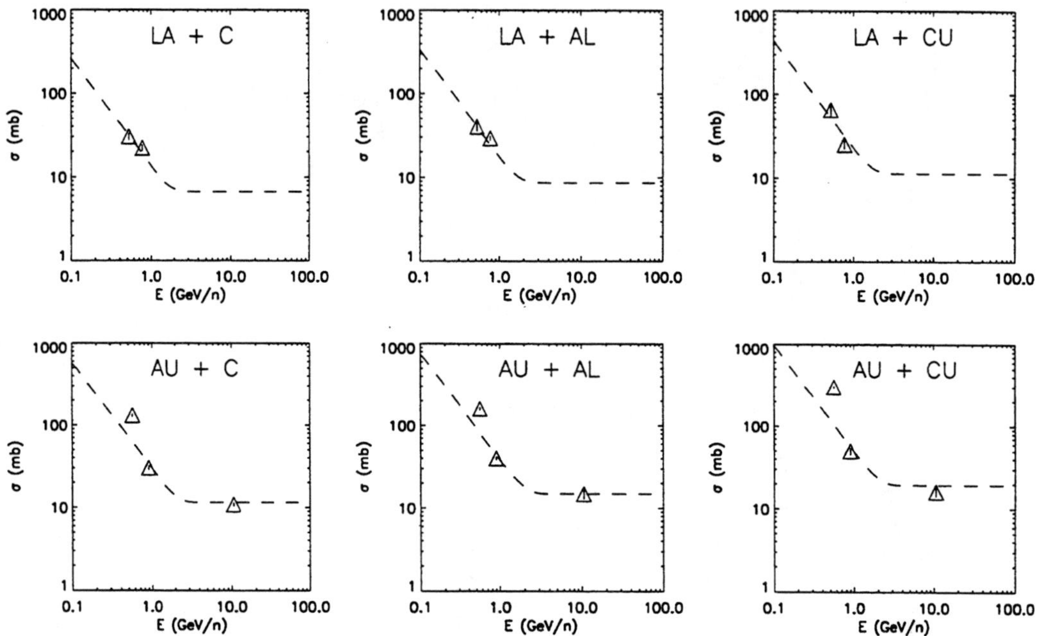

Fig. 8.— Charge-pickup cross sections of La and Au with C, Al and Cu. Dashed lines connect the calculated values. Data are from Nilsen et al. (1994) and Geer et al. (1995).

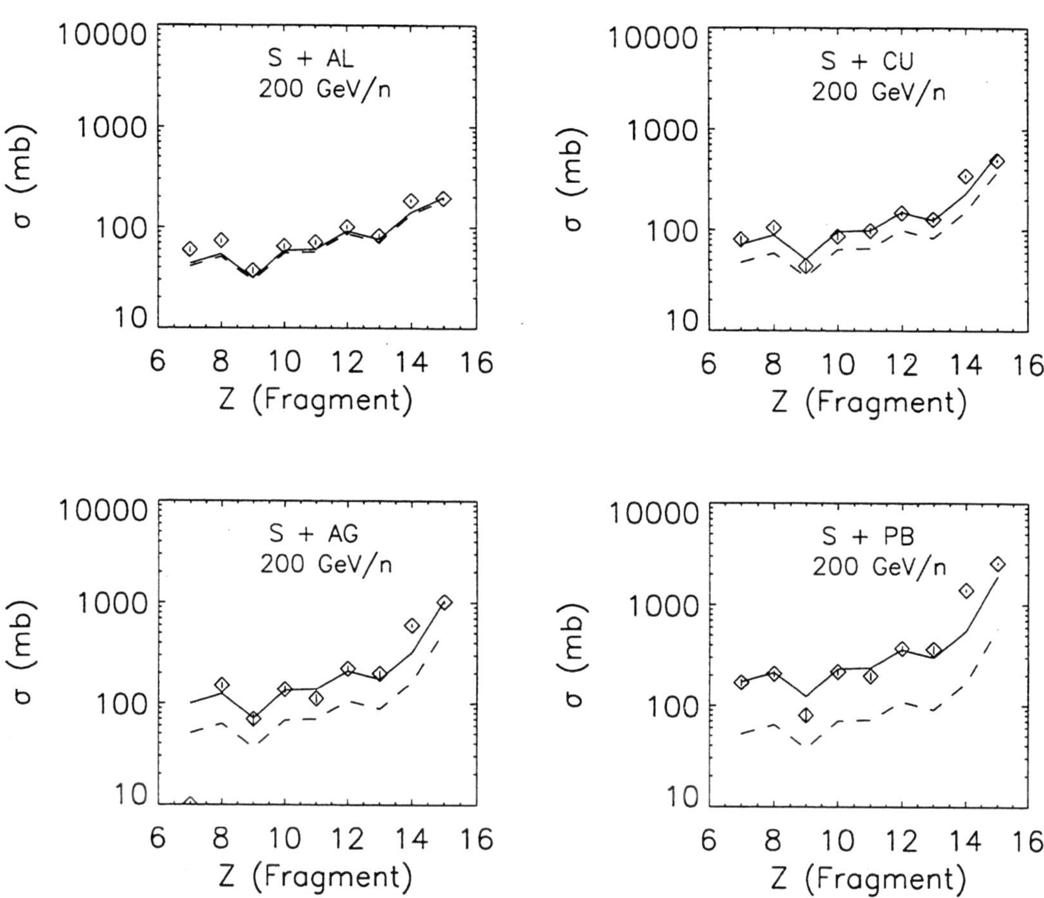

Fig. 9.— Fragmentation cross sections of S at 200 GeV/n with Al, Cu, Ag and Pb targets along with the calculated values before (dashed lines) and after (solid lines) correcting for electromagnetic dissociation. Data are from Brechtmann et al. (1988).

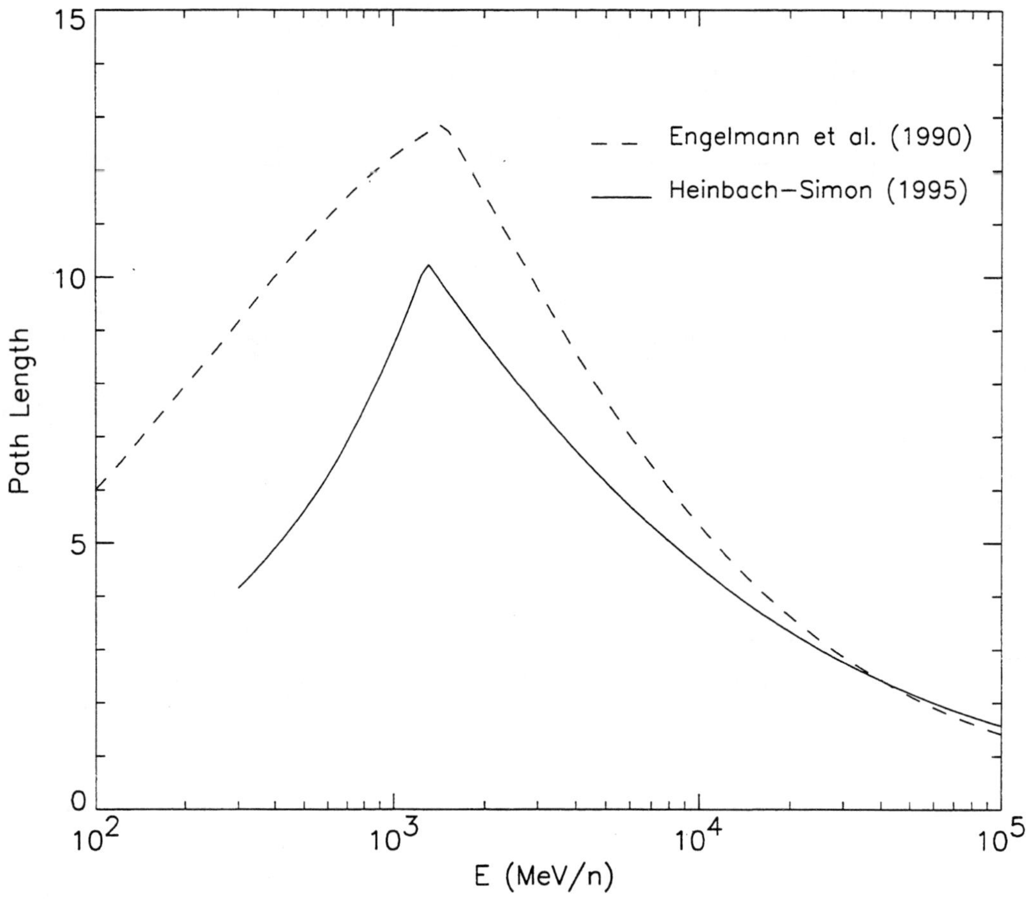

Fig. 10.— Energy dependence of Galactic escape path length. The curves shown are those of Engelmann et al. (1990) and Heinbach and Simon (1995).

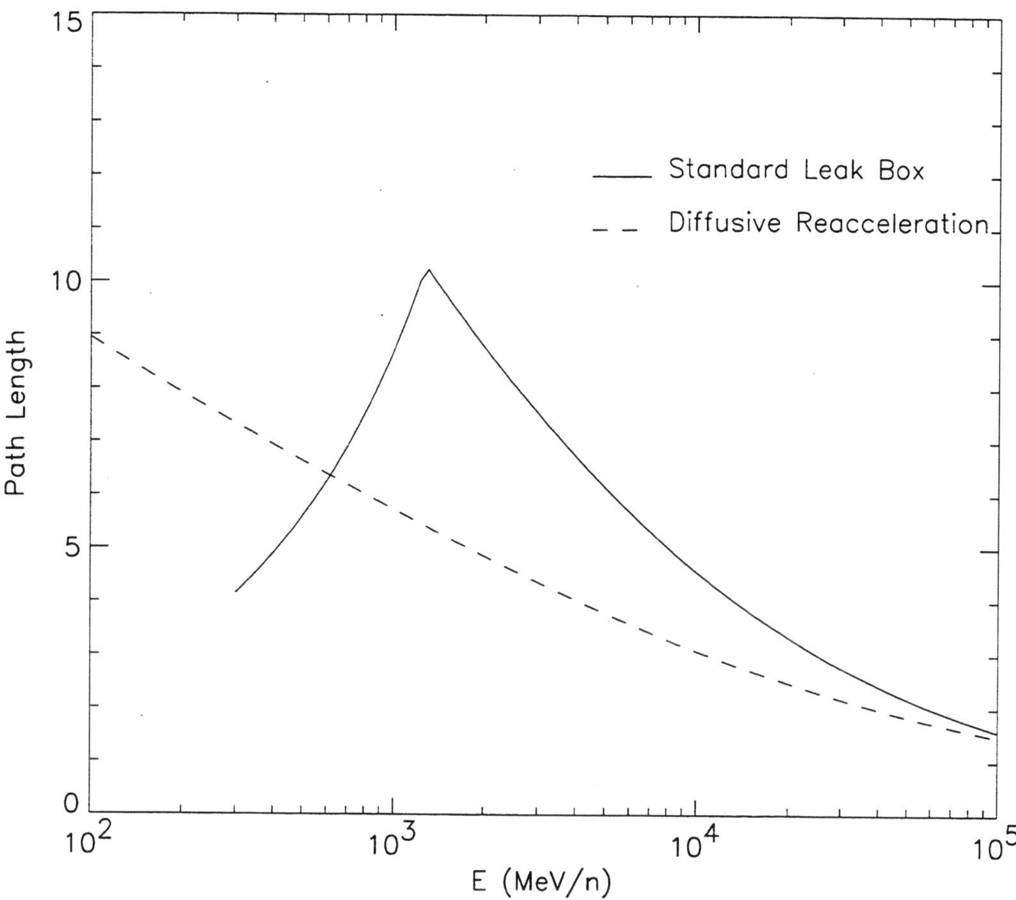

Fig. 11.— The energy dependence of galactic escape path length in the framework of the standard leaky box models vs. reacceleration with a strength of 0.6 for diffuse reacceleration. The curves shown are from Heinbach and Simon (1995).

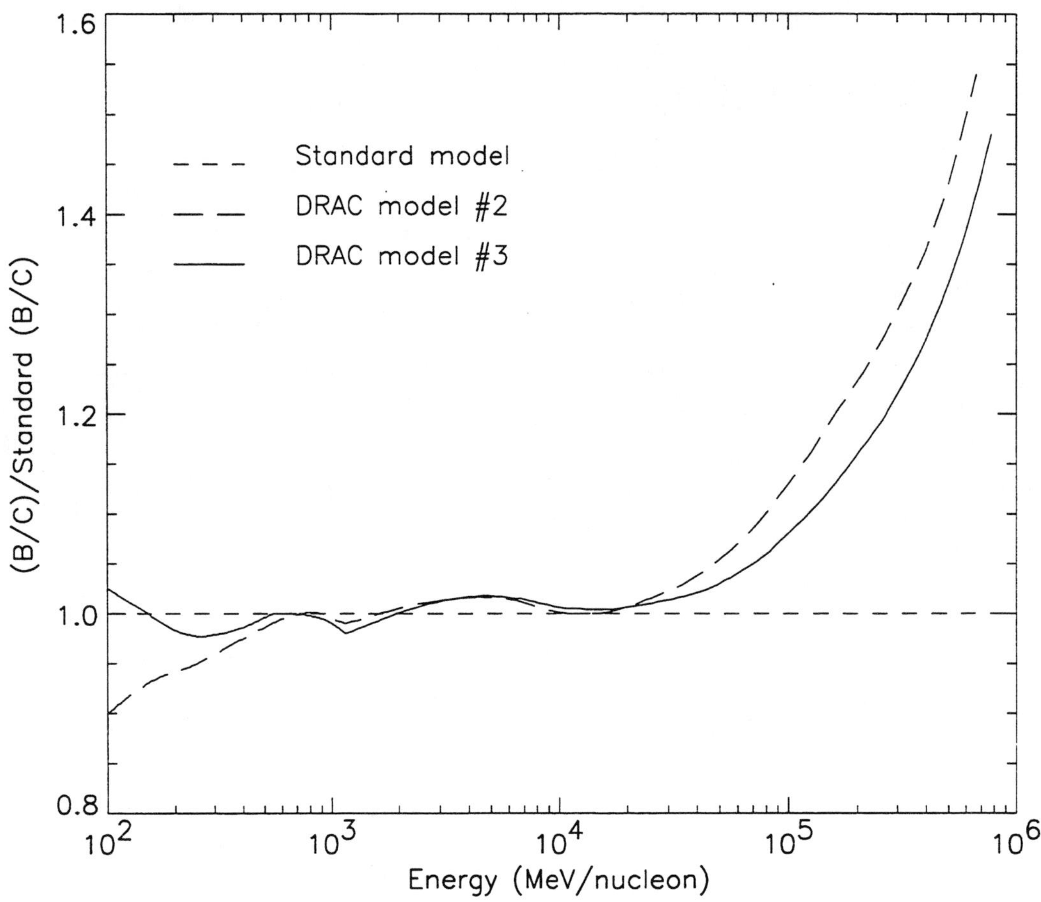

Fig. 12.— Comparison of B/C ratio predicted by the standard leaky box model to that by the distributed reacceleration models, due to Letaw, Silberberg and Tsao (1993). Results are shown as a normalized ratio to the standard model results.

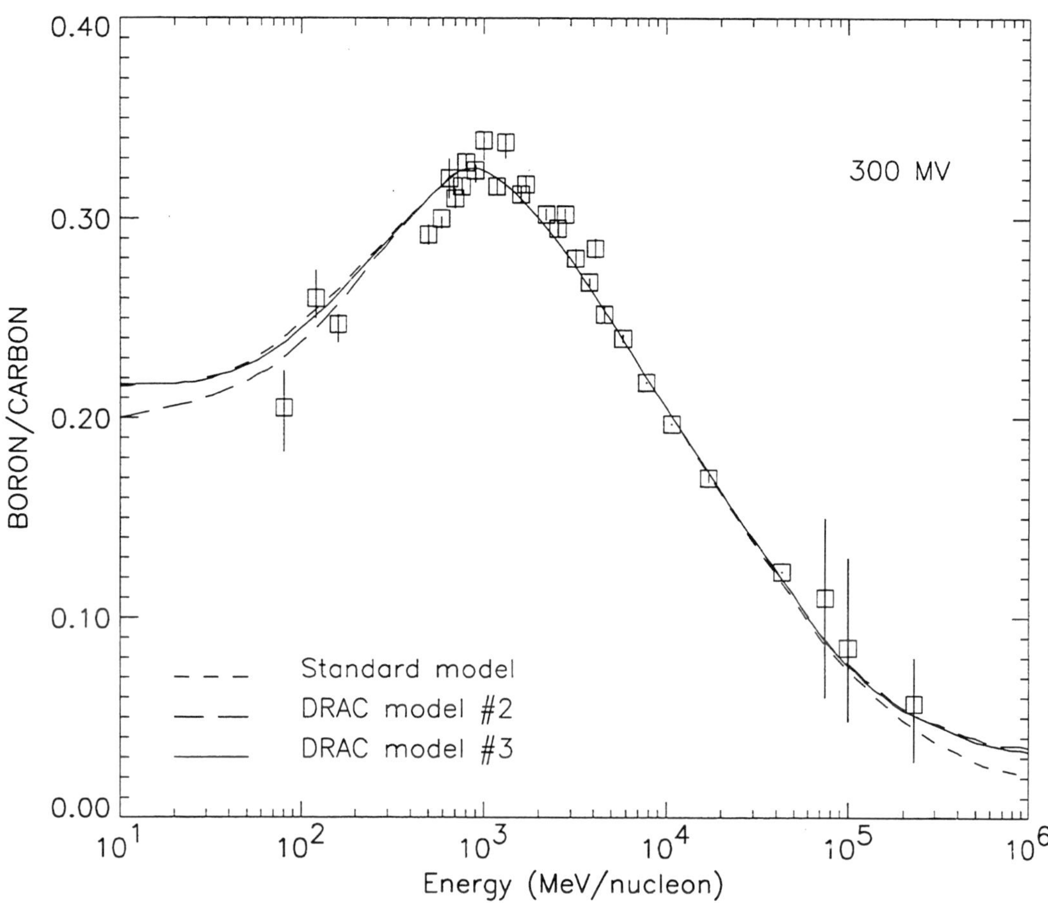

Fig. 13.— Same as Fig. 12, but without the normalization, and with data compiled from several observations. Model predictions are modulated by a potential of 300 MV in the force-field approximation.

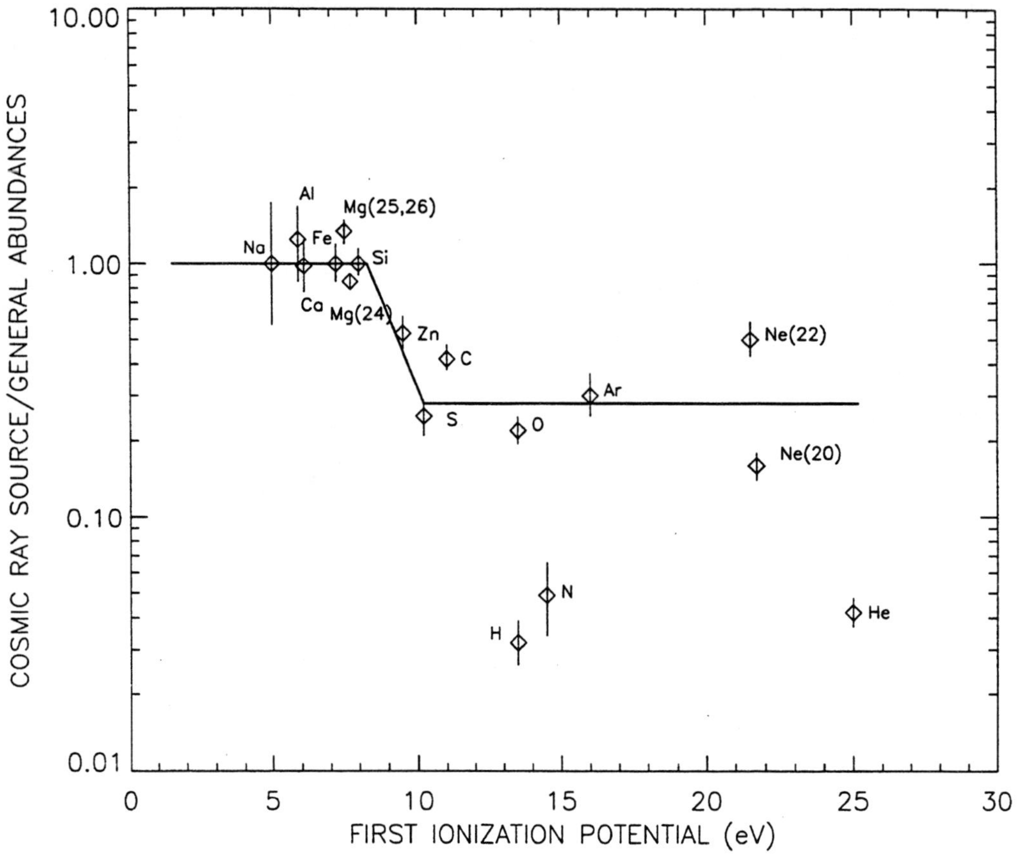

Fig. 14.— The ratio of cosmic-ray source abundances to the general abundances, from Silberberg and Tsao (1990). The calculated values of source abundances (not plotted to avoid cluttering) are within the shown error bars.

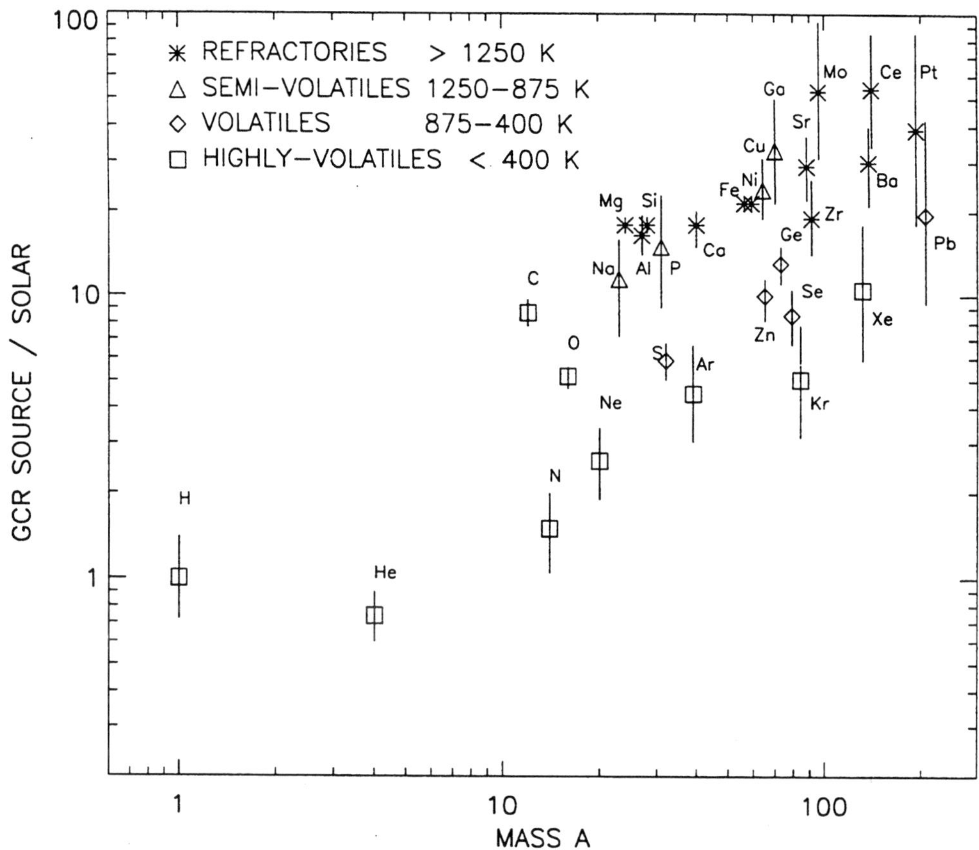

Fig. 15.— The cosmic-ray source abundances to solar (or General) abundances, as a function of atomic mass number A. These are shown for four classes of volatility. The data can be fitted by (CR/Sol) abundances $= a_i A^{b_i}$, $i = 1, 2, 3, 4$, (from Meyer et al. 1997).

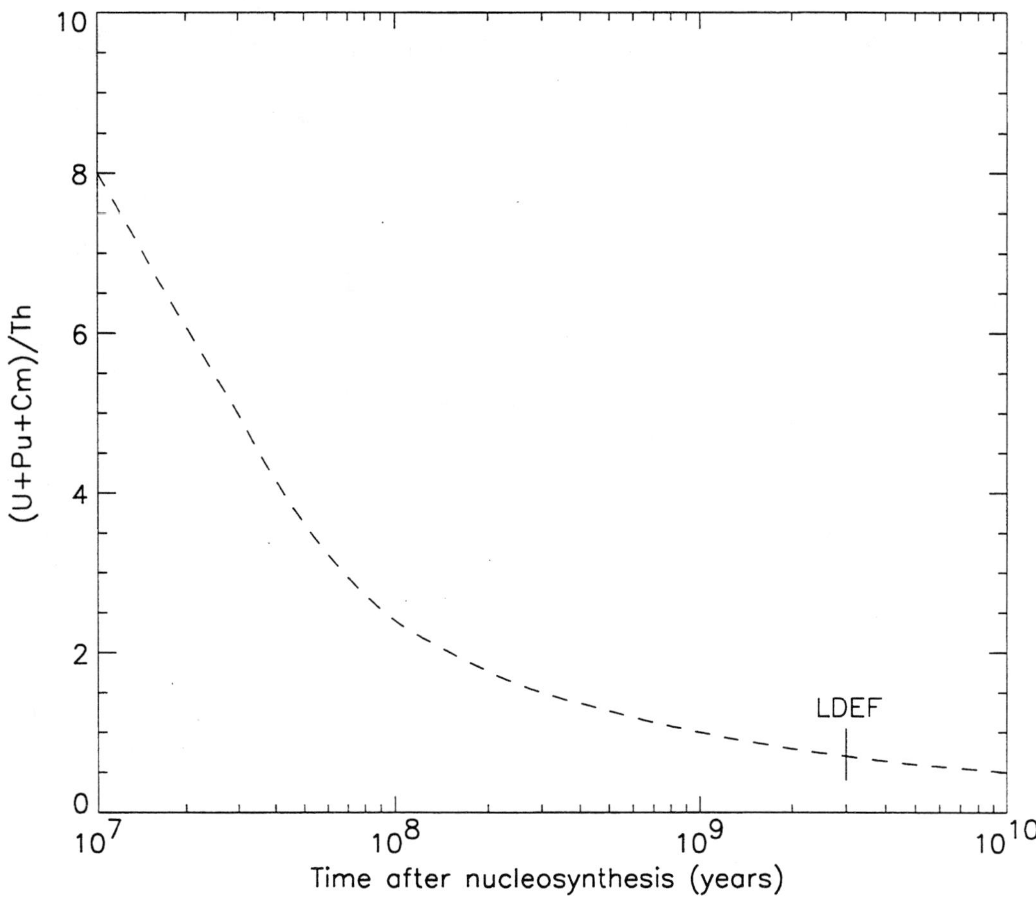

Fig. 16.— The ratio (U+Pu+Cm)/Th as a function of time after nucleosynthesis, based on Blake and Schramm (1974). If cosmic rays are injected by supernovae and accelerated by the shock waves of these supernovae, and propagated for 10^7 to 2×10^7 years before observation, the value is about 7. If the nucleosynthesized nuclei are condensed into stars, and spread out into the interstellar gas, the time is 10^9 to 10^{10} years ago, as supported by the LDEF observations. The presently observed galactic cosmic rays were accelerated out of this material about 10^7 to 2×10^7 years ago.

IV. Galactic cosmic rays: sources

Interpreting the Cosmic Ray Composition

Luke O'C Drury   Jean-Paul Meyer   Donald C Ellison

1   Introduction

Detailed composition measurements can be a very powerful means of tracing origins, a fact used regularly by forensic scientists and art historians. One of the main motivating factors for making detailed observations of cosmic rays was always the hope that a unique compositional signature could be found which pointed unambiguously to a particular source. This has proven much harder than expected, but we have now reached a point where it appears possible to begin to decipher the information contained in the compositional data; the key, we have discovered, is to read the data not in isolation, but in the context provided by our general astronomical knowledge and by recent developments in shock acceleration theory (Meyer, Drury and Ellison, 1997, 1998; Ellison, Drury and Meyer, 1997). In our view (not, it is only fair to warn the reader, yet universally accepted) the data show clearly that the Galactic cosmic ray particles originate predominantly from the gas and dust of the general interstellar medium.

2   What is the Composition?

Before attempting to interpret the data it is important to be clear about what exactly we are discussing. The raw measurements are the charge-resolved, and in some cases mass-resolved, differential energy spectra of the cosmic ray nuclei above the Earth's atmosphere. For instrumental and statistical reasons, good measurements with clean separation of the various species are only easily made for mildly relativistic nuclei. The measurements are affected by solar modulation at energies below a few GeV per nucleon. By correcting for these solar system effects we infer the local Galactic cosmic ray spectra which would be observed in the interstellar medium just outside the heliosphere. However it is clear that these in turn have been influenced, in varying degrees, by spallation nuclear reactions and other interactions during propagation through the interstellar medium. If we attempt to correct for these propagation effects we finally arrive at inferred *source spectra*. The relative fluxes of the various nuclear species at fixed energy per nucleon (equivalently, at fixed speed or Lorentz factor) in these demodulated and depropagated spectra constitute what is usually called the Galactic Cosmic Ray Source (GCRS) composition.

Quite a number of assumptions have already gone in at this stage. The heliospheric corrections are small above a few GeV per nucleon and probably uncontroversial. However the propagation corrections are clearly dependent on the propagation model used and often implicitly assume that there are distinct acceleration and propagation phases. Particularly with the current interest in so-called "reacceleration" models for propagation, it is not clear that such a sharp separation is justified. It should also be noted that most of the published data have been "depropagated" using the simple, but clearly unphysical, leaky-box model of Galactic cosmic ray confinement. By talking loosely about "the GCRS composition" without specifying precisely the energy per nucleon or rigidity at which the measurements were made we are also implicitly assuming that all the species have virtually identical spectra in energy per nucleon, an assumption which is approximately true for the main nuclear components in the range from 1 GeV to 1 TeV per nucleon (in fact, the data suggests that helium has a slightly flatter spectrum than hydrogen), but is certainly not true of the electrons[1]. It is worth noting that measuring at fixed kinetic energy per nucleon (which is the form traditionally used in experimental work) or fixed momentum per charge (ie rigidity, which is often used in theoretical work) give essentially equal relative abundances for all the heavy nuclei, but different values for the hydrogen abundance relative to the heavies.

3 Nuclear or Atomic Physics?

The obvious first thing to do is to compare the abundance pattern seen in the GCRS to the standard solar system pattern of abundances, which appears to characterise all undifferentiated bodies in the solar system including the sun itself. If corrections are made for the decay of long-lived radioactive nuclides, giving what is sometimes called the primordial solar-system or proto-solar abundance pattern, this appears to be close to the general local Galactic pattern of abundances (in as much as this can be determined); thus it has usually been taken as the base-line "standard" composition. However there is now increasing evidence that, both in the local interstellar medium (ISM) and in the surfaces of young B stars, the abundances of the heavy elements relative to hydrogen are systematically *lower* than in the sun by factors of order 1.5 to 2 (Snow and Witt, 1996, and references therein). In contrast, the solar system abundances *are* apparently typical of those in the local F and G type stars (Edvardsson et al, 1993; Andersson and Edvardsson, 1994). Bearing all this in mind we will continue to use the solar system abundances as reference values because they provide a well-determined set of values for all the elements which one might reasonably expect to be relevant to the local ISM, especially as regards the relative abundances of the heavy elements.

The GCRS and solar system abundances are compared in Fig. 1 (see table in Meyer, Drury and Ellison 1998). It is immediately obvious that the GCRS composition is disappointingly normal; all the elements are present, and the general pattern is strikingly similar in both the GCRS and the solar system. However there

---

[1]One often sees statements that the electron to proton ratio in the cosmic rays is 1 to 30 or 1 to 100. Any such statement is, however, meaningless if one does not specify how the comparison is made. The above applies to comparison at a fixed energy. If, by contrast, one compares at fixed Lorentz gamma factor the electrons are much more abundant than the protons (as pointed out by W Kundt)!

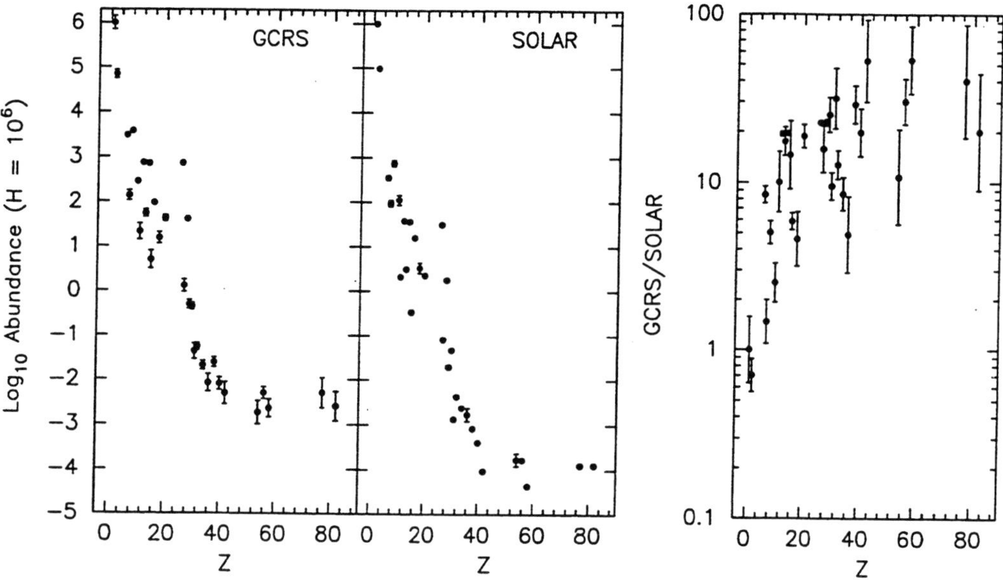

Figure 1: Plots of the GCRS and Solar abundances against atomic number for the major elements, both normalised to Hydrogen=10^6, and of the ratio GCRS/Solar. A numerical table of these values can be found in Meyer, Drury and Ellison 1998

are some significant differences, in particular Hydrogen is deficient in the GCRS, or the heavy elements are enhanced, by quite large factors relative to the solar system composition. For example iron (Fe) and silicon (Si) are about a factor 30 higher relative to hydrogen in the GCRS than in the solar system and this is a much larger factor than any uncertainty in the measurements. Note that adopting B star abundances would make this even more extreme and increase the overabundance of Fe and Si to between 45 and 60. The challenge is to interpret these slight (relative to the enormous variations between the individual elements), but clearly significant, differences between the two sets of abundances.

Now the heavy elements are known to be produced by nucleosynthesis in stars and, for many elements specifically in supernova explosions, and it has been suspected for a long time that cosmic rays are somehow linked to the supernova phenomenon (this was first suggested in the historic paper of Baade and Zwicky (1934) where they introduced the name supernova, and cogently argued for on energetic grounds by Ginzburg and Syrovatsky (1964) in their influential monograph). It is therefore very natural to seek to interpret the differences between the GCRS and local ISM (or solar system) abundances in terms of biases stemming from the nuclear physics associated with nucleosynthesis, perhaps during the slow core burning phase, but more likely during the rapid explosive phase. This effort was also stimulated by early reports suggesting high GCRS abundances of the ultra-heavy elements[2], including actinides, which are exclusively produced during the supernova explosion.

---

[2]The ultra-heavy elements are usually taken to be those with nuclear charge greater than 28, although the term is used very loosely

It is now clear that these attempts to interpret the abundance differences in terms of nucleosynthetic models are unconvincing. For example, Ne is depleted by about a factor 8 relative to Mg, Al and Na although all these elements are thought to be produced by C burning. Similarly S and Ar are depleted by factors of order 4 relative to Si and Ca although these elements are all produced by O and Si burning. No such large anomalies are found in supernova nucleosynthesis calculations, especially for elements produced in the same burning cycle (Woosley and Weaver, 1995; Timmes, Woosley and Weaver, 1995; Arnett, 1995). By contrast, Mg, Al (C burning), Si and Ca (O and Si burning) and Fe and Ni (e-process, i.e., explosive phase) are present in the GCRS in proportions within 20% of the solar values. However Mg, Al, Si and Ca are synthesised in core-collapse type II supernovae while Fe and Ni are predominantly produced in type Ia supernovae and the nucleosynthesis calculations of different supernova models typically yield deviations of these ratios by factors of order 2. In addition the ultra-heavy s-nuclei beyond $A = 90$, which are not produced in any type of supernova, are not underabundant relative to the above elements, or to the r-nuclei, and the general ultra-heavy abundances are not anomalously high. Further, with the exception of $^{22}Ne/^{20}Ne$ (and possibly $^{13}C/^{12}C$ and $^{18}O/^{16}O$) all isotopic ratios are consistent with solar values.

Remarkably, if the data are organised not by nuclear but by atomic properties some, though not *all*, of the differences can be accounted for. In particular if the ratio of the GCRS abundance to the solar system abundance is plotted against the first ionization potential (FIP) of each element a definite pattern, the so-called FIP effect, is evident (see Fig. 2). There is a large group of low-FIP elements which show a roughly constant and large enhancement of about 30 relative to H. At a FIP of order 10 eV there is a rather sharp break and the elements with larger FIP values show much smaller, but more scattered, enhancements.

Now of course the first ionization potential measures how easy it is to remove one electron from the outermost shell of electrons in the ground state of the neutral atom, and thus correlates strongly with chemistry. For example, elements with easily removed electrons tend to be metallic and chemically very reactive forming stable compounds which readily condense at high temperatures; elements with filled outer shells have very firmly attached electrons and are the inert gases which do not condense except under extreme laboratory conditions. So there exists, by and large, a relationship between the FIP of the various elements and their volatility, which is conveniently measured by the so-called condensation temperature[3]. It has long been known that the GCRS composition data can also be organised in terms of this condensation temperature; refractory (low-FIP) elements tend to be overabundant relative to volatile (high-FIP) ones. Fortunately, there are a few elements which do not follow the general FIP/volatility correlation and which, in principle, allow a distinction to be made between a FIP effect and a volatility effect in the GCRS composition. However these are not the easiest of elements to measure! Such data

---

[3]The exact definition of the condensation temperature is rather artificial. One imagines starting with a sample of solar composition gas at high temperature and a constant pressure of 10^{-4} atm and gradually lowering the temperature. The condensation temperature is the temperature at which 50% of the dominant solid compound formed by each element has condensed out of the gas phase.

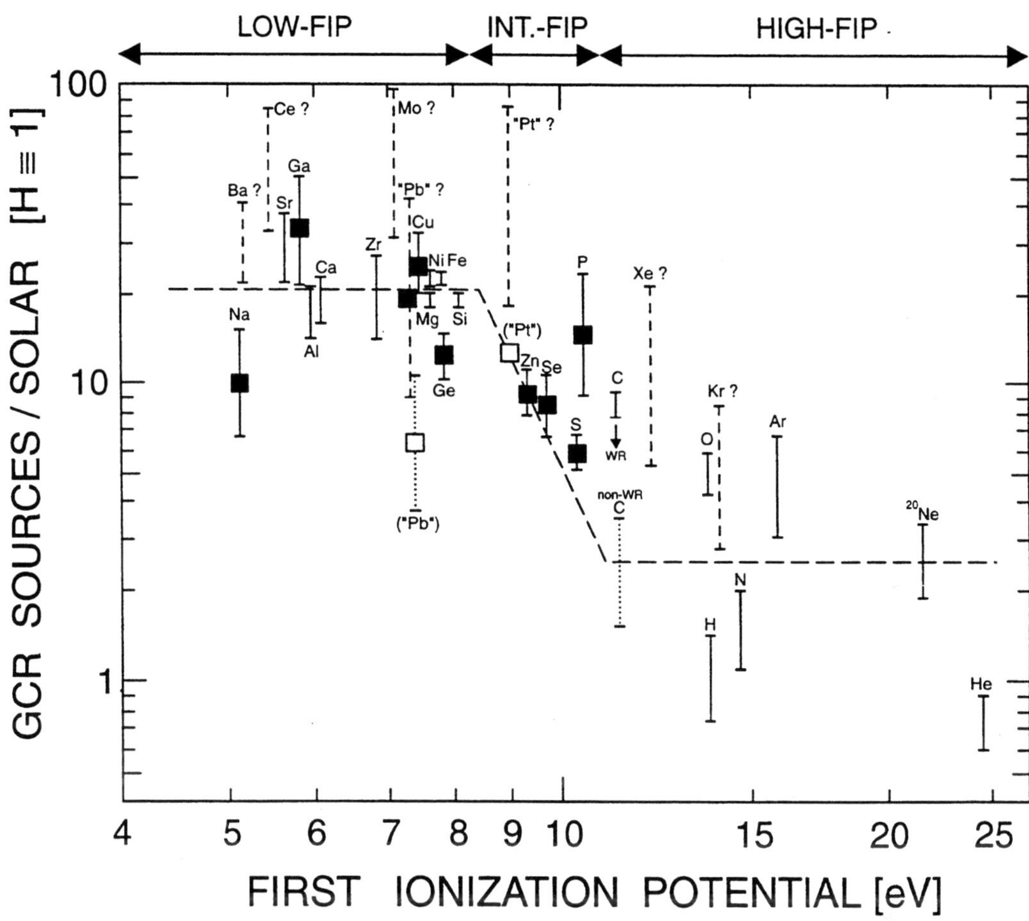

Figure 2: The GCRS to Solar abundance ratios plotted versus FIP. Solid squares denote those elements which can be used to distinguish between FIP and volatility. For detailed discussion see Meyer, Drury and Ellison 1998.

as is available tends to favour volatility rather than FIP as the better organising parameter (Meyer, Drury and Ellison 1997). However it is clear that neither FIP nor volatility alone completely accounts for the observations. In particular a simple two-step volatility or FIP bias does *not* account for the low relative abundances of H and He, the two most abundant elements! There must be some additional effect, parameter or process involved.

In Fig. 3 we sort the elements according to their volatility, and then plot their abundance enhancements versus the element mass A. This is a very interesting plot. It first shows that the refractory elements are globally enhanced relative to the volatile ones. Among the volatile elements the enhancements of all the inert gases and N appear to follow a smoothly increasing function of the mass (roughly $\propto A^{0.8}$); however, volatile H, C and O lie above this correlation. Among the refractories, by contrast, the enhancements are roughly independent of the mass[4].

In summary, the empirical evidence is that the GCRS composition is basically similar to that of the local ISM and has not been affected by specific nucleosynthetic processes (with the exception of the ^{22}Ne and associated ^{12}C excesses). But it does show clear signs of modification by factors depending on atomic physics or chemistry (an enhancement of low-FIP or, more probably, refractory elements) as well as on the element mass. It is very remarkable that the composition of fully-stripped relativistic nuclei should show traces of atomic physics effects with characteristic energies of only a few electron volts, and this is clearly a significant clue to the cosmic ray origin.

4   The solar coronal FIP effect

The FIP concept received a significant boost when it was discovered that this effect operates in the atmosphere of the sun and biases the composition of the solar corona, solar wind and solar energetic particles. By some not entirely understood mechanism, ionized heavy elements are preferentially lifted from the chromosphere into the corona, giving a coronal composition enhanced relative to the bulk solar composition in low FIP elements. This biased composition is then also seen in the solar wind and in solar accelerated particles.

This remarkable discovery prompted attempts to relate the origin of cosmic rays to the coronae of cool stars like the sun (Meyer, 1985). However it is certain that even if these stars are the source of the accelerated material, they cannot be the source of the energy needed for the acceleration. The only plausible known source of energy capable of driving the acceleration processes remains the mechanical explosion energy of supernovae. Thus this line of argument requires the dwarf stars to somehow inject large amounts of FIP biased but low energy ions (MeV) into the ISM which are later accelerated to the observed energies by passing supernova

---

[4]It is important to note that a tentative similar ordering of the data in terms of a combined FIP and mass effect would *not* order the data as satisfactorily. Specifically, the non-solar values of the GCRS abundance ratios between elements of similar FIP *and* mass, but widely different volatilities (Na/Mg, P/S, Ge/Fe, and Pb/Pt), cannot be interpreted in terms of a combined FIP and mass fractionation (Meyer, Drury and Ellison 1998).

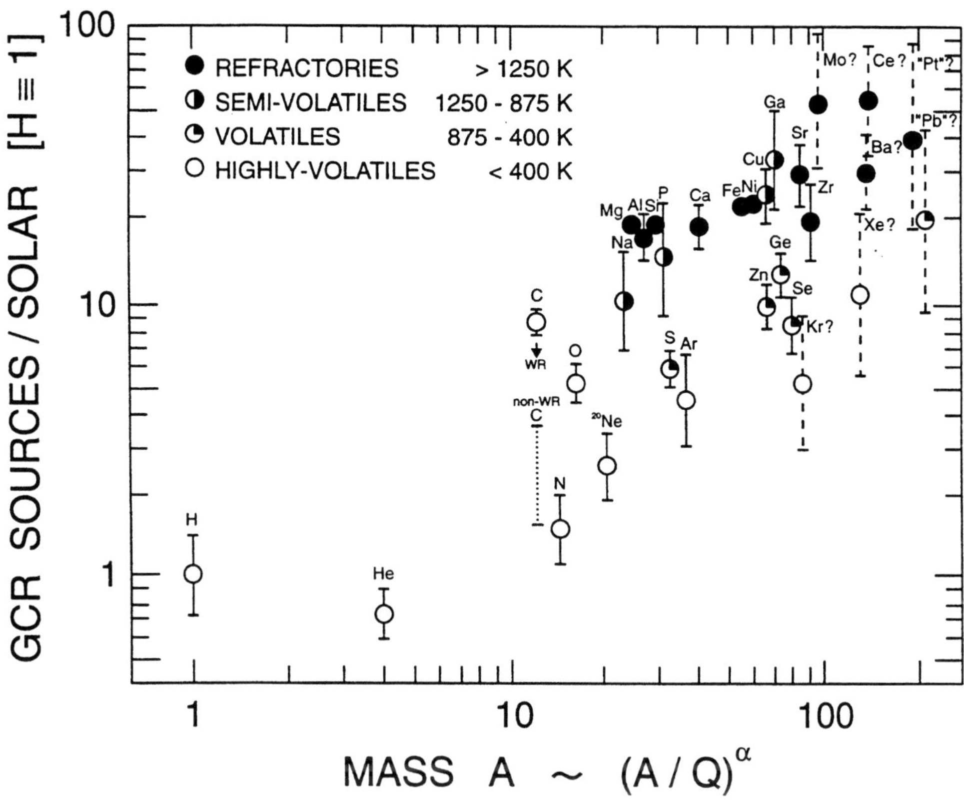

Figure 3: The GCRS to solar abundance ratios plotted against element atomic mass number *A* for four volatility classes.

shocks.

However there are problems in trying to make any such two-stage model work *quantitatively* (Epstein, 1981). The basic problem is that sub-relativistic ions suffer quite rapid ionization and Coulomb energy losses in the ISM even allowing for the effect of electron pick-up in screening the nuclear charge (Meyer, 1985). On the other hand, the mean time interval between passages of strong supernova shocks must be long in the general ISM (at least 10^7 y); otherwise the observed energy dependence of the secondary to primary ratios could not be accounted for[5]. In addition modern shock acceleration theory emphasises that the shock accelerates particles directly out of the "thermal" distribution. Not only is there no need for a separate pre-injected "seed" population: any such population, unless at rather high number density[6], will tend to be swamped by the ISM particles accelerated directly by the shock.

5   SNR acceleration from the ISM

As described elsewhere in this volume [x-refs here] much work has been done on diffusive shock acceleration applied to supernova remnants (SNRs) as a theoretical model for the origin of cosmic rays. However this has mainly concentrated on the spectrum and the total power and, until recently, the question of composition was largely ignored. It has been known for a long time that shocks are intrinsically efficient accelerators and that the resulting nonlinear modifications to shock acceleration (mainly the shock smoothing effect) tend to favour the acceleration of high rigidity over low rigidity species, that is of species with higher mass to charge ratios[7]. At a crude qualitative level this could be said to fit the observed enhancement of heavy elements over hydrogen (e.g., Ellison 1982); however the detailed pattern, and specifically the atomic physics correlations, are not accounted for. In addition it is well known that in most of the ISM the refractory elements are not in the gas-phase but are locked up in the solid state in interstellar dust grains. From UV absorption line studies, for example, it is known that the abundance of Fe in the gas phase is typically only 1% of its total local ISM abundance. If the SNR shock is accelerating ions from the ISM gas phase only, how can Fe be enhanced by a factor of at least 30 relative to hydrogen in the accelerated particles whereas in the gas phase flowing into the shock it is generally depleted by a factor of 100? One could of course suppose that the bulk of the acceleration occurs in a very hot phase of the

---

[5]These constraints would be significantly alleviated if GCR production were located in regions of active star formation containing many young low-mass stars with high levels of surface activity and a few massive stars to provide supernovae, all concentrated within a limited volume.

[6]A rough estimate is that the number density of MeV ions would need to be of order 10^{-6} that of the background plasma, which in turn would imply an energy density in the putative seed particles comparable to the thermal energy density of the plasma.

[7]This is often described as a more efficient acceleration of high rigidity species. However this can be rather misleading. What is meant is that the *steady state* differential velocity spectra of the higher rigidity species are less rapidly decreasing functions of velocity than those of the low rigidity species at the crucial low velocities close to the shock speed. However the low rigidity species have *shorter* acceleration time scales and are accelerated more rapidly, albeit to steeper spectra.

ISM where the grains are destroyed; however there the characteristic energies are far too high for few eV atomic physics effects to be important (and, in particular, to select the elements according to their FIP values); so, in any case, this would merely undo the grain depletion and not by itself generate an *overabundance* of the refractory/low-FIP elements.

The resolution of this problem, and, we believe, the key to interpreting the compositional data, is to recognise that dust grains in the ISM are charged and can therefore be shock accelerated. This is in fact quite an old idea. Epstein (1980) first suggested that charged dust grains could be accelerated, and that ions sputtered off the accelerated grains while the grains were in the upstream region would then be picked up and further accelerated by the shock[8]. With the advances in our understanding of shock acceleration it is now possible to calculate *quantitatively* and in some detail the process sketched out by Epstein. The full details can be found in Ellison, Drury and Meyer (1997); here we will concentrate on conveying the spirit of the calculations and the results.

The essential idea is to apply the modern theory of shock acceleration consistently to a SNR shock propagating in a dusty ISM. The refractory elements, such as Iron, Magnesium and Silicon, are known to be almost entirely condensed into small dust grains with a range of sizes extending from clusters of a few atoms to a maximum size of about 10^{-7} m (this size range is required to fit the UV, optical and IR data). These grains will be charged by a number of processes (secondary electron emission, photoelectric effect, plasma charging etc) to surface potentials of order 10 to 100 V (it is important to note that this is a standard part of ISM grain theory, not an assumption of our model; see, e.g. Spitzer 1978) implying mass to charge ratios for the larger grains of order 10^8 (and less for the smaller grains). This means that, relative to a shock moving at several hundred kilometers per second, their magnetic rigidity is *less* than that of a 10^{14} eV proton or electron. If the shock is capable of accelerating particles to these energies, the dust grains will inevitably also be scattered across the shock and accelerated.

In fact in the case of at least one remnant, that of the supernova of 1006, there is now direct observational evidence for the acceleration of electrons to these energies from the detection of X-ray synchrotron emission (Koyama et al, 1995) and inverse-Compton gamma-rays (Tanimori et al, 1998). The acceleration of protons to about 10^{14} eV, although not yet directly observed, is also required if SNRs are to provide the bulk of the Galactic cosmic ray population up to the "knee" energy. Thus it seems certain that at least some supernova remnant shocks are associated with magnetic field structures capable of scattering particles of rigidities up to 10^{14} V. The dust grains will have the same scattering mean free path as the ultra-relativistic protons and electrons of the same rigidity[9] but a much *lower* diffusion coefficient

---

[8]At about the same time Cesarsky and Bibring (1981) and Bibring and Cesarsky (1981) also discussed grain sputtering as a source of GCR material; however they did not consider grain acceleration and relied on downstream second order Fermi acceleration to accelerate sputtered ions.

[9]As this is a crucial aspect of the model it is worth discussing it in a little detail. The key point is that the trajectory a charged particle follows in a stationary magnetic field is determined only by the rigidity of the particle. Of course the time taken to traverse the trajectory will be different for particles of different velocities, but the path followed is the same for all particles of the same

because the grain velocity is only of order the shock speed and the diffusion coefficient is, within factors of order unity, just the product of the scattering mean free path and the particle speed.

The conventional picture is that the magnetic field structures themselves are generated by plasma instabilities driven by the accelerated proton pressure gradients in a bootstrap process which drives the mean free path down to a value of order the gyro-radius (Bohm scaling). For the subrelativistic dust grains this means that the effective diffusion coefficient for transport near the shock front rises as momentum, or velocity, to the second power (one from the increase in the mean free path and one from that in the speed). Thus as the grains are accelerated from an initial velocity of order the shock speed, the acceleration rate drops rapidly. At the same time the frictional drag on the dust particles resulting from collisions with atoms of the gas increases proportional to velocity. This sets a natural limit to the amount of grain acceleration determined by the balance between acceleration at the shock and frictional losses in the upstream and downstream regions. This process, which was not considered by Epstein, turns out to be crucial because it links the rate of gas collisions with the grain in the upstream region, and hence the amount of ion sputtering, to the acceleration rate of the shock. For any reasonable parameters the grains are only slightly accelerated, by about a factor ten in momentum or velocity, or one hundred in energy (corresponding to about 0.1 MeV per nucleon), but this is enough to produce a small amount of sputtering from the accelerated grains in the region ahead of the shock. We calculate that roughly 10^{-4} of the grain material will be sputtered in the upstream region, and give rise to secondary ions with velocities about ten times the shock speed. These ions will be picked up by the interstellar field and swept into the shock, which will then efficiently accelerate them to relativistic energies. Particles that are sputtered downstream from the shock are swept away from the shock and do not get accelerated by it. An important point is that the ions sputtered *upstream* are produced in association with and close enough to the shock to reach it before they have suffered serious energy losses (as noted above the energy loss times of subrelativistic ions are quite short).

In addition, of course, the same shock will directly accelerate ions out of the gas phase, but because these start down in the thermal distribution at velocities of order the shock speed there will be a strong rigidity dependent bias in the initial phase of the acceleration. Whatever the precise conditions of ionization, this will effectively result in a mass fractionation effect; in particular if, as plausible, we have a UV photoionized gas in which all species have ionization state one or two. Heavier species have larger mean free paths against scattering and thus sample more of the shock compression earlier in their acceleration than the lighter ions. This effect is not

---

rigidity (and charge sign). Thus, as long as the field varies on time scales longer than the time taken by a particle to traverse the scattering magnetic structures, the scattering mean free path depends only on the rigidity and will be exactly the same for a subrelativistic dust grain and a high energy proton. The time scale for variation in the field will be of order the length scale of the magnetic structure divided by the Alfvén speed, so basically all particles with velocities larger than the Alfvén speed see an essentially static field and will have scattering mean free paths which are determined only by their rigidity. The dust grains enter the downstream region with a velocity of order the shock velocity, and as the shock is super-Alfvénic, this condition is fulfilled for them.

easy to model analytically (x-ref Malkov; Berezhko, Yelshin and Ksenofontov, 1996) but has been simulated in Don Ellison's Monte-Carlo model of shock acceleration for many years. Where it has been possible to compare the Monte Carlo results with observations at the Earth's bow shock the agreement is generally excellent (eg Ellison, Moebius and Paschmann, 1990).

In terms of their contribution to the bulk composition of GCRs the most important SNR shocks are thought to be those associated with the larger older remnants nearing the end of their Sedov-like phase. The smaller faster shocks associated with young remnants certainly accelerate particles, but they process relatively small amounts of the ISM and the particles they accelerate (except at the highest energies) are trapped inside the remnant and subject to adiabatic losses as the SNR expands. In fact our results turn out not to be very sensitive to the assumed shock speed. We have considered two typical cases, a fast shock of velocity $2000\,\mathrm{km\,s^{-1}}$ and a slower older shock of velocity $400\,\mathrm{km\,s^{-1}}$. For both we have calculated the expected mass fractionation in the acceleration of the volatile element ions using Ellison's Monte Carlo code and consistently calculated the grain acceleration using the same code. We then used a simple approximation for the sputtering yields to estimate the flux of sputtered energetic ions of refractory elements into the shock, which we then accelerated to relativistic energies, again using the same Monte Carlo code. The results are shown in Fig. 4.

This is a very remarkable plot. We see at once that H, the inert gases and N show a clear mass fractionation effect which is well reproduced by the Monte Carlo code. In fact the agreement is best with the slower shock model, exactly as we expect on physical grounds. The refractory elements, which form the low-FIP group, fall exactly in the region where we predict accelerated sputtered ions from accelerated grains to lie (between the two horizontal lines given to roughly indicate errors in the model). It is important to note that we have not done any fitting to the data in this plot. We have simply taken known physics, a standard dusty ISM composition, and the Monte Carlo shock acceleration code and calculated *ab initio* what the composition of the accelerated particles should be at GeV per nucleon energies.

What we have done is to identify *two* distinct routes within the *one* shock acceleration mechanism whereby atoms of interstellar material can reach cosmic ray energies. The first route is the standard shock acceleration picture in which the shock-heated gas-phase ion distribution functions develop tails extending to very high energies and where high mass to charge ratio ions are preferentially accelerated. The second route involves modest acceleration of the interstellar grain population, sputtering of ions from the accelerated grains upstream of the shock, and subsequent acceleration of these sputtered ions at higher energies where the shock acceleration is already quasi-independent of rigidity resulting in little or no species dependent bias. The dynamical coupling between acceleration rate, frictional drag and sputtering links these two routes and shows that the second route, as measured by abundances at GeV energies, is about a factor 30 times as efficient as the first is for protons. This naturally explains the very similar enhancements of all the low-FIP refractory elements which we believe enter the cosmic ray population almost

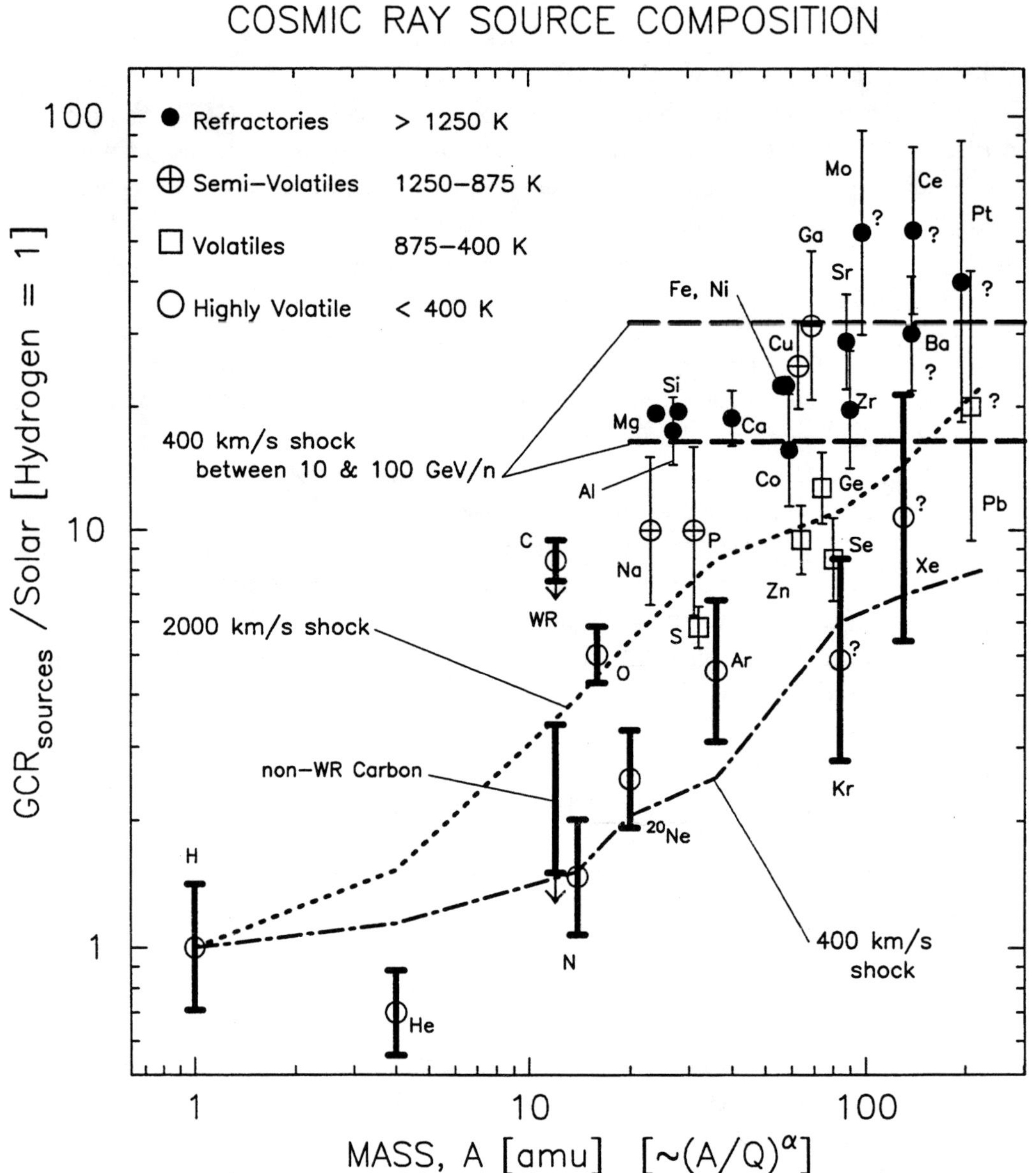

Figure 4: Comparison between the Monte Carlo model predictions and the observational data. The dotted and dot-dash lines indicate the mass-dependent fractionation of purely volatile species given by the code for two different shock speeds. The two dashed horizontal lines indicate the enhancement relative to hydrogen of sputtered ions from refractory grains predicted by the slower shock model.

exclusively through the second route, and therefore undergo the crucial first phases of the acceleration, not as individual ions, but as constituents of entire grains. The volatile elements, on the other hand, and in particular the dominant species H and He, enter almost exclusively as gas-phase ions through the first process and are subject to strong mass fractionation as predicted by the Monte Carlo model.

6   Carbon and Oxygen

That exceptions prove[10] rules is a principle going back to the scholastic philosophers; one of the best ways of testing the validity of any general principle is to look closely at the cases where it apparently fails. Clearly the two elements which appear exceptional in our interpretation of the data are carbon and oxygen. Both are clearly above the "volatile" curve, but also well below the "refractory" band. However both these elements are also special in at least two other respects. Firstly, although far from being completely condensed in the ISM, both are known to be important grain constituents. Secondly carbon, and to a lesser extent oxygen, are greatly enhanced in the winds from Wolf-Rayet (WR) stars[11].

Let us consider first the case of oxygen. The refractory metallic elements such as Mg, Al, Si, Ca and Fe which form grains do so by condensing as oxides (silicate minerals are known from infrared spectroscopy to be one of the main components of interstellar dust). It is generally estimated that this locks some 15% to 20% of the interstellar oxygen up in grains from which it will be preferentially accelerated in the same way as all other grain constituents. Relative to H, fig .. shows that O is enhanced by a factor of $f_O \approx 5$ whereas the neighbouring (in mass) volatile elements N and Ne are only enhanced by a factor $f_{\rm vol} \approx 2$. However sputtered grain material is enhanced by a factor $f_{\rm ref} \approx 20$ to 25. If a fraction x of the oxygen is in grains, then the resulting mixture of directly accelerated gas-phase "volatile" oxygen and sputtered "refractory" oxygen will give a net enhancement

$$x f_{\rm ref} + (1-x) f_{\rm vol} = f_O \qquad \text{so that} \qquad x = (f_O - f_{\rm vol})/(f_{\rm ref} - f_{\rm vol}). \qquad (1)$$

With the above observed values of the enhancement f factors we get a fraction x of order 0.15, perfectly consistent with the chemistry of silicate minerals. Recent depletion studies of the ISM suggest, if anything, rather higher fractions of interstellar oxygen in the dust component (Meyer, Jura and Cardelli, 1998), but this is hard to reconcile with the chemistry.

The other element which is known to form a significant interstellar grain component is carbon. Small graphite-like grains have long been a feature of dust models and are apparently required to reproduce the 2175 Å feature in the interstellar extinction curve. Although no generally accepted carrier has yet been identified for the unidentified interstellar absorption features, almost all suggestions involve substantial amounts of carbon. Carbon grains are observed to condense in the atmospheres

---

[10]In the original sense of *test* as preserved in the saying "the proof of the pudding is in the eating'

[11]Nitrogen is also enhanced in WN-type WR star winds, but only by factors of order 13. In contrast, carbon is enhanced by much larger factors of about 120 in the WC-type star winds. Oxygen can also be enhanced in the much rarer WO-type WR star winds.

of carbon stars from which they are ejected by radiation pressure into the ISM. In addition, Greenberg and his colleagues have argued that most interstellar grains acquire "mantles" of refractory organic deposits through condensation of organic compounds in molecular clouds and subsequent UV irradiation. In fact at present there is a "carbon crisis" in that the dust models all require more carbon in the dust than appears to be available (Snow and Witt, 1996; Mathis, 1996; Dwek, 1997). The best current estimates (Cardelli et al, 1996) suggest that the fraction of ISM carbon incorporated into dust grains is between 20% and 60%. The GCRS/solar system enhancement of about 9 relative to H would be compatible with a fraction of about 30% of the interstellar carbon in refractory grains, but in addition we expect a specific carbon excess from WR star nucleosynthesis.

The existence of such a component is indicated by the one firm isotopic anomaly in the GCRS, the well-established excess of ^{22}Ne. This strongly suggests a contribution from material contaminated by the winds from WR stars (van der Hucht and Williams, 1995). This is actually very natural. The most massive supernova progenitor stars are thought to evolve through a WR phase just prior to core collapse. The strong and fast WR wind will blow a circumstellar shell of material enriched in He burning products which will then be traversed by the subsequent SNR shock wave. This rather naturally accounts for a ^{22}Ne excess and must also contribute significant amounts of ^{12}C to the GCRS. In fact a large fraction of the ^{12}C should condense as grains in the C-rich WR star wind so that sputtering of these circumstellar grains may further enhance the importance of this contribution. This may not actually leave much room for a carbon enhancement from "ordinary" ISM grains! As for oxygen, the WR contribution of ^{16}O to the GCRS must be negligible in view of the observed lack of any associated excess of ^{25,26}Mg.

7　Conclusion

We have shown that standard shock acceleration applied to an ISM with the bulk composition of the local ISM, but where the refractory elements along with 15% of the oxygen and a significant fraction of the carbon are in dust grains, can replicate *all* the observed features of the GCRS abundance pattern and do so *quantitatively*, with the exception of the ^{22}Ne excess. This latter can be rather naturally explained in terms of an additional Wolf-Rayet wind component which must then also contribute to the C excess.

Our interpretation of the compositional data is based solely on calculable physical processes and standard astronomical inputs with essentially no adjustable parameters, yet manages to give a better match to the observations than any other interpretation we are aware of. This, we feel, argues strongly for the basic correctness of the underlying picture which locates the origin of bulk of Galactic cosmic rays in the processing of a dusty ISM by the strong blast waves driven by supernova explosions.

The exciting prospect is that we appear to be able to relate the GCRS composition to important astronomical questions about dust and the ISM. Obviously much more work needs to be done in refining the model (in particular the treatment of sputtering needs to be improved) and this needs to be related to our rapidly increasing knowledge of interstellar abundances and dust properties. However, we may ultimately be able to use cosmic ray composition studies to chemically analyze the interstellar dust!

8 Acknowledgements

The work of LD and JPM was supported by the TMR programme of the European Union under contract FMRX-CT98-0168 and that of DCE by the NASA Space Physics Theory Program. LD would like to thank the Service d'Astrophysique of the CEA/Saclay for their warm hospitality while working on the first draft of this article. Useful discussions on dust composition and ISM abundances with Renaud Papoular, Suzanne Madden and Anthony Jones are gratefully acknowledged.

References

Andersson, H. & Edvardsson, B. 1994 AA 290 590.

Arnett, D 1995, ARA&A 33 115

Baade, W. & Zwicky, F. 1934, Proc. Nat. Acad. Sci. U. S. 20 254.

Berezhko, E. G., Yelshin, V., & Ksenofontov, L. 1996, Sov. Phys. JETP, 82, 1,

Bibring, J-P. & Cesarsky, C. J. 1981, Proc 17th International Cosmic Ray Conference (Paris) 2 289.

Cesarsky, C. J. & Bibring, J-P. 1981, IAU Sym 94 "Origin of Cosmic Rays" ed G. Setti, G. Spada & A. Wolfendale (Dordrecht: Reidel) 361

Edvardsson, B., Andersen, J., Gustafsson, B., Lambert, D. L., Nissen, P. E. & Tomkin, J. 1993 AA 275 101.

Ellison, D. C., 1982, Ph.D. Thesis, The Catholic University of America.

Ellison, D. C., Drury, L. O'C. & Meyer, J-P. 1997, ApJ 487 197

Ellison, D. C., Moebius, E. & Paschmann, G. 1990 ApJ 352 376

Epstein, R. I. 1980, MNRAS 193 723

Epstein, R. I. 1981, IAU Sym 94 "Origin of Cosmic Rays" ed G. Setti, G. Spada & A. Wolfendale (Dordrecht: Reidel) 109

Ginzburg, V. L. & Syrovatsky, S. I. 1964, "The origin of cosmic rays", english translation by H. S. H. Massey incorporating revisions by the authors, ed. D. Ter Haar, (Pergamon: Oxford).

van der Hucht, K. A. & Williams, P. M. eds 1995, IAU Symposium 163 Wolf-Rayet Stars: Binaries, Colliding Winds, Evolution (Dordrecht: Kluwer)

Koyama, K. et al 1995 Nature 378 255

Meyer, D. M., Jura, M. & Cardelli, J. A. 1998 ApJ 493 222

Meyer, J-P. 1985, ApJS 57 173.

Meyer, J-P., Drury, L. O'C. & Ellison, D. C. 1997, ApJ 487 182

Meyer, J-P., Drury, L. O'C. & Ellison, D. C. 1998, ACE Workshop, Caltech, Jan. 1997, Space Sci. Rev., in press.

Spitzer, L. Jr., 1978, "Physical Processes in the Interstellar Medium", (Wiley: New York).

Snow, T. P. & Witt, A. N. 1996 ApJ 468 L65

Tanimori, T., Hayami, Y., Kamei, S., Dazeley, S. A., Edwards, P. G., Gunji, S., Hara, S., Hara, T., Holder, J., Kawachi, A., Kifune, T., Kita, R., Konishi, T., Masaike, A., Matsubara, Y., Matsuoka, T., Mizumoto, Y., Mori, M., Moriya, M., Muraishi, H., Muraki, Y., Naito, T., Nishijima, K., Oda, S., Ogio, S., Patterson, J. R., Roberts, M. D., Rowell, G. P., Sakurazawa, K., Sako, T., Sato, Y., Susukita, R., Suzuki, A., Suzuki, R., Tamura, T., Thornton, G. J., Yanagita, S., Yoshida, T., Yoshikoshi, T. 1998 ApJ 497 L25

Timmes, F. X., Woosley, S. E. & Weaver, T. A. 1995, ApJS 98 617

Woosley, S. E. & Weaver, T. A. 1995, ApJS 101 181

Understanding abundance anomalies in the Galactic cosmic rays

M. A. DuVernois

Department of Physics, The Pennsylvania State University
104 Davey Laboratory, University Park, PA 16802

ABSTRACT

The Galactic cosmic rays allow for a direct sampling of material
from outside of the solar neighborhood. This material is also a sample of
recent nucleosynthethic origin ($\sim 10^7$ years old) and could, in principle,
be dramatically different in composition than the standard (solar
system) abundances. Experimental data has, however, demonstrated
that the Galactic cosmic ray source (GCRS) composition is tantalizingly
similar to the standard composition. Beyond the similarities, the cosmic
ray source elements can be understood in terms of chemical properties:
first ionization potential (FIP) bias and volatility in particular. The
isotopic anomalies are examined in the context of supernova type Ia and
type II yields, nova yields, and Wolf-Rayet stars. The long-standing
GCRS neon abundance anomaly is difficult to explain in terms of
nucleosynthesis. Fractionation in the interstellar medium appears to be
a more likely solution.

1. Introduction

The Galactic cosmic rays, along with certain meteoritic inclusions, offer direct
samples of material from outside of the solar system. The abundance pattern of
the Galactic cosmic rays are interesting as an experimental data point for Galactic
chemical evolution in the current epoch. These measurements could then be
contrasted with the stellar abundance data at lower metallicities. The abundances
locked into the presolar nebula are well-measured both in the meteorites and in the
solar photosphere. This composition is presumed to be some average local Galactic
composition 4.55 Gyr ago. Since then, the Galaxy has undergone further chemical
evolution—metallicities have increased through stellar nuclear processing (Anders
and Grevesse 1989).

Naively then, the GCRS composition should be considerably enriched in
metals over the solar system material. In observations, the cosmic rays are a
mixture of primary (source) and secondary (spallogenic) particles. However, after

deconvolution of the effects of transport, the cosmic ray composition is similar to the solar composition. How these GCRS abundances differ from solar abundances, anomalies between the two, and how these differences likely arise is the focus of this work.

2. Cosmic ray elements

The elemental composition of the cosmic rays requires only modest instrument energy resolution ($\sim$3% for the iron group) and has a long history of measurements. Both satellite and balloon experiments have reported elemental abundances from tens of MeV/N up to a few GeV/N (e.g., Engelmann et al. 1989, DuVernois and Thayer 1996). For these experiments, observational abundance ratios are compared with theoretical results for varying source abundances. A simultaneous fit of all of the elements is required. See Figure 1 as an example of the Mg/Si experimental ratios from the above experiments, along with a propagated source abundance (DuVernois and Thayer 1996). Resolutions are sufficient to insure that errors are principally from the statistics of the observation and the systematics of the transport code rather than from misidentification of chemical species (Thayer 1997).

2.1. FIP and friends

It was observed in the late 1970s that the GCRS elemental abundance pattern was somewhat different from the typical solar photospheric abundances but was closer to the coronal abundances. In passing from the photosphere of the Sun into the corona, the solar ejecta undergoes a FIP fractionation. The detailed mechanism of this ion-neutral segregation is not completely understood, but observationally leads to a factor of 4-5 depletion of species with FIP$>$10 eV. The cosmic ray source composition relative to the solar system composition shows a similar break at 10 eV (Meyer 1985).

A FIP fractionation at 10 eV corresponds to an environment for the ion-neutral segregation of about 5000-8000K. This is consistent with the surface temperatures of F-M type stars. It is therefore possible that the cosmic ray preinjection particles represent a population akin to the solar energetic particles (SEP) which are extracted from the exterior of the sun, but from the corona rather than the photosphere. Low energy ions are capable of being efficiently accelerated by shocks, so this could represent a plausible injection mechanism for the Galactic cosmic rays.

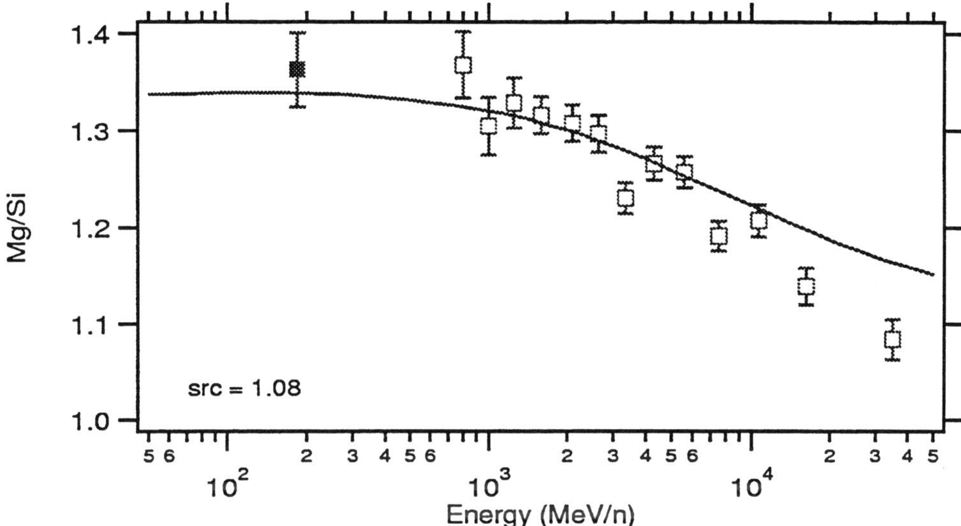

Fig. 1.— Cosmic ray observational data from Engelmann et al. 1989 and DuVernois and Thayer 1996 compared with standard cosmic ray propagation of the indicated source Mg/Si ratio. Fits such as this are required for all of the elements to assure a self-consistent propagation and determination of GCRS abundances.

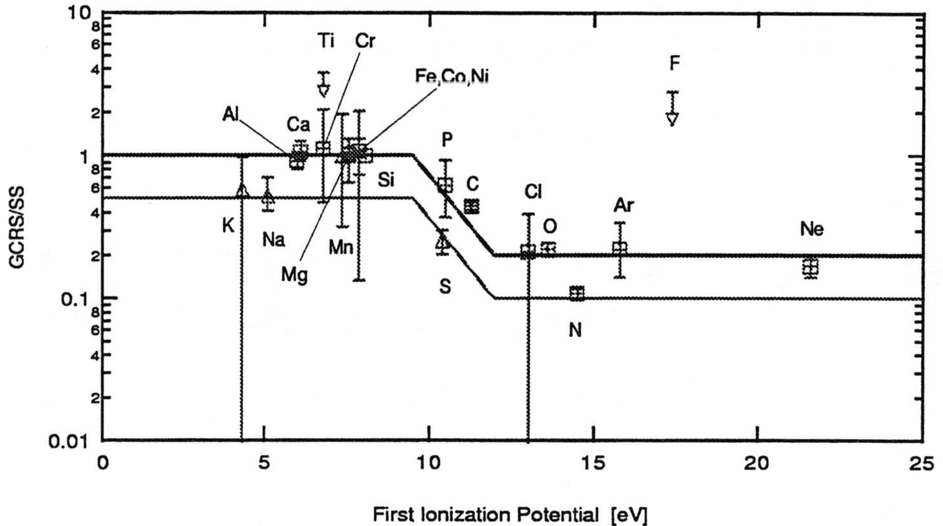

Fig. 2.— Cosmic ray source elemental abundances divided by the solar system (standard abundances) of Anders and Grevesse (1989) versus the first ionization potential. The data are from the *Ulysses* HET (DuVernois and Thayer 1996) with a slightly enlarged sample time. Triangles are volatile species (K, Na, Mn, & S). Ti and F data points are upper limits. The curves are to guide the eye and are described in the text.

In Figure 2 the Ulysses High Energy Telescope GCRS elemental composition relative to solar is shown (DuVernois and Thayer 1996; updated slightly). These data are from measurements at 50-350 MeV/N and propagated to the source using a standard Leaky Box model (Garcia-Munoz et al. 1987). The solid curve guides the eye for a source abundance which tracks photospheric-type FIP bias. The agreement with the data is quite good. A possible refinement is, however, the addition of a second curve (dashed line) which is a factor of two below the overall abundance curve. Note that the form of the break near 10 eV is neither a best-fit nor model-dictated. It is merely to guide the eye and suggestive of an exponential population gradient in FIP. This curve seems to correspond to the abundances of the volatile species (K, Na, Mn, and S) as well as nitrogen. A depletion of the volatiles and nitrogen by a factor of two over the coronal (FIP-biased) abundances would affect the stellar-wind-into-shock model mentioned above, or at least require a new component.

A bias based upon volatility would seem to imply some importance for the dust grains in the origin of cosmic rays (Meyer, Drury, and Ellison 1997). The model of Meyer et al. (1997) produces its elemental biasing completely in the dust grains and doesn't require a FIP-bias at all. If the FIP-bias is kept, perhaps volatile species could be depleted in the grains and found almost solely in the gaseous interstellar medium. A depletion of about a factor of two in the volatiles might then be interpreted as half of the cosmic rays coming from dust grains and half from the gas. All of them would receive FIP fractionation at an earlier time though.

Additional tests are easy to conceive of—there are many elements heavier than the iron-nickel group. These elements have been observed in a few experiments without a firm consensus being reached. The FIP fractionation appears to continue, but an additional bias, that of s-process versus r-process nucleosynthesis, crops up. Future experiments, such as the ultraheavy nuclei component of ACCESS, will address these elements to nearly the same precision as shown here for the lighter elements (Binns 1995).

It has been noted before that a low nitrogen abundance is very hard to understand from nucleosythethic arguments (Meyer, Drury, and Ellison 1997). Perhaps a chemical biasing in some manner in which nitrogen is similar to the volatiles could be responsible for its low cosmic ray abundance. We are led toward this since it is difficult to imagine a nucleosynthethic site which would underproduce N relative to solarlike C and O. In addition, a slight admixture of some exotic source to the overall solar-like source would be incapable of reducing the nitrogen abundance (Meyer, Drury, and Ellison 1997). What in dust grain physics would cause nitrogen to act as a volatile is unclear.

3. Cosmic ray isotopes

Largely because of the unknown structure of the chemical biases in the elemental abundance patterns of the cosmic rays, experiments to look directly at the isotopic abundances were built. Isotopic resolution also avoids the problem of the large correction factors in elemental ratios for the effects of solar modulation. Unfortunately, complete isotopic separation at iron requires better than 1% energy resolution over a dynamic range of nearly 40 thousand in energy loss (Simpson et al. 1992). Therefore isotopic measurements at the iron group are only now appearing with good isotopic separations (e.g., Connell and Simpson 1997, DuVernois 1997).

3.1. The historical trend

Observations of selected isotopic ratios relative to solar are shown in Figure 3. Most noticeable is the fact that there is a large scatter in the data, though in general not more than 3 standard deviations from the solar system abundances. The scatter between experiments is likely due to unanalyzed systematics in both the experimental components and the propagation algorithm. Some attempts have been made to characterize the propagation systematics (Thayer 1997, DuVernois 1997), but difficulty remains in comparing the work of different groups. However, there is another trend in the data, that of increasing agreement between the solar composition and the GCRS composition with new measurements (Thayer 1997). Some of the measurements in the iron-group are quite recent and have not been confirmed, but future experiments will confirm or refute these numbers within the next couple of years. Notable exceptions are the neon isotopic abundances—consistent with a large excess of ^{22}Ne (Garcia-Munoz, Simpson, and Wefel 1979).

3.2. Neon

This anomalous abundance led to several models and suggestions for a contribution to the GCRS from Wolf-Rayet stars. The outer layers of these extreme wind stars are enriched in neutron-rich isotopes of Ne, Mg, and Si in particular. A few percent contribution to the GCRS from Wolf-Rayet stars would be able to explain the neon anomaly. However, the predicted isotopic enhancements in Mg and Si have not been observed (see Fig. 3). Note that the Wolf-Rayet model used was not completely general and might be tunable somewhat to reduce the Mg and Si neutron isotopic enhancements.

The other suggested model (Supermetallicity) which fit the neon abundance

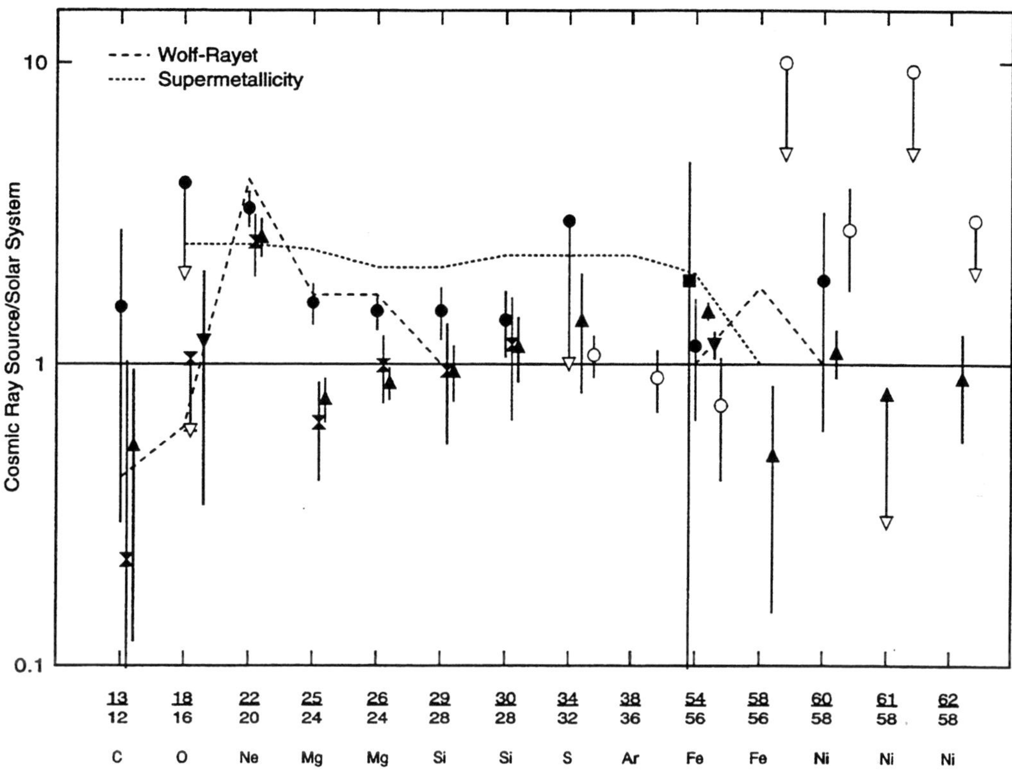

Fig. 3.— Galactic cosmic ray source isotopic ratios compared with the solar system (standard abundances) of Anders and Grevesse (1989). The plot is taken from Thayer (1996) and adapted from Wefel (1991). Data from various experiments and predictions from two models are shown. Wolf-Rayet predictions from Prantzos et al. (1985); supermetallicity predictions from Woosley and Weaver (1981); solid circles, Mewaldt (1989); solid square, Grove et al. (1990); triangle, Connell and Simpson (1993, 1995, 1997); hourglass, DuVernois et al. (1996); downward-pointing triangle, Lukasiak et al. (1995) & Webber et al. (1996); open circles, Leske (1993) & Leske and Wiedenbeck (1993).

pattern called for the GCRS to have its origin in extremely metal-rich stars (Woosley and Weaver 1981). By extrapolating the available stellar data, this would naturally give rise to enhanced neutron-rich species. Unfortunately for this model as well, the predictions are not borne out by the data for other species in the GCRS.

A more general approach is to find a combination of nucleosynthetic sources that produce the observed source abundances. In Table 1 the isotopic yields of a nova model, a type Ia supernova model, a type II supernova model, and a combined weighted supernova model are compared with the solar system abundances. Note that different units are used for the different models, however ratios can be directly constructed within columns. It is interesting to note that the weighted combination of type Ia and II supernovae leads to abundances quite similar to the solar system and to the GCRS. However, the supernova production of neon is only similar to that of the solar system and not to the GCRS.

Another scheme for enhancing the ^{22}Ne in the GCRS while not severely altering the other isotopic abundances might be to alter the relative contribution of novae in the GCRS to the solar material. Novae are currently thought to be relatively unimportant for the overall abundance pattern, but possibly important for a handful of isotopes (^{22}Ne being one such isotope). The bulk of the nova ^{22}Ne production comes from the decay of the radioactive ^{22}Na produced during the nuclear flash on the surface of the star. Current values for this production (see Table 1) of ^{20}Ne/^{22}(Na+Ne) from novae are greater even than the interated supernovae production. Therefore this suggestion is also unable to produce the observed neon abundances in the GCRS.

A different approach to understanding the GCRS neon abundances is to take a close look at solar-system neon. We've been considering this as the perfect reference composition—neglecting errors and biases. The standard abundances are in fact chosen to represent solar system material, but do vary from place to place within the solar system (Anders and Grevesse 1989). Neon shows some of the most extreme variations in isotopic abundance of any element. The various values of the ^{20}Ne/^{22}Ne ratio for sites in the heliosphere as well as the GCRS values are shown in Figure 3. Note in particular that there are neon abundances more 'out-of-whack' with the standard abundances in meteoritic inclusions than in the GCRS. Therefore a large degree of isotopic fractionation in neon occurred in the early solar system.

4.   Conclusions

For the GCRS elemental abundances, there appear to be two distinct fractionation processes. These are the well-known FIP bias, which reduces the

Table 1.   Isotopic yields from nucleosynthesis "engines."

Species	Nova[a]	SN Type Ia[b]	SN Type II[c]	SNIa + $\int$ SNII dM[d]	Solar[e]
^{16}O	0.5	1.43E-01	3.55E-01	1.80	2.37E+07
^{18}O	0.059	8.25E-10	1.35E-02	4.61E-03	4.76E+04
^{20}Ne	17	2.02E-03	2.08E-02	2.12E-01	3.20E+06
^{21}Ne	0.0056	8.46E-06	3.93E-05	1.08E-03	7.77E+03
^{22}Ne	0.00008	2.49E-03	1.25E-02	1.83E-02	2.34E+05
^{22}Na	0.554				
^{24}Mg	0.139	8.50E-03	3.16E-02	8.83E-02	8.48E+05
^{25}Mg	3.9	4.05E-05	2.55E-03	1.44E-02	1.07E+05
^{26}Mg	0.37	3.18E-05	2.03E-03	2.01E-02	1.18E+05
^{28}Si	3.2	1.50E-01	7.16E-02	1.05E-01	9.22E+05
^{29}Si	0.41	8.61E-04	3.25E-03	8.99E-03	4.67E+04
^{30}Si	1.74	1.74E-03	4.04E-03	8.05E-03	3.10E+04
^{32}S	2.9	8.41E-02	3.01E-02	3.84E-02	4.89E+05
^{33}S		4.50E-04	9.60E-05	1.78E-04	3.86E+03
^{34}S		1.90E-03	1.49E-03	2.62E-03	2.17E+04

[a] 1.35$M_\odot$ model, mass fraction x 100; Starrfield et al. 1996

[b] W7 model, synthesized mass ($M_\odot$); Nomoto, Thielemann, and Yokoi 1984, Tsujimoto et al. 1995

[c] 15$M_\odot$, synthesized mass ($M_\odot$); Tsujimoto et al. 1995

[d] Integrated over a Saltpeter Initial Mass Function (IMF) from 10–50$M_\odot$, synthesized mass ($M_\odot$); Tsujimoto et al. 1995

[e] Atoms/10^6 Si; Anders and Grevesse 1989

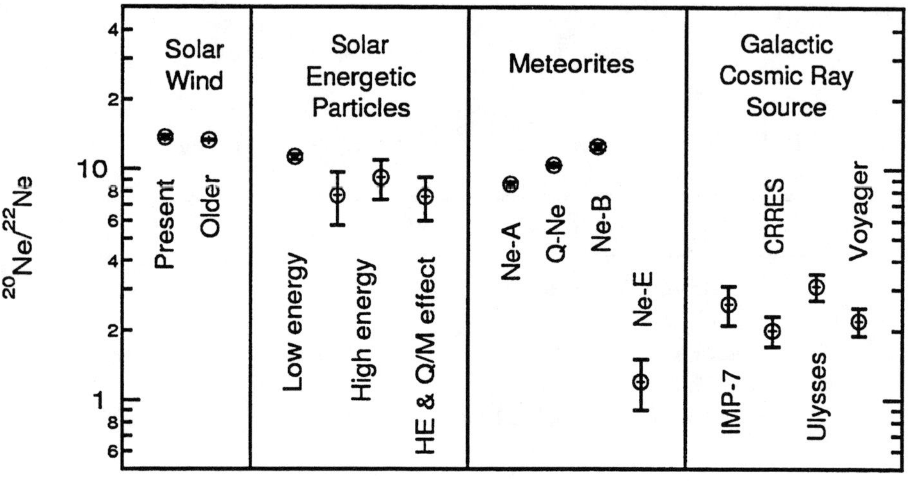

Fig. 4.— Neon isotopic abundances from various solar system sites and from the galactic cosmic rays. Solar wind: present day (Geiss 1973), older solar wind from lunar samples (Wieler, Baur, and Signer 1986). Solar energetic particles: low energy (Wieler, Baur, and Signer 1986), high energy (Dietrich and Simpson 1979; Mewaldt et al. 1984), with Q/M fractionation included (Mewaldt and Stone 1987). Meteorites: Ne-A (Tang and Anders 1988), Q-Ne (Anders 1988), Ne-B implantation (Podosek 1978), Ne-E (Srinivasan and Anders 1978). Galactic cosmic ray source: IMP-7 (Munoz, Simpson, and Wefel 1979), CRRES (DuVernois, Garcia-Munoz, Pyle, and Simpson 1993), Ulysses (Connell and Simpson 1993), Voyager (Lukasiak, Ferrando, McDonald, and Webber 1994).

abundance of species with FIP>10 eV by a factor of 5, and a second bias which reduces the abundances of volatile species (plus nitrogen) by a factor of about 2. This second effect is not seen in the solar coronal abundances and would appear to not be related to an ion-neutral segregation. Rather, it's an effect which appears more likely to involve grain, or molecule, formation and destruction. Both higher statistics and the inclusion of heavier elements (Z>28) could help confirm or deny this view.

The GCRS neon abundance seems to reflect a fractionation bias as well. With the other isotopes of the GCRS looking very solar-like, nucleosynthethic models falling short, and the solar abundance varying over the same order of magnitude within the early solar material, this remains the most likely scenario. Wolf-Rayet models can fit the observed neon abundances, but typically overproduce other neutron-rich isotopes—anomalies which are not seen.

Much of this work was performed while the author was at the University of Chicago, Laboratory for Astrophysics and Space Research. The author wishes to thank J. A. Simpson at Chicago and J. J. Beatty at Penn State for their support and J. W. Truran, J.-P. Meyer, B. Fields, M. R. Thayer, F. X. Timmes, and J. J. Connell for useful and encouraging suggestions for understanding cosmic ray and overall Galactic abundances.

REFERENCES

Anders, E. 1988, Meteorites and the Early Solar System, Kerridge, J. F., and Matthews, M. S., Univ. Arizona Press

Anders, E., and Grevesse, N. 1989, Geochim. Cosmochim. Acta, 53, 197

Binns, W. R. 1995, Adv. Sp. Res., 15, 6

Connell, J. J., and Simpson, J. A. 1993, Proc. 23rd International Cosmic Ray Conference, 1, 559

Connell, J. J., and Simpson, J. A. 1995, Proc. 24rd International Cosmic Ray Conference, 2, 602

Connell, J. J., and Simpson, J. A. 1997, ApJ, 475, L61

Dietrich, W. F., and Simpson, J. A. 1979, ApJ, 231, L91

DuVernois, M. A., Garcia-Munoz, M., Pyle, K. R., Simpson, J. A., and Thayer, M. R. 1996, ApJ, 466, 457

DuVernois, M. A. 1997, ApJ, 481, 241

DuVernois, M. A., and Thayer, M. R. 1996, ApJ, 465, 982

Engelmann, J. J. et al. 1989, A&A, 233, 96

Garcia-Munoz, M., Simpson, J. A., and Wefel, J. P. 1979, ApJ, 232, L95

Garcia-Munoz, M., Simpson, J. A., Guzik, T. G., Wefel, J. P., and Margolis, S. H. 1987, ApJS, 64, 269

Geiss, J. 1973, Proc. 13th International Cosmic Ray Conference, 5, 3375

Mewaldt, R. A., Spalding, J. D., and Stone, E. C. 1984, ApJ, 280, 892

Mewaldt, R. A., and Stone, E. C. 1987, Proc. 20th International Cosmic Ray Conference, 3, 255

Meyer, J. P. 1985, ApJS, 57, 173

Meyer, J. P., Drury, L. O'C., and Ellison, D. C. 1997, ApJ, in press

Nomoto, K., Thielemann, F.-K., and Yokoi, K. 1984, ApJ, 286, 644

Podosek, F. A. 1978, ARA&A, 16, 293

Simpson, J. A., et al. 1992, A&AS, 92, 365

Srinivasan, B., and Anders, E. 1978, Sci., 201, 51

Starrfield, S., Truran, J. W., Wiescher, M. C., and Sparks, W. M. 1996, MNRAS, in submission

Tang, M., and Anders, E. 1988, Geochim. Cosmochim. Acta, 52, 1235

Thayer, M. R. 1997, ApJ, 482, 792

Tsujimoto, T., Nomoto, K., Yoshii, Y., Hashimoto, M., Yanagida, S., and Thielemann, F.-K. 1995, MNRAS, 277, 945

Wieler, R., Baur, H., and Signer, P. 1986, Geochim. Cosmochim. Acta, 50, 1997

Source Abundances and Nuclear Cross Sections

C. Jake Waddington

School of Physics and Astronomy
University of Minnesota
wadd@physics.spa.umn.edu

Abstract

The elemental and isotopic abundances of the energetic nuclei in the cosmic radiation have been determined with steadily increasing precision. Some of these abundances are now known to within a few percent. The most precise determinations are generally limited to relatively low energy nuclei and to the lighter nuclei in the periodic table. The abundances measured in the solar system should allow us to deduce the abundances at the source of the cosmic radiation and hence better understand the nature of the source. All such deductions depend on evaluating the effects introduced by the propagation of these nuclei through the interstellar medium. Propagation calculations require a model of the transport mechanism and a knowledge of the changes introduced by nuclear interactions with the matter in the interstellar medium. At present, it appears that the most serious limitation to deducing source abundances arises from our lack of the appropriate nuclear cross sections for the production of nuclear fragments in these interactions. The limitations of our present knowledge of the nuclear parameters for the heaviest nuclei in the cosmic radiation will be discussed here, and the sensitivity of the deduced source abundances on the cross sections illustrated for a few examples.

Introduction

The ultra heavy (UH-) nuclei in the cosmic radiation, those with charge $Z \geq 30$, make up by number only some 1% of the multiply charged nuclei in the incident radiation. However, they include some two thirds of the periodic table, all those elements that require endothermic reactions to be created during nucleosynthesis. The elemental and isotopic abundances of these nuclei must contain very significant clues as to the nature of the processes that produce those nuclei that are present in the sources of the cosmic radiation. Understanding these processes will help to define the sources of the cosmic radiation and the acceleration mechanisms involved.

Since the discovery of the existence of these UH nuclei in the cosmic radiation, first as fossil tracks in meteoritic crystals by Walker, Fleischer and Price (1965), and then in the current cosmic rays by Fowler *et al.* (1967), continued measurements have slowly improved and refined our knowledge of the relative abundances in the incident radiation. Because to their much greater rarity, the abundances of the UH nuclei have not been determined with nearly the same precision as have the lighter nuclei. They have abundances that are $10^2 - 10^5$ times less than that of iron nuclei. Abundance measurements covering the whole range of charges culminated with the early 1980's HEAO (Binns *et al.* 1989), and Ariel (Fowler *et al.* 1987), spacecraft. These observations resulted in the elemental abundances of the more abundant even charged elements being relatively well defined for $30 \leq Z \leq 82$. Due a lack of charge resolution, those for the odd charged elements were not. More recent observations have concentrated on the abundances of the very heaviest elements, those with $Z \geq 70$. The exposure for 6 years of plastic detectors on the Long Duration Exposure Facility (LDEF) by Keane *et al.* (1997) has increased the numbers of $Z \geq 70$ nuclei observed, although with poor charge resolution. The TEK exposure of etchable glass detectors on the Mir space station by Westphal *et al.* (1998), has reported results on those nuclei with $Z \geq 74$ with significantly better charge resolution than hitherto. Essentially nothing is known yet about the isotopic abundances of any of the UH nuclei.

In order to relate the observed abundances to those at the source it is necessary to adjust them for the effects introduced as the nuclei propagate from the source. The nuclei diffuse through the matter and fields of the interstellar medium (ISM) until they reach the solar system, are brought to rest by energy losses, are destroyed by interactions, or escape from the galaxy. They can lose energy due to ionization losses, or encounter shock waves causing them to gain energy (re-acceleration). Many of them will undergo nuclear interactions with the matter, breaking them up into lighter fragments.

The energies of the nuclei are further reduced as they penetrate the heliosphere and enter the solar system, by an amount that depends on the solar activity cycle. Solar modulation results in the nuclei losing between 200 to 800 MeV/nucleon as they penetrate into the vicinity of the Earth. For high energy nuclei, > 5 GeV/nucleon, these energy changes have little effect, but the best measurements of charge are typically made on nuclei with lower energies. The energy loss at any one time in the solar cycle has a spread that depends on the incident energy (Goldstein, Fisk and Ramaty, 1970) while the mean loss depends on the level of solar activity. This energy loss is particularly significant for cosmic ray nuclei with $E \leq 500$ MeV/nucleon, which are just those for which it is possible to determine the isotopic abundances of the lighter nuclei. A consequence of these energy losses, is that the energies that the nuclei have in the near ISM are not well defined with respect to the energies at which they are observed. This energy spread, combined with the uncertain possibility of reacceleration in the ISM, and the dependence of the rate of ionization energy loss on the energy of the nuclei, means that the energies at which nuclear interactions occur in the ISM are uncertain. This uncertainty is particularly serious in the energy range where the cross sections depend on the energy. For the heaviest UH nuclei, $Z \geq 70$, the cross sections appear to vary at energies of up to 4.0 GeV/nucleon. At higher energies it appears that the regime of "limiting fragmentation" is reached and the cross sections can be regarded as constant. For lighter nuclei limiting fragmentation is reached at lower energies, although for $_{47}$Ag or $_{36}$Kr nuclei, it has not been reached below 1.3 GeV/nucleon (Nilsen *et al.*, 1995).

Propagation

The propagation of the relatively abundant elements lighter than nickel can be characterized by leaky box models in which the nuclei are assumed to diffuse through the ISM. It can be assumed that these models should be equally applicable to the propagation of the heavier nuclei. The nuclei will have exponential probabilities of interacting or of escaping from the confinement volume, the galaxy. Escape lengths, expressed in terms of matter traversed, ranging between 5 and 8 g.cm^{-2} have been derived from studies of the lighter nuclei. As expected, the probability of escape increases as the rigidity, or energy, increases, (e.g. Garcia-Munoz *et al.* 1987) so the escape lengths decrease. These typical escape lengths can be compared with the mean free paths for interaction of very heavy nuclei, such as those with $Z \approx 80$ of ≈ 1.1 g.cm^{-2}, much less than the escape lengths for moderate energy nuclei. Hence, the propagation of the heaviest nuclei will be dominated by interactions, rather than by escape. Furthermore, since the lower energy highly charged nuclei experience large energy losses due to ionization, they undergo rapid energy changes while propagating. This can introduce significant uncertainty because of the energy dependence of the interaction probabilities at the energies where observations are typically made.

In order to study the effects introduced by these energy dependent cross sections, it is convenient to use a relatively simple, albeit somewhat naive, weighted slab approach based on the methodology of the leaky box model. In this calculation, nuclei with a certain initial energy are propagated through many thin slabs of material. In each slab the energy is calculated from that in the previous slab and the effects of any interactions are based on the cross sections assumed for that energy. The output from each slab is weighted by the probability that the nuclei would escape while in that slab. Nested leaky box models assume that initially nuclei cannot escape until they have traversed a finite amount of matter, a "truncation" length. Leaky box models, nested or not, have been remarkably successful in representing the data on the lighter nuclei (Garcia-Munoz *et al.* 1987), although their physical significance has been questioned by Jones *et al.* (1990) and others. Nevertheless, they provide a suitable framework to study the effects of these energy dependent cross sections. A recent study by Ptuskin *et al.* (1996, 97) has shown that even when energy losses are included, weighted slab models give results that generally agree with "exact" solutions to better than 10%, justifying the use of such models, given the other large uncertainties. By using a weighted slab approach it is possible to take account of the varying cross sections and to include the effects of energy losses in a computationally simple manner.

Various assumptions usually have to made in specific calculations of propagation. Although helium probably makes up some 10% by number of the ISM, it is generally assumed in propagation calculations that the medium is pure atomic hydrogen. Little is known about the cross sections in helium, even for light nuclei. If they resemble those in carbon more than those in hydrogen, then the $\approx 20\%$ of interactions occurring in helium will be less energy dependent. If a fraction of the hydrogen is ionized, then the energy loss calculations should be adjusted for the greater losses in the ionized medium, (e.g. Soutoul, Ferrando and Webber, 1990; Thayer 1995).

The populations of nuclei in each slab are weighted by an exponential path-length model, with weights, w, at a path length of x, of a form that allows for the possibility of truncation.

$$x < T \qquad w = 0$$

$$x \geq T \qquad w = \exp\left(\frac{-(x - T)}{L}\right) \qquad \text{Eq. 1}$$

for various assumed values of the escape length, L and truncation, T. These weighted populations are then summed.

These sums for each source element can then be weighted by an assumed source composition, and the remaining primary nuclei plus the created and surviving fragments of each element summed to predict the resulting composition spectra that would be observed. This procedure can then be repeated for various initial energies and weighted by an assumed source energy spectrum. The resulting predicted charge spectra can then be compared with actual observations. If a match is found then the composition at the source of the cosmic radiation can be assumed to be that of the initially assumed source spectrum, within the uncertainties.

Fragmentation of the Heaviest Nuclei

A few of the cross sections needed for these calculations can be measured experimentally using beams of energetic nuclei from accelerators. Most have to be inferred by some form of scaling from those measured. Many of the detectors used in studies of the interactions of energetic heavy nuclei in targets at accelerators cannot look at the individual fragments produced. Instead, they only measure a sum of the squares of the charges of all the fragments that emerge. Fig. 1 shows a plot of the signals from Cherenkov counters looking at interactions of 4.0 GeV/nucleon $_{79}$Au nuclei incident on a polyethylene target (Cummings *et al.* 1998). It can be seen that when the fragment has a small charge change, ΔZ, from that of the incident nucleus, which implies the production of a single fragment with a large charge, there are clear, well resolved, peaks for each charge. Such collisions are presumably peripheral in nature. As ΔZ increases, the resolution is degraded, due the smearing effects of multi-fragmentation. For the most central collisions, with no dominant single fragment, there is no charge resolution and the distribution becomes continuous.

                                                                    C. Jake Waddington

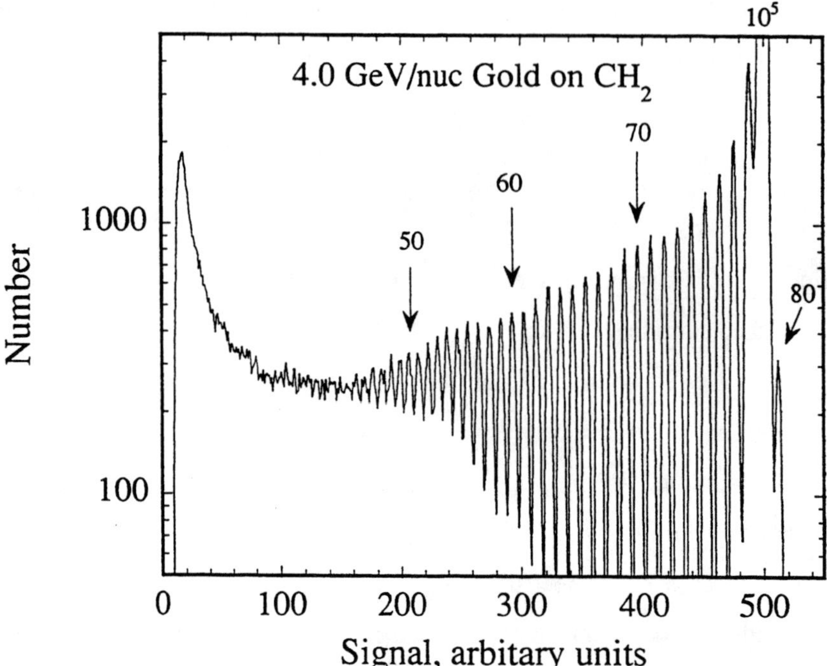

Fig. 1. A plot of the signals from Cherenkov counters resulting from the interactions of 4.0 GeV/nucleon Au nuclei interacting in a polyethylene target (Cummings *et al.* 1998)

The proportion of interactions that do not result in the emission of a single well defined fragment is considerable. In the example of Fig. 1 some 37% of all interactions of the $Z = 79$ projectiles produce no fragment with $Z \geq 49$, which can be compared with the 34% that produce fragments with $78 \geq Z \geq 70$. Multi-fragmentation of 10.6 GeV/nucleon Au nuclei has been studied using nuclear emulsions (Cherry *et al.*,1996, 1998), who find that with heavy target nuclei as many as 20 $Z \geq 2$ fragments can be generated in a single interaction. Even when the target is a hydrogen nucleus, some 37% of all interactions result in the production of between 3 and 6 $Z \geq 2$ fragments.

Fortunately, the cross sections for producing any one particular fragment element with large ΔZ are typically an order of magnitude less than those for small ΔZ. Compilations of natural cosmic abundances all show that in general the elemental abundances decrease as the charge increases, e.g. (Grevesse and Anders, 1988). Hence, fragments with large ΔZ normally can not significantly change the observed cosmic ray abundances. Thus $_{82}$Pb fragmenting into $_{78}$Pt, with $\Delta Z = 4$, can significantly change the observed abundance of Pt, but Pb fragmenting into $_{52}$Te, with $\Delta Z = 30$, will not significantly change the observed abundance of Te. There can be exceptions for certain rare odd charged nuclei where there are low abundances of the slightly heavier elements. For example, $_{43}$Tc, which can be stable in the cosmic radiation, but not in the source, will have more fragments from the $_{50}$Ba and heavier elements than from the rare $_{44}$Ru to $_{49}$In nuclei.

The magnitude of the individual charge peaks of fragments from bombardment of a target, Fig. 1, can be used to determine the cross sections for the production of each fragment element. It is necessary to measure the incident intensity of the projectile nuclei and the target thickness. In order to allow for background effects an exposure to projectiles of equal energy but with no target present is also needed. In addition, unless the target used is very thin, it is necessary to allow for the effects of secondary interactions within the target, (Binns *et al* 1989, Cummings *et al.*, 1990, Waddington *et al* 1994, Nilsen *et al* 1994, Geer *et al* 1995, Cummings *et al* 1998). It should be noted that the largest cross sections for a fragment, those for $Z = 78$, $\Delta Z = -1$, are still less than 10% of the total charge changing cross section. Fig. 2 also shows the large variations of the cross sections with energy when hydrogen is the target, Waddington *et al.* (1998)

The cross sections measured at accelerators do not take account of the radioactive decays which would occur during the millions of years of propagation in the ISM. Based on the studies of Silberberg and Tsao (1990), (Silberberg *et al.*,1997a,b) it appears that these neutron rich very heavy nuclei preferentially lose neutrons in the peripheral interactions that produce the heavy fragments, see also Cummings *et al.* (1990). As a consequence these fragments tend to β decay, or K-capture, to lower charge nuclei, and the even/odd charge ratios tend to be increased. The resulting cross sections for the particular example of 2.0 GeV/nucleon gold nuclei, are illustrated in Fig. 3. It can been seen that the result is to enhance the production of nuclei with even charges. Whether the nuclei can capture K electrons in flight is energy dependent and hence the probabilities of K-decay are uncertain. However, the effect of allowing or forbidding K-decay is generally rather minor when considering the most significant abundance ratios of groups of nuclei, but will be significant for observations with better charge resolution, such as those expected from TREK.

Current measurements of cross sections for very heavy nuclei have been limited to just one heavy projectile element, gold, and have not had isotopic resolution. Propagation

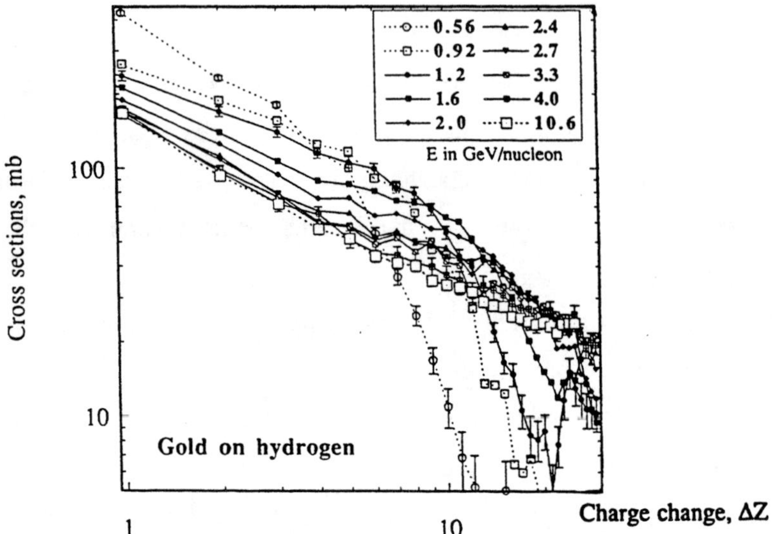

Fig. 2. Measured cross sections of gold nuclei fragmenting in a hydrogen target as a function of ΔZ for a range of energies between 0.56 and 10.6 GeV/nucleon. Uncertainties are shown for some of the values. See text for references.

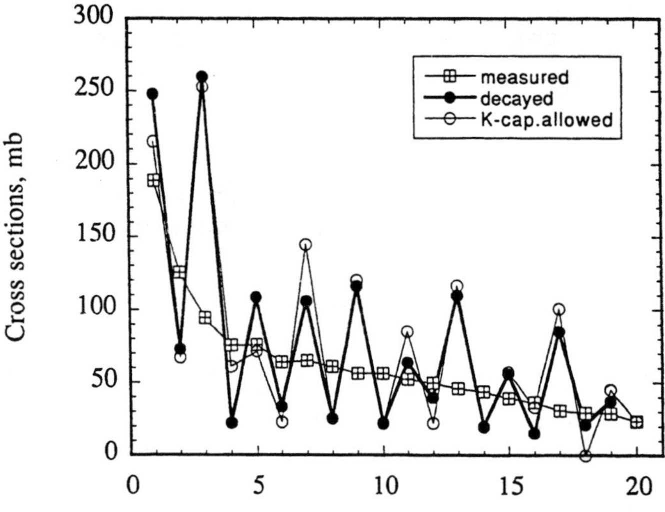

Fig. 3. Effects of decay on the cross sections of gold nuclei in a hydrogen target at 2.0 GeV/nucleon. Values are shown for the measured cross sections as well as the values when all decays of fully stripped nuclei are allowed. Also shown are the cross sections when electron pickup allows all K captures to occur.

calculations have to use cross sections deduced from those measured. The cross sections for elements other than gold have to be scaled from those measured by assuming that they depend only on the mass number, typically as $(A/197)^{2/3}$. These elemental cross

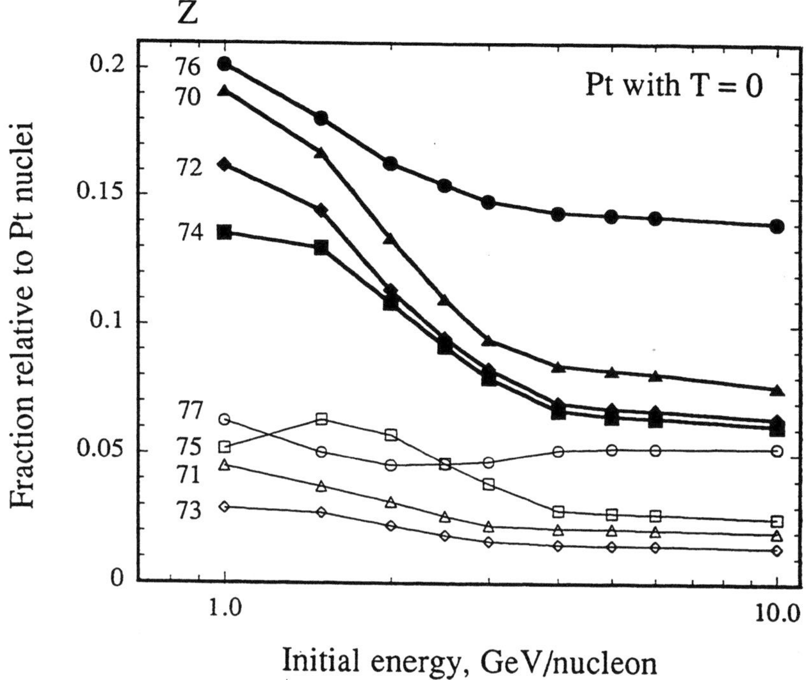

Fig. 4. The relative production of fragments from propagation of an incident beam of Pt nuclei as a function of the initial energy. Fragments with even charges are shown as heavy points, those with odd charges as light points. Propagation assumes no truncation and an escape length of 5.5 g.cm^{-2}. Points are connected to guide the eye.

sections are then corrected for the effects of decay by using the isotopic yields predicted from semi empirical relations derived from studies of proton bombardment, Silberberg and Tsao. These corrections have to take account of the stripped nature of the nuclei in the cosmic radiation, which affects K-capture decays, as well as the existence of bound state β-decay isotopes, Takahashi *et al.* (1987). These corrections are energy dependent, since these very highly charged nuclei can capture the occasional K-shell electron while propagating, even at quite high energies. Here also, the poorly defined energies of the nuclei introduce uncertainties.

Propagation Examples

The sensitivity of propagation calculations on the uncertainties in the cross sections can be illustrated by looking at a few simple examples. Consider the propagation from a

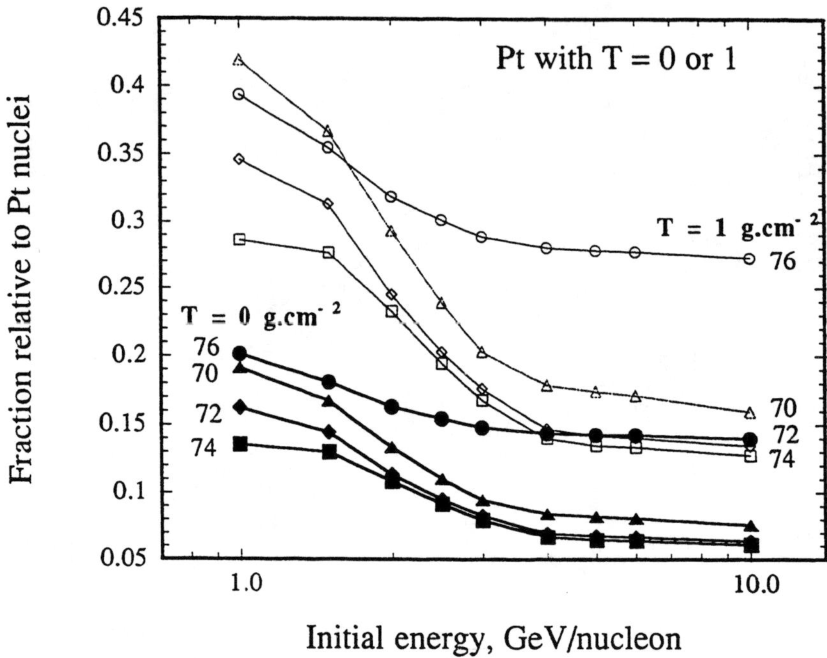

Fig. 5. The effects of 1.0 g.cm^{-2} of truncation during propagation on the even charged fragments from an incident beam of Pt nuclei as a function of the initial energy. The relative production of the even charged fragments with no truncation from Fig. 4 are shown by the solid points, those with T = 1.0 g.cm^{-2} by the open points. Points are connected to guide the eye.

source of pure $_{78}$Pt nuclei. In this case the cross sections used were scaled from those of the gold nuclei and values interpolated between the measurement energies. The predicted yields of fragments at a number of initial energies are shown in Fig. 4. Here it is assumed that T = 0 g.cm^{-2} and that L = 5.5 g.cm^{-2}. Notice that the yield of the even charged fragments decreases sharply between 1.0 GeV/nucleon and $\approx$ 6.0 GeV/nucleon. Those fragments with even charges are significantly more abundant than their neighboring odd charged nuclei. The effects of even a small amount of truncation can be considerable. Considering only the even charged fragments, Fig. 5, introducing 1.0 g.cm^{2} of truncation increases the fraction of fragments by a factor of two. Reacceleration can also introduce significant changes, (Waddington 1998).

These nuclei lose energy rapidly due to ionization losses, but also have very short interaction mean free paths. The energy spectra of Pt nuclei with an initial energy of 2.0 GeV/nucleon and of the resulting fragments are shown in Fig. 6 after propagation. It can

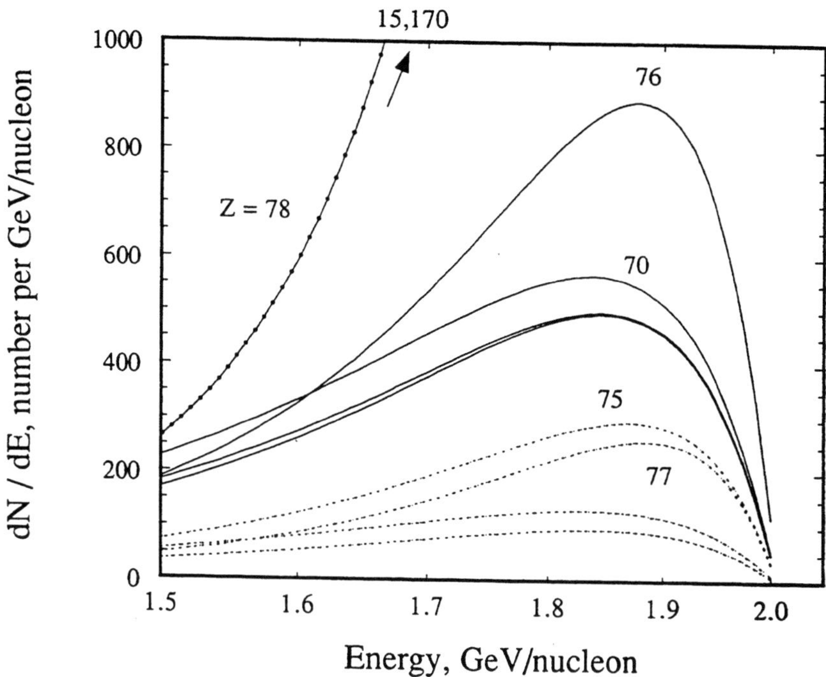

Fig. 6. The expected energy spectra of Pt nuclei and the resulting fragments from the propagation of an initial beam of 2.0 GeV/nucleon Pt nuclei. Spectra of even charge fragments are shown as solid lines, those of odd charged nuclei dashed.

be seen that even at such a low initial energy the majority of the fragments have energies just one or two hundred MeV/nucleon less than at the source. Of course, the effects of solar modulation will be to reduce these energies further before they are detected. Introducing truncation, to first order, introduces a sharp cutoff at about 1.87 GeV/nucleon, although there is also a small shift in the relative yields of the fragments, due to those initial Pt nuclei which were unable to escape during the truncation period.

Conclusions

Observations of the abundances and energy spectra of the very heavy elements in the cosmic radiation can be expected to become more precise during the next few years. This will increase the need for better determinations of the nuclear parameters that are required in any propagation calculation. The current measurements of the cross sections from a gold beam have provided an indication of the probable energy dependence, but they can hardly be adequate to address precision calculations of the propagation of the wide range of isotopes with $Z \geq 70$. There have to be two main concerns.

Firstly, there are no comprehensive measurements of the production of isotopes. A knowledge of these cross sections is essential in order to take account of the effects of decay. At present these effects have to be estimated from the predictions of the semi empirical relations of Silberberg and Tsao (1997), which are based on a few cross sections of isotopes produced by proton bombardment. In general, these semi empirical relations have not been too successful in predicting the results from new measurements.

Secondly, the only heavy nuclei used to measure cross sections of these very heavy nuclei have been those of gold, a rare element in the cosmic radiation. It is not clear that the cross sections for even nearby elements should be simply scaled from those from gold. It is unlikely that it will be possible to make extensive measurements of cross sections from other beams anytime in the near future. What is needed is a predictive code based on the physics of nuclear fragmentation. Such codes do exist but at present give results that are less successful at predicting measured values than are the semi empirical relations. For example, there is a purely nuclear theory abrasion-ablation code by Gaimard and Schmidt, 1991, and the ISAPACE model (Blaich *et al.*, 1992), which couples an intranuclear cascade with an evaporation code. These have been compared (Waddington, 1998) with the results found for light nuclei, such as ^{32}S, and found to give rather worse results than the more empirical models. Hopefully, the new results obtained with gold nuclei, as well as those obtained for somewhat lighter nuclei such as Ag and Kr, (Nilsen *et al.* 1995) will provide enough additional data to permit these models to be improved.

With new observational results and improved nuclear parameters it should become possible over the next few years to obtain a much better definition of the abundances of these very heavy nuclei in the source of the cosmic radiation.

References

Binns, W. R., Garrard, T. L., Gibner, P. S., Israel, M. H., Kertzman, M. P., Klarmann, J., Newport, B. J., Stone, E. C., and Waddington, C. J., 1989a, Ap. J., **346**, 997, and references therein.

Binns, W. R, Cummings, J. R.,. Garrard, T. L,. Israel, M. H, Klarmann, J., Stone, E. C. and Waddington, C. J., 1989b, Phys. Rev., **C39**, 1785

Blaich, T., Beremann-Blaich, M, Fowler, M.M., Wilhelmy, J. B., Britt, H. C., Fields, D. J., Hansen, L. F., Namboodiri, M.N. and Sangster, T. C., 1992, Phys. Rev., **C45**, 689

Cherry, M. L., Dabrowska, J A., Deines-Jones, P., Holynski,. Jones, W. V., Kolganova, E. D., Olszweski, A., Sengupta, K., Skorodko, T. Yu , Szarska, M.,Waddington, C. J., Wefel, J. P., Wilczynska, B., W. Wolter, W., and Wosiek, B.,1995, Phys .Rev., **C52**, 2652-2662.

Cherry, M. L., Dabrowska, J A., Deines-Jones, P., Holynski, R., Nilsen,, B.S., Olszweski, A., Szarska, Trzupek, A.,Waddington, C. J., Wefel, J. P., Wilczynska, B., Wilczynski, H., Wolter, W., and Wosiek, B., Wozniak, K., 1998, To be published in Eur.Phys.J. C

Cummings, J. R., Waddington, C. J., Nilsen B. S. and Garrard, T. L., 1998, submitted to Phys. Rev.

Cummings, J. R., Binns, W. R., Garrard, T. L., Israel, M. H., Klarmann, J., Stone, E. C., and Waddington, C. J., 1990, Phys. Rev., **C42**, 2508 and 2530

Gaimard J. J. and. Schmidt, K. H., 1991, Nuclear Phys., **A531**, 709

Geer, L. Y., Klarmann J., Nilsen, B. S., Waddington, C. J., Binns, W. R., Cummings, J. R. and. Garrard, T. L., 1995, Phys. Rev., **C52**, 334-345

Fowler, P. H., Adams, R. A., Cowen V. G. and Kidd, J. M., 1967, Proc. Roy. Soc. **A.301**, 39

Fowler, P. H., Walker, R. N. F., Masheder, M. R. W., Moses, R. T., Worley A.and Gay, A. M. , 1987, Ap. J., **314**, 739, (1987)

Garcia-Munoz, M., Simpson, J. A., Guzik T. G., Wefel, J. P. & Margolis, S. H. 1987, Ap. J Suppl. **64,** 269

Grevesse, N. and Anders, E., 1988, AIP Conf. Proc. **183**, p. 1, Cosmic Abundances of Matter, ed. C. J. Waddington

Goldstein, M. L., Fisk L. A. and Ramaty, R., 1970, Phys. Rev. Lett., **25**, 832

Jones, F. C., Crawford, H. J., Engelage, J., Guzik, T. G., Mitchell, J. W., Waddington, C. J., Webber, W. R. & Wefel, J. P., 1990, Proc.21st Int. Cosmic Ray Conf., Adelaide, Australia, **3**, 333

Keane, A. J., Thompson, A., O'Sullivan, D.,. Drury , L. O. C. and Wenzel, K-P., 1997, 25th ICRC, Durban, **3**, 361

Nilsen, B. S., Waddington, C. J., Binns, W. R., Cummings, J. R.,. Garrard T. L. and Klarmann, J., 1995, Phys. Rev., **C52**, 3277

Ptuskin, V. S., Jones, F. C., and Ormes, J. F., 1996, ApJ, **465**, 362

Ptuskin, V. S., Jones, F. C., Ormes, J. F. and Soutoul, A., 1997, ICRC, Durban, **4**, 261

Silberberg R.. and Tsao, C. H., 1990, Phys. Rep., **191**, 351 (see also URL http://www.spdsch.phys.lsu.edu)

Silberberg, R., Tsao, C. H., Barghouty, A. F.and Shapiro, M. M., 1997a, 25th ICRC, Durban, **4**, 321

Silberberg, R., Tsao, C. H., Barghouty, A. F., Shapiro, M. M. and. Adams, J. H., 1997b, *25th ICRC*, Durban, **4**, 325

Soutoul, A., Ferrando P. & Webber,W. R., 1990, Proc 21st Int. Cosmic Ray Conf., Adelaide, Australia, **3**, 337

Takahashi, G. J., Boyd, R. N., Mathews, G. J., Yokoi, K., 1987, Phys. Rev., **C36**, 1522.

Thayer, M. R., 1995, Proc.24th Int. Cosmic Ray Conf., Rome, Italy , **3**, 124

Waddington, C. J., Binns, W. R., Cummings, J. R., Garrard, T. L., Gauld, B. W., Geer , L. Y., Klarmann J. and Nilsen, B.S., 1994, Nucl. Phys., **A566**, 427c

Waddington, C. J., 1996, New Astro. Rev., to be published.

Walker, R. M., Fleischer R. L. and. Price, P. B, 1965, Proc. ICRC, London, **2,** p 1086, see also. Fleischer, R. L., Price, P. B.,. Walker, R. M., Maurette M. and Morgan, G., 1967, J. Geophys. Res., **72**, 355

Westphal, A. J., Price, P. B., Weaver B. A. and Afanasiev, V. G., 1998 Nature, to be published.

Early Galactic Li, Be and B: Implications on Cosmic Ray Origin

R. Ramaty

Laboratory for High Energy Astrophysics, Goddard Space Flight Center, Greenbelt, MD 20771; ramaty@gsfc.nasa.gov

and

R. E. Lingenfelter

Center for Astrophysics and Space Sciences, University of California San Diego, La Jolla, CA 92093-0424; rlingenfelter@ucsd.edu

ABSTRACT

Be abundances of old, low metallicity halo stars have major implications on cosmic-ray origin, requiring acceleration out of fresh supernova ejecta. The observed, essentially constant Be/Fe fixes the Be production per SNII, allowing the determination of the energy supplied to cosmic rays per SNII. The results rule out acceleration out of the metal-poor ISM, and favor Be production at all epochs of Galactic evolution by cosmic rays having the same spectrum and source composition as those at the current epoch. Individual supernova acceleration of its own nucleosynthetic products and the collective acceleration by SN shocks of ejecta-enriched matter in the interiors of superbubbles have been proposed for such origin. The supernova acceleration efficiency is about 2% for the refractory metals and 10% for all the cosmic rays.

1. Introduction

It has been known for almost three decades that cosmic-ray interactions in the interstellar medium (hereafter ISM) have an important role in producing the Galactic inventories of the light elements Li, Be and B, hereafter LiBeB (Reeves, Fowler & Hoyle 1970; for reviews see also Reeves 1994 and Ramaty, Kozlovsky & Lingenfelter 1998a). However, not all the LiBeB isotopes are cosmic-ray produced. About 10% of the ^{7}Li inventory results from nucleosynthesis in the Big Bang (e.g. Spite & Spite 1993), and nucleosynthesis in a variety of Galactic objects, including core collapse supernovae (SNIIs, Woosley & Weaver 1995), novae (Hernanz et al. 1996) and giant stars (Plez, Smith & Lambert 1993; Wallerstein & Morrell 1994), can produce a large fraction of the remaining 90%. Core collapse supernovae also contribute to ^{11}B production via ^{12}C spallation by neutrinos during the explosion (Woosley et al. 1990; Woosley & Weaver 1995). But essentially all of the ^{6}Li, Be and ^{10}B, and about 50% of the ^{11}B are cosmic-ray produced.

Starting about a decade ago, observations with ground based telescopes led to Be abundance determinations in old, low metallicity halo stars (see Vangioni-Flam et al. 1998 for a recent compilation of the data). These stars, with Fe-to-H abundance ratios as low as 10^{-3} of the solar value, were born in the early Galaxy and still preserve fossil records of conditions that existed in that epoch of Galactic evolution. Thus, as the Be is entirely cosmic-ray produced, the early Galactic data extend the time scale of cosmic-ray research from the 10^7 year mean age of the current epoch cosmic rays to the more than 10 billion year age of the Galaxy. In particular, the fact that the average early Galactic ISM, unlike the ISM of the current epoch, was almost totally devoid of metals (C and heavier atoms), provides new clues on the nature of the source material out of which cosmic rays are accelerated. As we shall see, the early Galactic Be data strongly suggests that the cosmic rays are accelerated out of fresh supernova ejecta rather than the average ISM, because if they had been accelerated out of the ISM, the Be in the early Galaxy would have been highly underproduced. To allow the direct acceleration of supernova ejecta before they mix into the average ISM, two related cosmic-ray origin scenarios were developed recently. The first considers the individual supernova shock acceleration of its own nucleosynthetic products (Lingenfelter, Ramaty & Kozlovsky 1998), while the second addresses the collective acceleration by successive SN shocks of ejecta-enriched matter in the interiors of superbubbles (Higdon, Lingenfelter & Ramaty 1998). Freshly formed dust grains in supernovae play a central role in both models, as we shall see.

2. LiBeB Origin

Li and Be abundances for various stars as a function of their metallicity are shown in Fig. 1. The metallicity of a star is defined in terms of its Fe abundance, [Fe/H]$\equiv$log(Fe/H)$-$log(Fe/H)$_\odot$, where Fe/H is the Fe abundance by number relative to H and (Fe/H)$_\odot$ is the solar system value. [Fe/H] increases with time, reflecting the accumulating production of supernova nucleosynthesis, and thus provides a convenient, but nonlinear, representation of elapsed time since the formation of the Galaxy. Studies of Galactic chemical evolution (e.g. Ramaty, Lingenfelter & Kozlovsky 1998b; see also Pagel 1997 and references therein) have provided information on the age-metallicity relation. The halo phase of our Galaxy, for which [Fe/H] $\lesssim -1$, corresponds to a period of $\sim 10^9$ years preceding the formation of the Galactic disk. LiBeB observations of the early Galaxy (e.g. Molaro et al. 1997; Hobbs & Thorburn 1997; Duncan et al. 1997; Garcia Lopez et

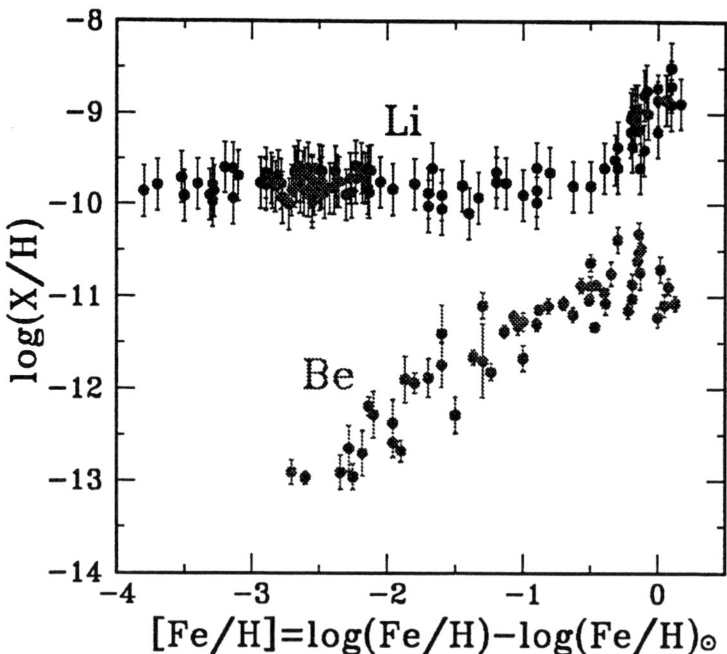

Fig. 1.— Li and Be abundances for stars of various ages as a function of their Fe abundance. Data compilation by M. Lemoine for Li and Vangioni-Flam et al. (1998) for Be.

al. 1998) are very challenging because their spectral lines are weak and they are usually blended with interfering lines from other more abundant elements. The observations and their interpretation therefore require large telescopes and very efficient, high resolution detectors. While the Li and Be lines can be observed from the ground, the B lines require observations from space (with the Hubble Space Telescope, see Duncan et al. 1997).

The flat portion of the Li evolution (Fig. 1), usually referred to as the Spite plateau (Spite & Spite 1993), is generally believed (see Reeves 1994) to represent the ^{7}Li abundance resulting from nucleosynthesis in the Big Bang. The subsequent increase in Li/H is due to nucleosynthesis in the various Galactic objects mentioned in the Introduction. Unlike Li/H, Be/H increases with increasing [Fe/H] at all metallicities, implying a Galactic origin for Be even at the lowest metallicities where $\log(\mathrm{Be/H}) \simeq -13$. Indeed, the maximum contribution of Big Bang nucleosynthesis is insignificant, $\log(\mathrm{Be/H})_{\mathrm{BBNS}} \simeq -15$ (Orito et al. 1997).

The implications of the Be data for cosmic-ray origin become more obvious (Ramaty et al. 1998a) when $\log(\mathrm{Be/Fe})$, rather than $\log(\mathrm{Be/H})$ is considered (Fig. 2). Here the horizontal line provides (Vangioni-Flam et al. 1998) the best

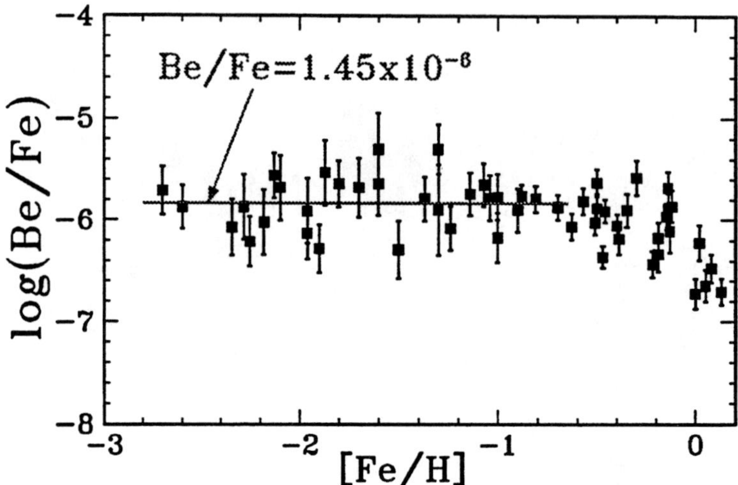

Fig. 2.— Observed Be-Fe abundance ratio as a function of [Fe/H]; data compilation Vangioni-Flam et al. (1998). The best fit, for [Fe/H] $\lesssim$ -1, implies that $2.3 \times 10^{-8} M_\odot$ of Be are produced per average SNII. The decrease at higher [Fe/H] is due to contributions of Type Ia supernovae which make Fe but very little Be.

fit to the data for [Fe/H]< -1. Fe production in this epoch (Timmes, Woosley & Weaver 1995) is dominated by SNIIs, which result from massive ($>10M_\odot$), short lived ($\lesssim$ 30 Myr) stellar progenitors with an IMF (initial mass function) averaged Fe yield per SNII of $\sim$0.1$M_\odot$, essentially independent of metallicity (Woosley & Weaver 1995). The decrease of Be/Fe for [Fe/H] $\gtrsim -1$ most likely results from the additional Fe production in Type Ia supernovae (Timmes et al. 1995). These come from more slowly evolving white dwarf systems and are less frequent than the SNIIs, but because they do not produce neutron star remnants, they eject much more Fe per supernova and thus account for about half of the present Fe production.

The essentially constant observed Be-to-Fe abundance ratio for [Fe/H] $\lesssim -1$, coupled with the fact that Fe in this epoch is produced only by SNIIs with a yield independent of [Fe/H], requires an essentially constant Be production per SNII,

$$Q_{Be} \simeq 0.1 \times 1.45 \times 10^{-6} \times (9/56) = 2.3 \times 10^{-8} M_\odot \, , \tag{1}$$

independent of [Fe/H]. This is in conflict with cosmic-ray acceleration out of the ISM because in that case the composition of the cosmic rays (particularly C/H and O/H) would evolve in proportion to that of the ISM and Q_{Be} would increase with [Fe/H], contrary to the requirements of the data.

Independent evidence against cosmic ray acceleration purely out of the ISM is provided by energetics. The energy in cosmic rays per SNII, W_{SNII}, needed to produce the required amount of Be depends on the composition of both the ISM and the cosmic rays, on the energy spectrum of the cosmic rays, and on the cosmic-ray escape length from the Galaxy, X_{esc} measured in g cm^{-2} (Ramaty, Kozlovsky, Lingenfelter & Reeves 1997). In Fig. 3 (from Ramaty et al. 1998b) we show W_{SNII}, as a function of [Fe/H] for two values of X_{esc}, and for two Galactic cosmic-ray origin models: a proposed CRS model for which the cosmic rays at all [Fe/H] have the same source composition and spectrum as the current epoch cosmic rays, and the ISM model for which the cosmic rays are accelerated out of a metallicity dependent interstellar medium with an energy spectrum that is also identical to that of the current epoch cosmic rays. The ambient ISM composition for both models is solar, scaled with $10^{[Fe/H]}$, except that O/H is allowed to increase by a factor of 3 for [Fe/H]< -1 to allow for the well known increase of O/H at low metallicities relative to its solar value (see Pagel 1997). To take into account recent shock acceleration results (Ellison, Drury & Meyer 1997), the cosmic-ray C/H and O/H relative to the corresponding metallicity dependent ISM values are also increased by factors of 1.5 and 2. We see that the ISM model requires that W_{SN}, the cosmic-ray energy per SNII, not only be metallicity dependent, which is unlikely, but also untenably large, exceeding the total available ejecta kinetic energy ($\sim 1.5 \times 10^{51}$ erg, Woosley & Weaver 1995) when [Fe/H]< -2. This reinforces the previous conclusion that cosmic rays accelerated out of the average, metal poor ISM cannot be responsible for the Be production in the early Galaxy.

For the CRS model, on the other hand, W_{SNII} is essentially constant (Fig. 3), equal to the very reasonable value of $\sim 10^{50}$ erg/SNII, practically the same as the energy supplied per supernova to the current epoch cosmic rays (Lingenfelter 1992). That these two energies are consistent, led to a different cosmic-ray paradigm, direct acceleration out of fresh supernova ejecta, at least for the refractory metals (Ramaty et al. 1998a; Lingenfelter, Ramaty & Kozlovsky 1998; Higdon, Lingenfelter & Ramaty 1998). The constancy of Be/Fe is a straightforward consequence of such a model and clearly such a cosmic-ray origin provides the simplest explanation for the origin of Be throughout the entire evolutionary history of our Galaxy. Moreover, as we shall see (§3), acceleration of fresh ejecta, rather than average ISM material, is to be expected since the hot phase of the interstellar medium, where shock acceleration is most efficient (Axford 1981), is probably highly enriched in fresh gas and dust from the same supernovae whose shocks accelerate the cosmic rays (Higdon et al. 1998). There is also another, more complex scenario for Be production by a possible, separate low energy cosmic-ray (LECR) component, also accelerated from fresh nucleosynthetic matter (Cassé, Lehoucq & Vangioni-Flam 1995; Ramaty, Kozlovsky & Lingenfelter 1996.) Such LECRs might allow the acceleration of the standard Galactic cosmic rays out of

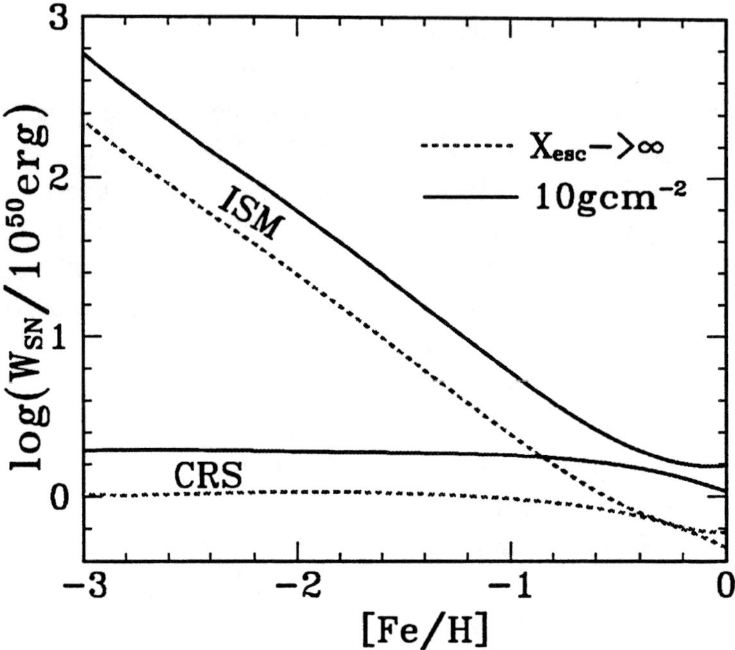

Fig. 3.— Energy in cosmic rays per SNII required to produce $2.3\times10^{-8}M_\odot$ of Be. The cosmic ray source composition is metallicity independent for the CRS model and metallicity scaled for the ISM model. $X_{esc}\simeq 10$ g cm^{-2} is the approximate current epoch cosmic-ray escape length; in the early Galaxy it could have been different, depending on the density and magnetic structure of the early Galactic halo. When $X_{esc}\rightarrow\infty$, the cosmic rays are trapped in the halo until they are either stopped by Coulomb collisions or destroyed by nuclear reactions; this choice of escape length yields the lowest W_{SN} for the given Be production.

the ISM at all epochs of Galactic evolution, including the current one, but the nuclear gamma-ray lines that would provide the only evidence for their existence have not yet been seen. We consider these issues in §3, but before that we briefly discuss the B and ^{6}Li data.

Observations (Duncan et al. 1997; Garcia Lopez 1998) with the Hubble Space Telescope of the B abundance show (Fig. 4) that the B-to-Be abundance ratio is also essentially independent of [Fe/H], implying a common origin for these two elements. It has often been pointed out (e.g. Reeves 1994) that there is a problem with a pure cosmic-ray origin for B in that its isotopic ratio, ^{11}B/^{10}B=4.05±0.2 measured in meteorites (Chaussidon & Robert 1995) and ^{11}B/^{10}B =3.4±0.7 in the interstellar medium (Lambert et al. 1998), exceeds the calculated ratio (2 to 2.5) for production by the Galactic cosmic rays (Ramaty et al. 1997). However,

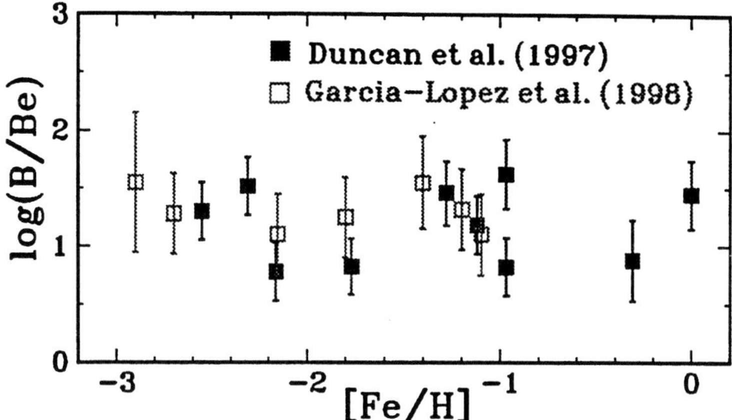

Fig. 4.— Observations of the B-to-Be ratio as a function of [Fe/H]. The fact that this ratio does not vary much with metallicity implies a common origin for these two elements. SNII accelerated cosmic rays produce both B and Be and additional B comes from C spallation by neutrinos in SNIIs.

significant ^{11}B production is also expected from ^{12}C spallation by neutrinos in SNIIs (Woosley et al. 1990). In fact, if $\sim 30\%$ of the ^{11}B is from neutrinos the observed ^{11}B/^{10}B can be explained, and since both the neutrino and cosmic-ray induced spallation processes are related to such supernovae, the constancy of the B-to-Be ratio is assured. The neutrinos mostly make ^{11}B and not ^{10}B because their temperature is not high enough for interactions above the higher threshold energy for ^{10}B production. The required additional ^{11}B production per SNII (Ramaty et al. 1997), about $(2\text{-}7)\times10^{-7}$ M$_\odot$, is consistent with the supernova calculations (Woosley & Weaver 1995).

The CRS model predicts (Ramaty et al. 1997) an abundance ratio ^{6}Li/Be=5±0.5, essentially independent of [Fe/H] and consistent with the meteoritic value of 5.8 (Anders & Grevesse 1989). But at low metallicities ([Fe/H]$\simeq$-2.3), values of ^{6}Li/Be as high as 60 are reported (Smith, Lambert & Nissen 1998) which, even though still subject to large uncertainties, appear inconsistent with CRS production alone and would imply an additional source which dominated ^{6}Li production at early times. Since significant ^{6}Li can be produced by the α-α reaction ^{4}He(α,pn)^{6}Li whose cross section is very large below about 100 MeV/nucleon, ^{6}Li/Be is strongly dependent on both the cosmic-ray composition (He/CNO) and energy spectrum, suggesting two possible additional sources. LECRs might account for these early Galactic ^{6}Li data, but if ^{6}Li/Be were to remain constant as a function of [Fe/H], as would be the case if the LECR component were also responsible for producing the Be at all metallicities, then ^{6}Li/Be would be inconsistent with the meteoritic value (at [Fe/H]=0), and the total cosmic-ray produced Li abundance (which includes ^{7}Li with ^{7}Li/^{6}Li=1.5) would significantly exceed the Li/H data around [Fe/H]$\simeq -1$ (see Fig. 1). Alternatively, there could have been significant ^{6}Li production via the α-α reaction by possible pre-Galactic cosmic rays consisting almost entirely of primordial protons and α particles that would produce no Be. Further studies of this very exciting possibility, however, must await improvements in the reliability of the early Galactic ^{6}Li measurements, which are very difficult. The contribution of Big Bang nucleosynthesis to Galactic ^{6}Li again is very small (Nollett, Lemoine & Schramm 1997).

3.   Cosmic-Ray Origin

Motivated by the need to accelerate the cosmic rays out of fresh supernova ejecta, two related CRS scenarios were developed. The first considers the individual supernova shock acceleration of its own nucleosynthetic products (Lingenfelter et al. 1998), while the second addresses the collective acceleration by successive SN shocks of ejecta-enriched matter in the interiors of superbubbles (Higdon et al. 1998). In the individual supernova model, freshly formed high velocity grains in the slowing ejecta reach the forward supernova shock which then accelerates the grain erosion products. The superbubble model emphasizes the fact that the bulk of the SNIIs occur in the cores of supernova generated superbubbles where the ambient matter is likely to be dominated by fresh supernova ejecta. In both scenarios, grain erosion products play a central role. They provide an explanation for the observed cosmic-ray enrichment of the highly refractory Mg, Al, Si, Ca, Fe and Ni relative to the highly volatile H, He, N, Ne, Ar, an idea developed in detail previously for the ISM model (Meyer, Drury & Ellison 1997). In both the individual supernova and superbubble scenarios, the accelerated C and O originate from grains, O from oxides (MgSiO$_3$, Fe$_3$O$_4$, Al$_2$O$_3$, CaO) and C mainly from graphite. As shown previously (Lingenfelter et al. 1998) such an origin for the C and O can explain the problematic C-to-O ratio in the cosmic rays which exceeds the corresponding solar ratio by about a factor of 2.

The similarity of the cosmic-ray source and solar abundance ratios of refractory elements, mainly Mg, Al, Si, Ca, relative to Fe, has been mentioned (Meyer et al. 1997) as an argument against the supernova ejecta origin for the cosmic rays. That this is not the case was demonstrated by Lingenfelter et al. (1998), the principal reason being that the combined contributions from SNIIs and Type Ia supernovae are responsible for both the cosmic-ray source and solar abundances of these refractories.

The presence of s-elements in the cosmic rays has also been used as an argument against cosmic-ray acceleration out of supernova ejecta (e.g. Meyer et al. 1997), since such elements are not synthesized in the supernova explosions. However, this does not contradict acceleration out of the ejecta because s-elements are present in supernova ejecta along with the other much more abundant products of pre-supernova burning. The s-process elements are made in the cores of stars and can be ejected both in supernova explosions and in strong stellar winds (e.g. Arnould & Takahashi 1993). In fact, the observations of SN1987A (e.g. Mazzali, Lucy & Butler 1992 and references therein) show very significant overabundances of the prominent s-process products, Sr and Ba, relative to Fe by perhaps as much as an order of magnitude compared to solar values. Only a fraction of such supernova ejecta could account for the much more modest Sr and Ba overabundances of about 1.5, required for cosmic-ray source ratios (e.g. Binns et al. 1989). Moreover, the enrichment of r-process nuclei in the cosmic rays, especially the strong Pt peak (Waddington 1996), provides direct support for a supernova ejecta origin. The r-process elements are thought to be made just above the newly formed neutron star (e.g. Cardall & Fuller 1997) in core collapse supernovae. Thus, both r- and s-process elements are blown off in the supernova ejecta along with the products of explosive burning and other products of earlier burning, and all the refractories condense in the ejecta.

Finally, electron-capture decay nuclei, such as ^{59}Ni (decaying into ^{59}Co with mean life of 1.1×10^5 yrs), can give a measure of the time between nucleosynthesis and acceleration, since such decay is suppressed once the nuclei are accelerated and fully ionized. Preliminary ACE data on the ^{59}Ni-^{59}Co ratio (Wiedenbeck et al. 1998) show that ^{59}Ni has decayed, suggesting delayed acceleration. This result favors the superbubble model for which the mean time between successive supernova explosions, $\sim 3 \times 10^5$ yr, ensures that each supernova shock will on average accelerate ejecta accumulated from many previous supernovae on time scales clearly exceeding the ^{59}Ni mean life.

4. Nuclear Gamma-Ray Line Emission

The possible existence of a distinct low energy component of cosmic rays which could not be observed in the inner solar system because of solar modulation, is a topic of major interest for cosmic-ray research. Evidence for a strong enough LECR component that could produce significant amounts of LiBeB could only come from nuclear gamma-ray line data (e.g. Ramaty, Kozlovsky & Lingenfelter 1979). Indeed, as mentioned above, following the reported (Bloemen et al. 1994) detection with COMPTEL/CGRO of C and O gamma-ray lines from the Orion star formation region, Cassé et al. (1995) suggested that the LECRs postulated to exist in the Orion region might be responsible for the Be (and B) production in the early Galaxy. The motivation for this idea was the indication (based on the reported spectrum of the line emission) that the LECRs in Orion are enriched in C and O relative to protons and α particles (see Ramaty 1996 for review and Ramaty, Kozlovsky & Lingenfelter 1996 for extensive calculations of LiBeB production by LECRs). It was proposed (Bykov 1995; Ramaty et al. 1996; Parizot, Cassé & Vangioni-Flam 1997) that such enriched LECRs might be accelerated out of metal-rich winds of massive stars and the ejecta of supernovae from massive star progenitors ($>60M_\odot$) which explode within the bubble around the star formation region due to their very short lifetimes. These arguments led to the suggestion (Vangioni-Flam et al. 1996; Vangioni-Flam, Cassé & Ramaty 1997) that the composition of the LECRs could be independent of Galactic metallicity, thus allowing them to reproduce the constancy of Be/Fe in the early Galaxy. The problem with this model is its energetic inefficiency, mostly because it relies on $>60M_\odot$ SNII progenitors which are much less numerous, but not significantly more energetic than the progenitors of all the SNIIs ($>10M_\odot$) which accelerate the cosmic rays in the CRS model. Moreover, the Orion gamma-ray line data have recently been withdrawn by the COMPTEL team (private communication, V. Schönfelder, 1998). The planned high resolution observations of the Galaxy-wide nuclear gamma-ray line emission by the INTEGRAL mission should better define the possible contribution of LECRs to Galactic Be production.

5.   Conclusions

We have seen how recent atomic spectroscopy observations of light element abundances in old halo stars have brought exciting new insights to the question of the origin of the cosmic rays, a problem that so far has been investigated mainly by in situ cosmic-ray observations. The cosmic rays in the early Galaxy, or at least their C and O, must have been accelerated from freshly nucleosynthesized matter rather than from the then extremely metal poor interstellar medium. This strongly suggests that the present epoch cosmic rays are also accelerated from fresh material, unlike in most current models. We have outlined two recently proposed scenarios for cosmic-ray acceleration from supernova ejecta that could account for such an origin.

REFERENCES

Anders, E., & Grevesse, N. 1989, Geochim. et Cosmochim. Acta, 53, 197

Arnould, M., & Takahashi, K. 1993, in *Origin and Evolution of the Elements*, N. Prantzos, et al. eds. (Cambridge) 395

Axford, W. I. 1981, Annals N.Y. Acad. Sci., 375, 297

Binns, W. R. et al. 1989, ApJ, 346, 997

Bloemen, H. et al. 1994, A&A, 281, L5

Bykov, A. M. 1995, Space Sci. Rev., 74, 397

Cardall, C. Y., & Fuller, G. M. 1997, ApJ, 486, L111

Cassé, M., Lehoucq, R., & Vangioni-Flam, E. 1995, Nature, 373, 318

Chaussidon, M., & Robert, F. 1995, Nature, 374, 337

Duncan, D. K. et al., 1997, ApJ, 488, 338

Ellison, D.C., Drury, L.O'C., & Meyer, J-P. 1997, ApJ, 487, 197

Garcia Lopez, R., Lambert, D. L., Edvardsson, B., Gustafsson, B., Kiselman, D., & Rebolo, R. 1998, ApJ, 500, 241

Hernanz, M., José, J., Coc, A., & Isern, J. 1996, ApJ, 465, L27

Higdon, H. C., Lingenfelter, R. E., & Ramaty, R., 1998, ApJ (Letters), submitted

Hobbs, L. M., & Thorburn, J. A. 1997, ApJ, 491, 772

Lambert, D. L. et al. 1998, ApJ, 494, 614

Lingenfelter, R. E. 1992, in *Astronomy & Astrophysics Encyclopedia*, S. Maran ed. (New York: Van Nostrand), 139

Lingenfelter, R. E., Ramaty, R., & Kozlovsky, B. 1998, ApJ, 500, L153

Mazzali, P. A., Lucy, L. B., & Butler, K. 1992, A&A, 258, 399

Meyer, J-P., Drury, L. O'C., & Ellison, D. C. 1997, ApJ, 487, 182

Molaro, P., Bonifacio, P., Castelli, F., & Pasquini, L. 1997, A&A, 319, 593

Nollett, K. M., Lemoine, M., & Schramm, D. N. 1997, Phys. Rev. C, 56, 1144

Orito, M., Kajino, T., Boyd, R. N., & Mathews, G. J. 1997, ApJ, 488, 515

Pagel, B. E. J. 1997, *Nucleosyn. and Chem. Evolution of Galaxies* (Cambridge)

Parizot, E. M. G., Cassé, M., & Vangioni-Flam, E. 1997, A&A, 328,107

Plez, B., Smith, V. V., & Lambert, D. L. 1993, ApJ, 418, 812

Ramaty, R. 1996, A&A (Suppl.), 120, C373

Ramaty, R., Kozlovsky, B., & Lingenfelter, R. E. 1979, ApJS, 40, 487

Ramaty, R., Kozlovsky, B., Lingenfelter, R. E. 1996, ApJ, 456, 525

Ramaty, R., Kozlovsky, B., & Lingenfelter, R.E. 1998a, Phys. Today, 51, No. 4, 30

Ramaty, R., Kozlovsky, B., & Lingenfelter, R. E. 1998b, Nuclei In the Cosmos V, in press, (astro-ph/9809211)

Ramaty, R., Lingenfelter, R. E., Kozlovsky, B., & Reeves, H. 1997, ApJ, 488, 730

Reeves H. 1994, Revs. Mod. Phys. 66, 193

Reeves, H., Fowler, W.A., & Hoyle, F. 1970, Nature Phys. Sci, 226, 727

Smith, V. V., Lambert, D. L., & Nissen, P. E. 1998, ApJ, in press

Spite, F., & Spite, M. 1993, in *Origin and Evolution of the Elements*, eds. N. Prantzos et al. (Cambridge), 201

Timmes, F. X., Woosley, S. E., & Weaver, T. A. 1995, ApJS, 98, 617

Vangioni-Flam, E., Cassé. M., Fields, B. D., & Olive, K. A. 1996, A&A, 468, 199

Vangioni-Flam, E., Cassé. M., & Ramaty, R. 1997, in *The Transparent Universe*, eds. C. Winkler et al., ESA SP-382, p. 123

Vangioni-Flam, E., Ramaty, R., Olive, K., & Casse, M. 1998, A&A, 337, 714

Waddington, C.J. 1996, ApJ, 470, 1218

Wallerstein, G., & Morell, O. 1994, A&A, 281, L37

Wiedenbeck, M. et al. 1998, ACE News 17

Woosley, S. E., Hartmann, D. H., Hoffman, R. D., & Haxton, W. C. 1990, ApJ, 356, 272

Woosley, S.E., & Weaver, T.A. 1995, ApJS, 101, 181

Cosmic Ray Source Composition and its Relation to the Interstellar Matter

Kunitomo Sakurai

*Institute of Physics, Kanagawa University, Rokkakubashi,Yokohama
221-8686,Japan*

ABSTRACT

In this article, several topics as related to the source composition of cosmic rays are considered in some detail. They are:

(1) Observed nature of the source composition of cosmic rays and its relation to the origin of cosmic rays,

(2) Source matter of cosmic rays and its possible mechanism of forming the source composition of cosmic rays,

(3) Possible processes on the condensation of the elements into the source matter of cosmic rays,

(4) The behavior of cosmic ray nuclei as being controlled either by the first ionization potential or the condensation temperature for each nucleus in its atomic state, and

(5) Causal relation of the acceleration of cosmic rays with the formation of cosmic ray source matter.

Some additional discussion will be given while considering the topics as described above.

1. Introduction

It is now believed that most of cosmic rays are born somewhere in this Milky Way Galaxy, except for the highest energy cosmic rays, but there remain many problems unsolved as yet to understand the origin of cosmic rays and its related subjects. Since the understanding on the origin of cosmic rays may give us some insight crucial in understanding such high energy phenomena as nonthermal γ –ray, X–ray and radio emissions associated with supernova explosions and many of other energetic events in the interstellar space, it is necessary to pursue where and how these cosmic rays are accelerated to high energy and how their chemical composition is formed in the regions where they are born.

One of the key issues to understand where cosmic rays are born necessarily comes from the observed data on their source composition which has been composed from the matter mixed up from those which were originated from gases ejected from supernova explosions, stellar winds released from various types of stars and other sources. It is thus necessary to analyze the observed data on the cosmic ray composition in the space near the earth to deduce the source composition of cosmic rays by referring to the theoretical treatment on the propagation of cosmic rays in the interstellar space, because the observed composition of cosmic rays are contaminated with the secondary components of various nuclei which have been produced while propagating in the space between their source regions and the earth's neighborhood.

At present, the source composition of cosmic rays has been known in greater detail as the results of the analyses of the observed data on the cosmic ray composition (e.g., Simpson, 1983; Webber, 1997). It thus follows that this composition can be taken into account in understanding such a problem as how and where it has been formed in the galactic space. The mechanism of the formation of this composition may be causally related to various physical processes as observed in this space, because cosmic rays are mostly accelerated from the matter provided from high energy phenomena, such as supernova explosions taking place there.

2. Observed Data on the Source Composition of Cosmic Rays

Many observations have been performed to obtain the data on the chemical composition of cosmic rays for a wide range of particle energy by using satellites and spacecraft launched into the nearby space and beyond. Since the observed result of this composition consist of both the primary and the secondary components of cosmic rays, it is necessary to theoretically estimate how much this secondary component contribute to this composition as the results of the disintegration of primary cosmic rays due to their interaction with the gases ambient in the interstellar space before cosmic rays reach the points where their observations are being made. In doing so, the mechanism on how cosmic rays propagate in the interstellar space while interacting with hydrogens and other elements ambient in this space has to be worked out theoretically based on reasonable assumptions on their interactions just mentioned.

By the extensive efforts by Shapiro, Silverberg and others(e.g., Silverberg et al., 1973, 1983), at present, it can be said that the propagation mechanism of cosmic rays and its relation to their disintegration in the interstellar space has been reasonably well understood. So, the data currently available on the source composition of cosmic rays can be referred to be reasonable and then applied to search for the regions where this source composition is formed. Such regions, of course, must supply the matter from which cosmic rays are accelerated, though some selective mechanism seems to work out to form the source composition of cosmic rays.

According to the observed data currently available on the source composition of cosmic rays, it is clear that the most heavy nuclei, identified mainly as refractory and siderophile, are relatively enriched in this composition as compared with the chemical composition of the solar photosphere as obtained from spectroscopic observations. However, it is noted that, as shown in Fig.1, the cause of the relative enrichment of these heavy nuclei is causally related to the first ionization potentials of these nuclei in their atomic states(e.g., Cassé and Goret, 1978). Volatile elements, such as helium, neon and others are less abundant in the source composition than in the solar photosphere.

Many of heavy nuclei relatively enriched in this source composition are identified as refractory and siderophile as specified above. Furthermore, their first ionization potentials in their atomic states are relatively lower as compared with those for volatile elements. Since the elemental volatility is necessarily related to the condensation temperature of the elements in their atomic states(Sakurai and Ito, 1990), it may be interesting to look for the relation between the source composition of cosmic rays and this temperature for each nucleus in its atomic state(Sakurai, 1995, 1997). This relation is shown in Fig.2, in which the data on the condensation temperature for each element have been taken from Wasson(1985). Most of the elements relatively enriched in the source composition of cosmic rays are well confined in the area where the condensation temperature is about equal to or higher than 1000K. It is suggested from Fig.2 that both of volatile and low-volatile elements must have been lost effectively from the regions where cosmic ray source matter is formed.

The result as indicated in Fig.2 suggests that such matter as identified with cosmic ray sources must have passed through states with temperature lower than 1000K while being formed somewhere in the interstellar space. In this cooling process, volatile elements may have been lost more efficiently as compared with refractory and siderophile elements as will be discussed in the next section. It is here remarked that such matter enriched with refractory and siderophile elements could be identified as the matter similar to carbonaceous C_2 chondrites or materials like these.

In order to search for the regions in which the source composition of cosmic rays is formed in the interstellar space, the observational data on the isotopic composition of cosmic rays are also useful to find some idea about the formation of cosmic ray source matter. In particular, the cosmic ray nuclei which decay by K—electron capture as indicated in Table 1, may give an important information about the forming process of cosmic ray source matter(e.g., Schramm, 1992). From the observations of these nuclei, we can obtain the information of the behavior of these nuclei in the interstellar space after ejected from the explosions of supernovae identified as type II. Some discussion will be made on these nuclei when we consider the acceleration process of cosmic rays in the section 4.

3. The Nature of the Source Composition of Cosmic Rays

Some years ago, it was pointed out(Israel,1985) that the source composition of cosmic rays is similar to the chemical composition of carbonaceous C_2 chondrites, from which volatile elements have already been efficiently lost, so that they are less abundant than those found in the solar photospheric abundances. Thus, those chondrites are overabundant of non–volatile element as refractory and siderophile ones as compared with the solar abundances. The existence of the complexes enriched with Ca and Al, being called Ca–Al inclusions, inside these chondrites is reflected upon the chemical composition of these chondrites.

The observed results of the interstellar depletion for various elements indicate that both refractory and siderophile elements are highly depleted in comparison with volatile and low–volatile elements(e.g., Spitzer, 1985; Savage and Sembach,

1996). This means that the former elements are efficiently condensed into dusts or grains, as aggregates of dusts, in the interstellar space in comparison with the latter elements during the processes of the formation of these dusts or grains there after all of these elements were released from supernova explosions or stellar winds from red supergiant or Wolf–Rayet stars. It can be, therefore, said that the mechanism of forming such dusts or grains may be similar to that of the formation of cosmic ray source matter, the chemical composition of these dusts or grains, being responsible for the interstellar depletion, seems well coincident with that of this source matter, from which cosmic rays are necessarily born. Volatile elements may have been lost efficiently through the condensation process.

Since the elemental volatility for each corresponding element, the condensation temperature for various elements must be somehow dependent on this potential(Sakurai, 1995). When this potential is plotted against this temperature for each element, the result shown in Fig.3 is necessarily obtained for the elements found in the source composition of cosmic rays. This result strongly suggests that both refractory and siderophile elements are quite efficiently accumulated into the interstellar dusts and grains, which may be identified with the source matter of cosmic rays because of their relatively higher condensation temperature.

When we consider the relative abundances of ultra–heavy nuclei(Z $\geq$ 30) in the sources matter of cosmic rays, it becomes clear that these nuclei must have been supplied from the debris released from the explosions of supernovae classified as type II(Sakurai,1995). These supernovae are identified as the explosions of massive stars which are enriched with these nuclei(Cardelli,1994).

When the source composition of cosmic rays relative to the chemical composition of the solar photosphere is compared with the interstellar depletion with respect to each element, both refactory and siderophile elements are concentrated to a domain as seen in Fig.4, which is separated from other elemental groups mainly composed of volatile elements. Two elements, Ca and Al, seem to reflect upon the chemical composition of Ca–Al inclusions. These results seem to support that the source matter of cosmic rays is made from those which have lost volatile elements efficiently.

4. Cosmic Rays Acceleration as Related to the Formation of the Source Matter of Cosmic Rays

Some consideration will be given on the acceleration process of cosmic rays which is closely related to the formation of the source composition of cosmic rays as obtained from the analyses of the observed data on their nature for the last few decades. As shown in the last section, both refractory and siderophile elements, including ultra–heavy ones, are relatively enriched in the source composition as compared with those found in the chemical composition of the solar photosphere, which is now thought of as that which is recognized as the representative to the cosmic chemical abundances(Sakurai,1995).

Such enrichment of refractory and siderophile elements in the source composition of cosmic rays gives us a clue to search for the matter whose composition is abundant with these elements. Since the condensation temperature of them is higher than 1000K or so (e.g., Wasson, 1975), they must have been concentrated into dust and grains, from the material originally supplied from supernova explosions and stellar winds, in the interstellar space. They are concentration thus must be reflected upon in the observed data on the interstellar depletion of elements, because this depletion of elements is causally connected with the condensation of elements into dust in the interstellar space. Thus the data on this depletion can be considered in identifying the matter of cosmic ray sources.

If we compare this source composition with the chemical composition of dusts and grains being prevailed in the interstellar space as observed in the phenomena of the interstellar depletion of the elements(Savage and Sembach, 1996), the cosmic ray sources should have been formed from these dusts and grains after formed as the results of their condensation process while making their temperature decrease to 1000K or less as estimated from the results shown in Fig.2. Grains cited here may be identified with aggregates of dust material and so their chemical composition is almost the same as that of dusts. It is thus concluded that the materials contained in dust and grains could be identified as the source matter of cosmic rays(Sakurai, 1989, 1991).

While drifting in the interstellar space after released from supernova explosions or the end products of stellar winds, they sometimes become charged due to the ionization resulting from their interaction with shock waves and other disturbances associated with supernova explosions. These charged dusts and grains may be accelerated by the action of these shock waves propagating in the interstellar space(e.g., Epstein, 1980, Meyer et al., 1997, Ellison et al., 1997). Sometimes, they may be disintegrated into charged atoms, which tend to be accelerated more efficiently into higher energy.

In order for those dusts and grains to be formed from the gases ejected from supernova explosions or stellar winds from red supergiant or Wolf–Rayet stars, these ejected gases must condense initially into these dusts while being cooled off in the neighborhood of the sites of supernova explosions or in the space not far from those stars. This means that cosmic rays are not generated directly from the ionized gases released from supernova explosions as the results of the acceleration of the gases expanding from the sites of supernova explosions. These gases, at first, have to become cool enough for dusts to grow, by losing their own heat, immediately after released into the nearby space from supernova explosions.

There is some evidence to suggest that cosmic ray particles are not directly accelerated from hot gases just released from supernova explosions. For instance, none of the radioactive nuclei decaying by K–electron capture such as ^{57}Co, ^{55}Fe, ^{44}Ti and others, all of which are short lived because of decay constants, have been detected by the direct observation by satellites and spacecraft, despite that these nuclei must have been produced by the rapid process of nucleosynthesis associated with supernova explosions of type II. Another evidence is that according to the observations of nonthermal radio emissions from the sites of supernova explosions, the beginning of these emissions is usually delayed for a few years or more after these explosions(e.g., Brown and Marscher, 1978; Weiler et al., 1986). The lack of these emissions indicates that the acceleration of cosmic rays is not at work in the neighborhood of the sites of supernova explosions for a few years up to a decade or more after supernova explosions. Therefore, there seems to exist the period ample enough for various elements released from supernova explosions to cool off and then condense to form dusts in the space not far from these sites.

According to the optical observations of the supernova 1987a, the light curve after its explosion is well explained by taking into account the radioactive decay of ^{56}Co, which seems to have been synthesized enormously in gases released from the explosion(e.g., Arnett et a., 1989). This means that almost all of ^{56}Co nuclei have never been accelerated to cosmic rays, since these nuclei of cosmic ray energy must be mostly fully ionized so that they are never able to decay by capturing K–electrons.

However, there may be some claim by referring to such observations that powerful X–ray emissions from expanding shell ejected from the explosion of supernova SN 1006 are a direct evidence for cosmic rays to have been accelerated in this shell, though these emissions are associated with the acceleration of relativistic electrons(Koyama et al., 1995). For such a claim, we can argue that these emissions are not directly accompanied by the explosion of supernova SN 1006, because this supernova exploded almost 1000 years ago. Thus these observations cannot be considered as the evidence that the acceleration of cosmic rays takes place in the expanding gas clouds immediately after supernova explosions.

5. Future Problems

The source composition of cosmic rays under consideration is limited to the particles in the medium energy range(1GeV per nucleon at most) . Although the direct observational data on these particles in the energy much higher than 1 GeV per nucleon, say 10^{14} eV, are lacked now because of the difficulty to develop the observing technique, it is not clear whether or not such nature as the excess of heavy nuclei in the source composition of cosmic rays as found in particles of medium energy is maintained in such high energy region. However, since the degree of this excess does not seem to be the same between the medium and higher energies cosmic rays, it is urgently required to directly confirm whether the source composition of cosmic rays is highly dependent on particle energy per nucleon or not. In the case of solar cosmic rays, the excess of heavy nuclei in them is clearly seen in the particles of relatively low energy (less than 100 MeV per nucleon) as compared with those of GeV energy. If such a nature could be applied to cosmic rays of galactic origin, it is necessary for us to reconsider the acceleration process of these cosmic rays.

When we consider the propagation mechanism of cosmic rays in the interstellar space after released from their acceleration regions, it is necessary to take into account possible controlling effects of the galactic magnetic field for cosmic rays to traverse the interstellar space. The trapping of cosmic rays in the galactic disk due to the effect is much more efficient to heavy nuclei such as Fe's and beyond in comparison with light and medium nuclei, because their gyration radii tend to become shorter with their atomic numbers. Thus, there is a possibility that, in the higher energy range as 10^{14} eV or more, heavy nuclei become much more abundant as compared with cosmic rays of the energy as 1 GeV or less. The problem on how the source composition of cosmic rays changes with particle energy still remains as one of the future studies.

Detailed mechanism on how the source composition of cosmic rays is enriched with refractory and siderophile elements as shown in Fig.2 has not been found as yet, though an idea to interpret this fact has been proposed in this article based on the efficient condensation of these elements into the source matter of cosmic rays.

References

Arnett, W.D., Bahcall, J.N.Kirshner, R.P. & Woosley, S.E. 1989, Ann. Rev. Astron. Astrophys, 27, 629.

Brown, R.L. & Marscher, A.P. 1978, Ap, J. 220, 467.

Cardelli, J.A. 1994, Science 265, 209.

Ellison, D.C., Drury, L. O'C. & Meyer, J.P. 1997, Ap. J. 487, 197.

Epstein, R.I. 1980, Mon. Not. RAS. 193,723.

Israel, M.H., 1985, Annals New York Acad. Sci. 470, 188.

Koyama, K. et al. 1995, Nature 378, 255.

Meyer, J.P., Drury, L.O'C. & Ellison, D.C. 1997, Ap, J. 487, 183.

Sakurai, K. 1989, Adv. Space Res. 12, 149.

Sakurai, K. 1995, Adv. Space Res. 15, 35.

Sakurai, K. 1997, Nuclear Phys. A621, 56c.

Sakurai, K. & Ito, K. 1990, in Neutrinos in Cosmic Ray Physics and Astrophysics, ed. K. Sakurai (Tokyo, Terra Sci. Pub.), 201.

Savage, B.D. & Sembach, K.R. 1996, Ann. Rev. Astron. Astrophys., 34, 279.

Schramm, D.N. 1992, in Essays in Nuclear Astrophysics, ed. C.J. Barnes et al. (Cambridge, Cambridge Univ.), 325.

Silverberg, R. & Tsao, C.H. 1973, Ap. J. Suppl. 25,315.

Silverbrg, R. Tsao, C.H. & Letaw, J.R. 1983, in Composition and Origin of Cosmic Rays, ed. M.M.Shapiro (Dordrecht, Kluwers), 231.

Simpson, J.A. 1983, Ann. Rev. Nuclear Part. Phys. 33, 232.

Spitzer, L. 1985, Ap. J. Lett. 290, L21.

Wasson, J.L. 1985, Meteorites: Their Record of Early Solar–System History(New York, Freeman).

Webber, W.R. 1997, Space Sci. Rev. 81. 107.

Weiler, K.W., Sramek, R.A., Pangia, N., van der Hulst, J.M. & Sulvati, M. 1986, Ap. J. 301, 790.

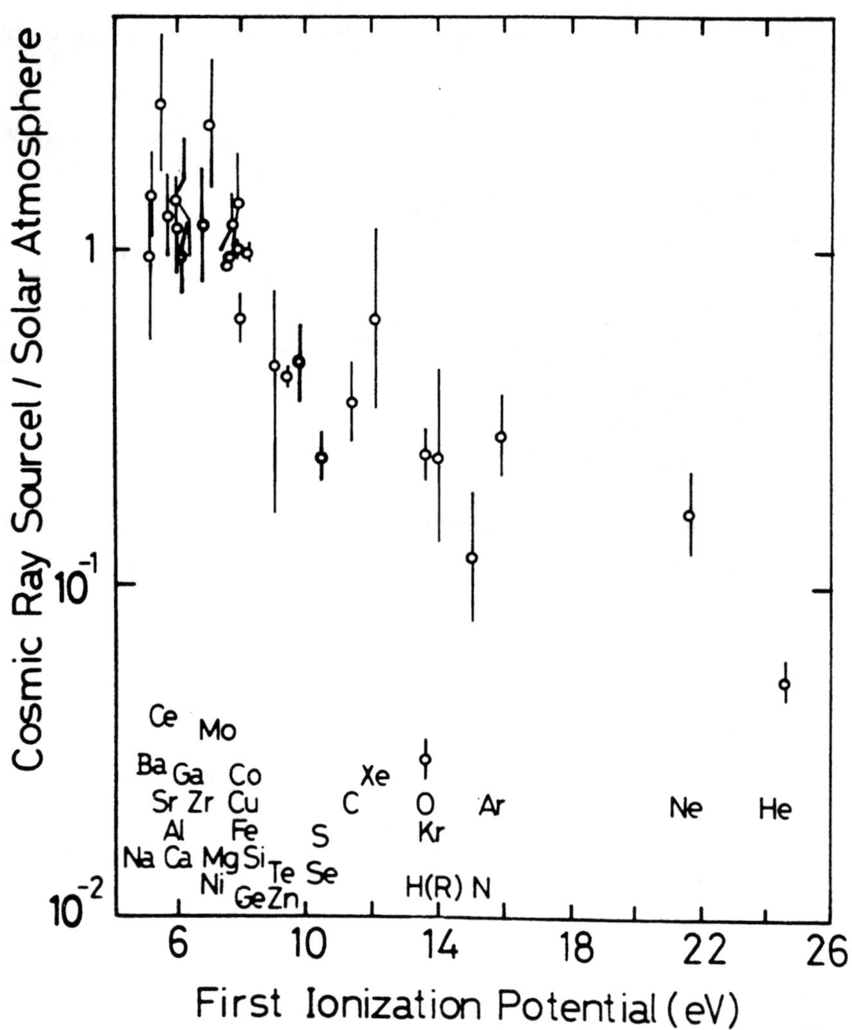

Fig.1   The chemical composition of cosmic rays in their sources relative to that of the solar photosphere as expressed with first ionization potentials of the elements.

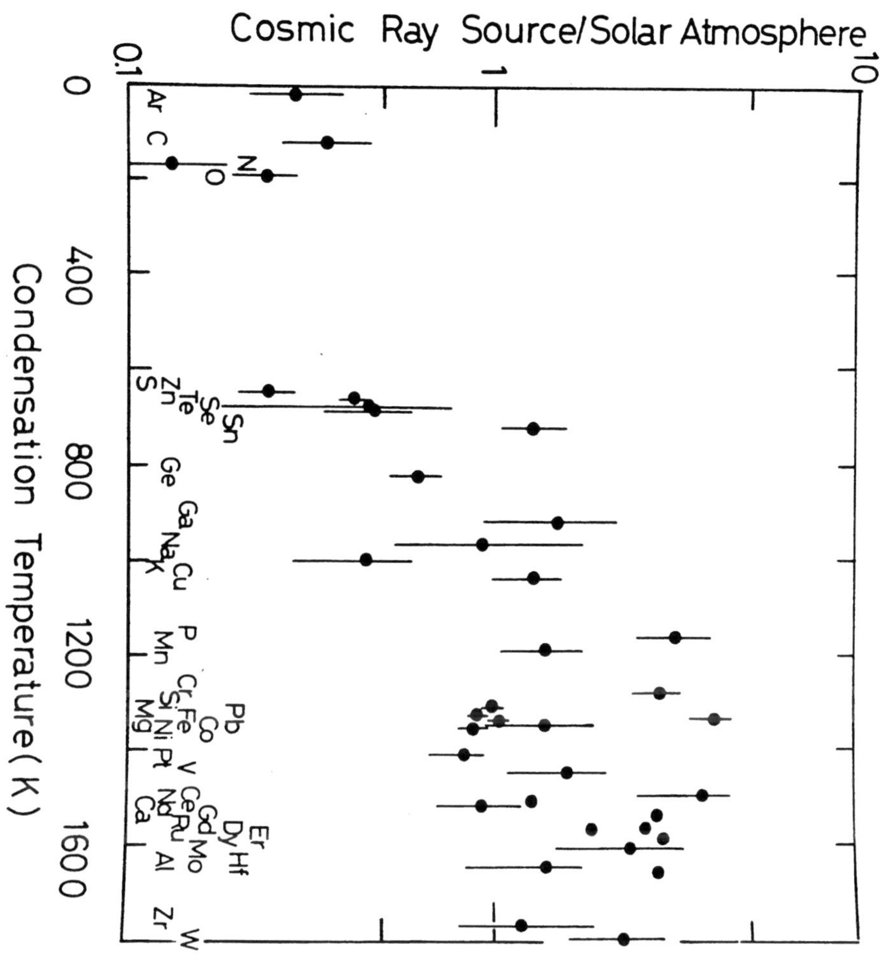

Fig.2   The chemical composition of cosmic rays in their sources as compared with that of the solar photosphere as a formation of the condensation temperature of the elements. Most of ultra–heavy nuclei(U.H.) are refractory.

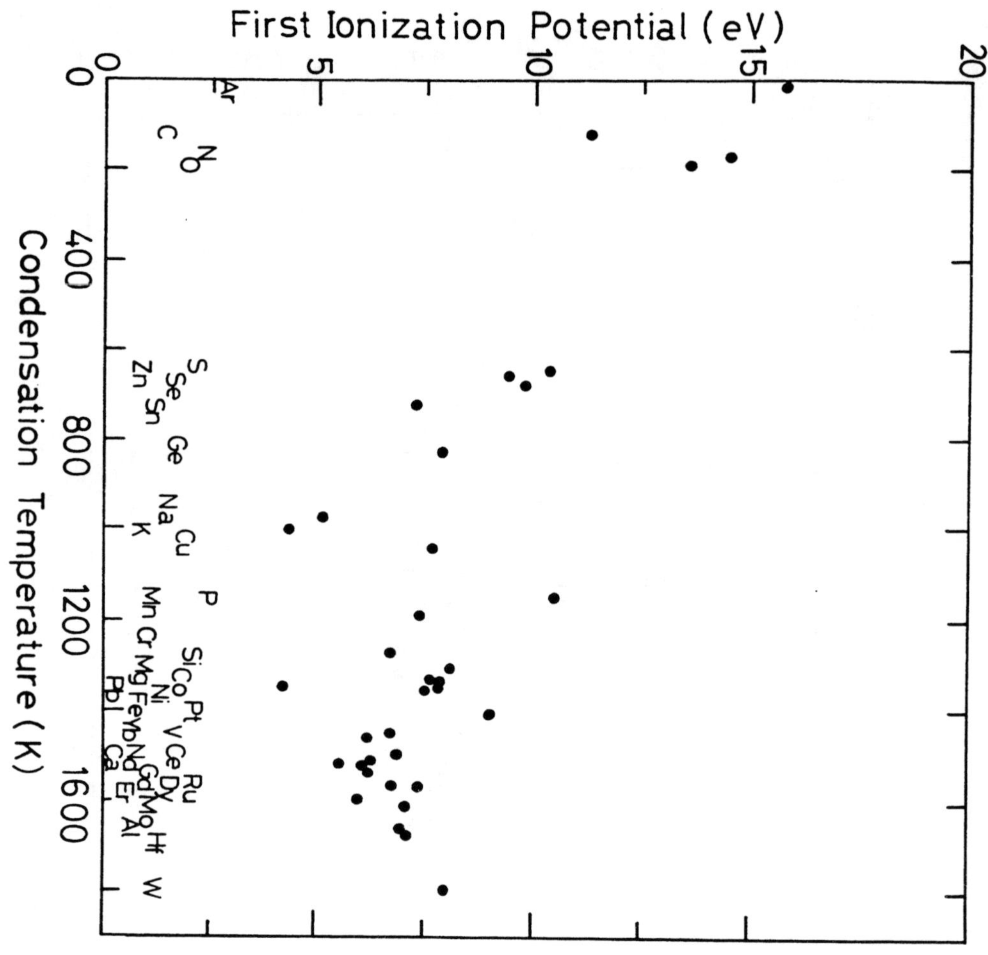

Fig.3   The relation between the first ionization potential and the condensation temperature of the elements(Sakurai, 1995).

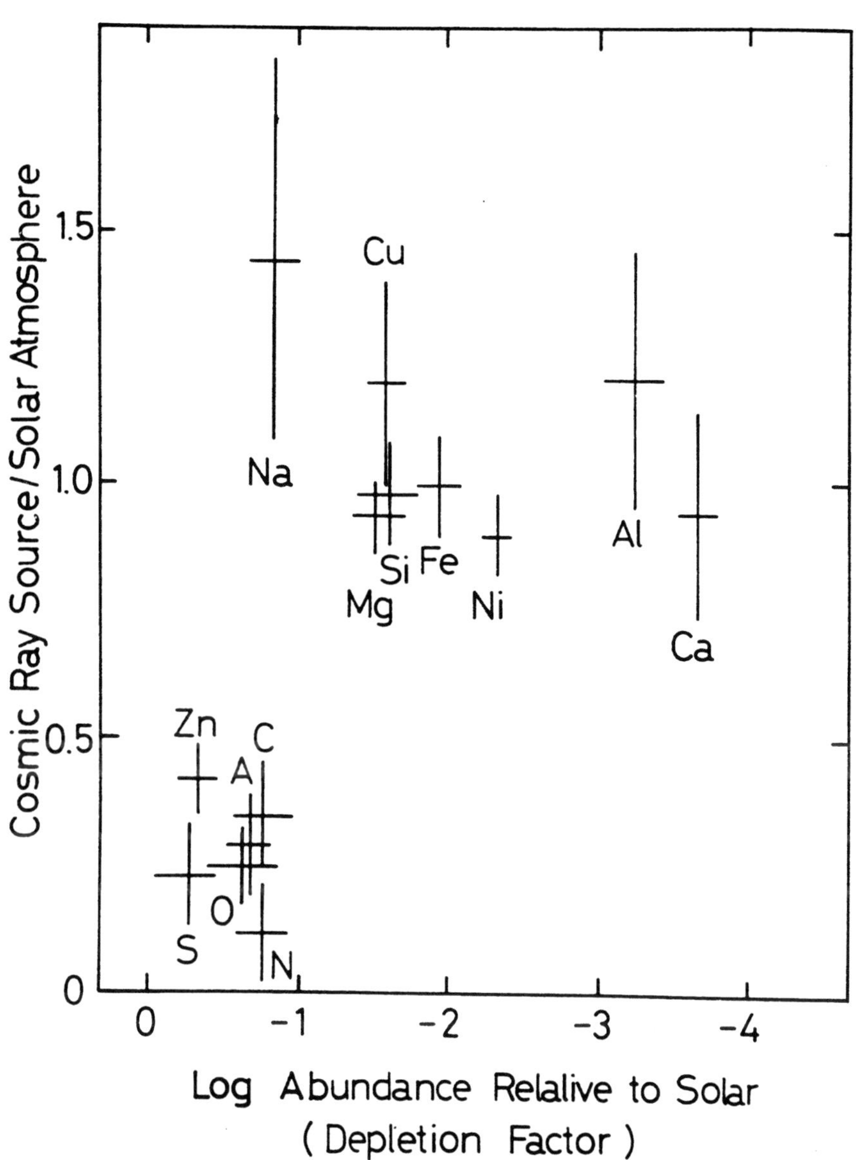

Fig.4   The relation of the interstellar depletion to the chemical composition of cosmic rays in their sources relative to that of the solar photosphere.

Table 1 Isotopic Elements Decaying by K–Electron Capture(Webber, 1997).

Decay Process	Effective Lifetime 350–500–650 MeV per nucleous
$^{37}\text{Ar} \rightarrow {}^{37}\text{Cl}$	40–75–90
$^{44}\text{Ti} \rightarrow {}^{44}\text{Ca}$	18–30–48
$^{49}\text{V} \rightarrow {}^{49}\text{Ti}$	12–22–36
$^{51}\text{Cr} \rightarrow {}^{51}\text{V}$	9–18–30
$^{54}\text{Mn} \rightarrow {}^{54}\text{Cr}$	7.5–13–24
$^{55}\text{Fe} \rightarrow {}^{55}\text{Mn}$	6–10–20
$^{57}\text{Co} \rightarrow {}^{57}\text{Fe}$	4.5–7.5–16

Ultra-Heavy (Z≥30) Galactic Cosmic-Ray Nuclei and the ACCESS Charge (Z) Identification Module (ZIM)

W.R. Binns[1], J.H. Adams[2], Jr., L.M. Barbier[3], E.R. Christian[3], J.R. Cummings[1], G.A. DeNolfo[4], P.L. Hink[1], M.H. Israel[1], J.F. Krizmanic[3], R.A. Leske[4], W. Menn[5], R.A. Mewaldt[4], J.W. Mitchell[3], S.M. Schindler[4], M. Simon[5], S.H. Sposato[1], C.J. Waddington[6] , M.E. Wiedenbeck[7]

[1] Department of Physics and McDonnell Center for the Space Sciences, Campus Box 1105, Washington University, St. Louis, MO. 63130, [2] Space Sciences Division, Code 7654, Naval Research Laboratory, Washington DC, 20375, [3] Code 661, NASA/Goddard Space Flight Center, Greenbelt MD 20771, [4] Space Radiation Laboratory, Caltech, M/C 220-22, Pasadena, CA 91125, [5] Physics Department, University of Siegen, Adolph Reichwein Street, 57068 Siegen, Germany, [6] School of Physics and Astronomy, University of Minnesota, 116 Church Street S.E., Minneapolis, MN. 55455, [7] Jet Propulsion Laboratory, 4800 Oak Grove Drive, Pasadena, CA 91109
email: wrb@howdy.wustl.edu

Abstract. The measurement of all individual element abundances in the galactic cosmic rays over the full charge spectrum $1 \leq Z \leq 92$ is one of the key goals of cosmic ray studies. It is believed that they may include a sample of galactic matter that is a accelerated out of the dust and/or gas in the interstellar medium (ISM) and/or material that is freshly synthesized and accelerated directly in supernovae (SN). In this paper we describe some of the science objectives that can be achieved by a program to complete the determination of these element abundances. A detector that we are presently studying as part of the ACCESS mission for the International Space Station is described. This detector is the ACCESS Charge (Z) Identification Module (ZIM). It utilizes silicon dE/dx detectors, aerogel and acrylic Cherenkov counters, and a scintillating fiber detector which serves as both a hodoscope and a time-of-flight counter.

1 Introduction

There is great interest in determining the abundances of galactic cosmic rays over the full charge spectrum since they may include a sample of galactic matter that is accelerated out of the dust and/or gas in the interstellar medium (ISM), as well as material that is freshly synthesized and accelerated directly in supernovae (SN). Elemental abundances of individual nuclei measured so far ($1 \leq Z \leq 28$ and even-Z nuclei for $Z \leq 60$), when corrected for propagation through the ISM, are quite similar to those of the solar system with the striking exception that the elements exhibit a fractionation relative to the solar system (SS) abundances that can be ordered by the elements' first-ionization potential (FIP) (e.g., Ferrando, 1993; Binns *et al.*1989). Nuclei with high FIP are generally depleted with respect to those with low FIP. A similar fractionation occurs for solar energetic particles (SEP), most samples of the solar wind, and the solar corona. On the sun, there is strong evidence that nuclei with low FIP are preferentially selected from an ion-neutral gas in the solar atmosphere (chromosphere and photosphere), where $T \sim 10^4 K$, and transported into the corona where some are accelerated and injected into interplanetary space (Stone, 1988). This almost certainly occurs on other stars possessing a relatively low temperature chromosphere which can preferentially inject low-FIP particles into the corona (F through M stars). Essentially all samples of accelerated cosmic matter appear to exhibit this ordering. However, a fractionation of elements based on volatility rather than FIP produces a very similar ordering since FIP and volatility are strongly correlated for all but a few elements (Epstein, 1980; Bibring and Cezarsky, 1981; Sakurai, 1995; Meyer *et al.*, 1997). In addition, several of the most abundant elements (H, He, C, and O) fit the general FIP picture rather poorly. Resolution of the FIP vs. volatility question is quite crucial since it directly addresses the origin of cosmic rays, one of the primary goals of galactic cosmic-ray investigations. The final resolution of this question must depend upon a study of the ultra-heavy (UH) cosmic-ray nuclei, those with $Z \geq 30$, since only a few elements have FIP and volatility that are not correlated.

2 Scientific objectives of UH elemental abundance studies

Current Status of UH Observations

Since the early 1980's, the definitive data on the UH elemental composition for $Z>30$ has been the combination of results from HEAO-3 and Ariel-6 (Binns, *et al.*, 1989; Fowler, *et al.*, 1987). Those two experiments had sufficient charge resolution to determine abundances of even-Z elements for $Z \leq 60$, but not the less abundant adjacent odd-Z elements. For higher-Z nuclei the combination of inadequate resolution and low numbers of particles collected limited the measurements to abundances of groups of several elements.

The Trek large detector, which consisted of a large area of glass track-etch detectors exposed outside of *Mir*, has recently reported new results by Westphal, *et al.*, (1998). For the narrow charge range $76 \leq Z \leq 82$, they have achieved the best resolution to date, ~0.45 cu. They have cleanly resolved $_{82}$Pb from the $_{76}$Os-$_{78}$Pt group for the first time and the resolution for the other even-Z nuclei in this range is sufficient to obtain improved abundance estimates over the HEAO-3/Ariel-6 measurements. However, these abundances are not normalized within the TREK experiment itself to the more abundant elements of lower Z owing to the relatively narrow band of charges over which the glass track detectors are sensitive. Ratios such as Pb/Fe would be likely to help distinguish FIP from volatility models.

Results from LDEF have much greater statistical weight than those from Trek, including 40 or more actinides compared with 4 from Trek, but the charge resolution is considerably poorer. The LDEF report (Keane, *et al.,* 1997) showed a single broad peak centered in the $_{76}$Os-$_{78}$Pt region with a barely discernible shoulder near $_{82}$Pb.

Cosmic-Ray Source Models (FIP vs Volatility)

The cosmic-ray source (CRS) elemental abundances, which must be inferred from direct observations by correcting for the secondary component that originates in interstellar fragmentation of heavier nuclei, are generally similar to the abundances of elements in the solar system (SS); however, there appears to be preferential acceleration, governed by atomic properties of the elements, that results in deviations of CRS from SS. One way of ordering the variations of the CRS/SS ratio is by the first-ionization-potential (FIP) of the elements. Figure 1 plots this ratio for elements with $Z \leq 32$ (Ferrando, 1993). Elements with FIP>10.5 eV have a ratio that is approximately one-fourth to one-eighth that of elements with FIP<8.5 eV. Elements with intermediate FIP have intermediate CRS/SS. UH elements have a similar ordering, insofar as they are known, although the measurement and cross-section uncertainties are greater. It is useful to plot the ratio of the GCR abundances measured at earth to SS abundances propagated to earth. In Figure 2 (Binns, et al., 1989) the top panel is a plot of this ratio for propagated SS abundances with no FIP enhancement. The bottom panel is the same thing but with a FIP enhancement. This clear correlation over the full range of measured elements suggests injection of material into the cosmic-ray accelerator by stellar flares, similar to solar flares, or from a partially ionized gas with temperature of the order of 10^4K. Since the accelerator is likely to work only on ionized atoms, the elements with lower ionization potentials would be preferentially accelerated.

Meyer *et al.*, (1997) argue that the correlation is better understood by considering the volatility of the elements and that the apparent dependence upon FIP is coincidental. In this model elements of high condensation temperature, which are likely to form refractory grains, are preferentially accelerated because interstellar grains themselves are accelerated by shocks. Ionized atoms sputtered off those grains have substantial energy which results in preferential injection of those refractory nuclei. Figure 3 is a plot of the CRS/SS ratio vs. condensation temperature taken from Meyer *et al.*, (1997). This shows that refractories and semi-volatiles have roughly the same ratio within statistics while volatiles have a reduced ratio. The highly volatile elements shown at the extreme right of the figure are plotted vs. mass. Mass appears to order these nuclei very nicely.

On the other hand, Ramaty *et al.*, (1997) suggest a different origin for the grains. From the observed constancy of Be/Fe in stars over a wide range of metalicities, coupled with the required supernovae energy input to maintain that constancy if cosmic rays are accelerated out of the interstellar medium, they infer that cosmic-ray acceleration is not solely from the interstellar medium, but must be primarily freshly nucleosynthesized matter. This argument would suggest that interstellar grains might not be significant sources of cosmic rays. However, they argue that high velocity grains formed in a supernova could be the injection source for cosmic rays. There is evidence for such grains in SN1987a

Shapiro (1997a and b) has recently proposed a FIP model based on solar-like coronal mass ejection of supra-thermal ions from dMe and dKe dwarf stars.

Most of the elements with low FIP are refractory, so it is difficult to distinguish between the FIP and volatility models. Meyer, *et al.* (1997) and Binns (1995) point out that there are several UH elements that break the degeneracy between low-FIP and refractory characteristics. Elements that are volatile or semi-volatile, but have low FIP, and for which it should be possible to infer source abundances, are $_{32}$Ge, $_{37}$Rb, $_{50}$Sn, $_{55}$Cs, $_{82}$Pb, and $_{83}$Bi. For these elements source abundances have been established only for Ge, Sn, (and the "Pb group"). The ZIM instrument now under study as described below will collect and resolve a significant number of each of these elements. While the observed number of Rb and Cs will be affected by fragmentation of the more abundant element just above ($_{38}$Sr and $_{56}$Ba), the source abundances are large enough that there would be a factor of at least 1.5 greater abundance of these elements expected in the cosmic rays (relative to adjacent elements) if the source were controlled by FIP than if it were controlled by volatility.

Source Models -- r-Process and s-Process

The elemental abundances of Z>30 nuclei in the solar system are understood as being mainly a mixture of nuclei created in two neutron-capture nucleosynthesis processes, r-process and s-process. These processes occur in very different astrophysical settings. The r-process is believed to occur primarily in supernova while the s-process occurs in red giant stars. A key question is whether the galactic cosmic rays originate in a similar mix of material. The HEAO-3 and Ariel-6 data indicated a mix very similar to the solar system for Z<60 but an apparent enhancement of the cosmic ray r-process nuclei relative to the solar system for Z>60. This observed enhancement was seen as an excess of about a factor of two in the interval $62 \leq Z \leq 80$; however, because of the limited charge resolution and low numbers of events in that interval of HEAO data, it was not possible to cleanly resolve less abundant secondaries from primaries nor to verify that individual elements had the relative abundances expected of an r-process source.

In addition, it turns out that the only r-process elements for which abundances were determined were elements of intermediate or high FIP (Te, Xe, and the Os-Ir-Pt group), so if indeed FIP does govern source fractionation for $Z \leq 30$ and s-process elements, it is not known at present, whether the r-process elements are subject to the same fractionation (*i.e.*, whether they are accelerated in the same source). Cs is an r-process element with very low FIP, so its measurement would help resolve this issue. If, however, volatility governs source fractionation, then the measured overabundance of the Os-Ir-Pt group, which was interpreted as an r-process enhancement in the HEAO-3 data, would be reduced. The ZIM instrument, with its ability to resolve odd-Z elements in the UH charge range for the first time, would clarify the r/s mixture in the comic rays and thus provide a vital clue to the nature of the cosmic-ray sources.

Galactic Propagation and Nearby Sources

The amount of interstellar material traversed by cosmic rays between their acceleration and their observation at earth is indicated by the abundance of cosmic-ray secondaries, nuclei that have been created by spallation in the ISM and are generally rare relative to heavier elements. Whereas the "light" secondaries such as Li, Be, and B probe the long pathlength end of the distribution of material traversed, an instrument that measures UH secondaries will probe the short-pathlength end, providing information about short pathlengths (a fraction of a g/cm^2). For plausible assumptions about the diffusion mean-free-path of cosmic rays in the galaxy (Ptuskin & Soutoul, 1990), 0.5 g/cm^2 corresponds to a distance of roughly 150 parsecs (and a transport time for a relativistic particle of about a half million years).

The observed abundances of light secondaries (Li, Be, B) and Fe secondaries (elements $21 \leq Z \leq 25$) can be explained by an exponential distribution of path lengths in the interstellar medium with a mean of about 8 g/cm^2 that varies somewhat with energy. The Fe secondary abundances are consistent with a 1 g/cm^2 truncation of the exponential at short pathlengths, but those data are not very sensitive to short pathlengths since the interaction mean free path for Fe in interstellar gas is 3 g/cm^2 and for C, N, O, the lighter nuclei mainly responsible for the production of Li, Be, and B in the GCRs, it is about 7 to 9 g/cm^2,

Secondaries of the heavier elements provide a much more sensitive test of pathlength truncation because the interaction mean free path of nuclei with $50 \leq Z \leq 56$ is about 1.6 g/cm^2, and for $76 \leq Z \leq 83$ is about 1.2 g/cm^2. To determine the pathlength distribution cleanly it is necessary to have elements that are almost entirely secondary in the observed cosmic rays. Among the UH cosmic rays, it is primarily the odd-Z elements that are so underabundant that those observed in the cosmic rays can be identified as predominantly secondaries. Hence a definitive answer about the influence of truncation (and thus about possible near-by sources) requires cleanly resolved odd-Z elements from the more abundant adjacent even-Z elements. It will be particularly interesting to probe the short path lengths since the CRIS experiment on the ACE spacecraft is returning data that includes new cosmic-ray clocks (radioactive nuclei) that probe shorter times than do previously measured clocks.

Energy Spectra

The energy spectra of the UH cosmic rays were measured by HEAO and Ariel over a limited energy range. Low-energy (several hundred MeV/nucleon) and high-rigidity (>5GV) HEAO data sets (Binns, *et al.*, 1989) suggested somewhat different compositions, but the differences were of low statistical significance. It is important to determine whether the UH composition does indeed vary with energy. Similarly the HEAO data suggested, again with limited statistical significance, that the UH secondary/primary ratio has a slightly softer energy spectrum than the secondary/primary ratio at Z<26. Such a difference would suggest a different galactic confinement or a different source distribution; or it could be a reflection of the energy dependence of the fragmentation cross sections.

The ZIM instrument will have energy resolution (from an Aerogel Cherenkov counter) up to about 10 GeV/nucleon which corresponds to a total energy near ~1 TeV for $Z\sim50$. The numbers of particles will be sufficient to study spectra up to this energy for individual elements of $Z\leq38$. About 2000 $_{38}$Sr of all energies and about 1000 of the group $50\leq Z\leq56$ should be detected by ZIM in a 4 year mission on the International Space Station. About 5% of those particles will have energies exceeding 10 GeV/nucleon.

The ZIM Instrument

The instrument that we are currently studying for flight as the charge module (ZIM) for ACCESS (Figure 4) uses silicon dE/dx detector arrays, two Cherenkov counters with radiators of different refractive index for velocity measurement, and a coded scintillating fiber hodoscope for trajectory determination and time-of-flight (TOF). (For a more detailed discussion of how this combination of detectors provides unambiguous charge determination see Binns *et al.*, (1997). A cross-section drawing of the baseline instrument is shown in Figure 1. The outside dimensions of the detector are 2.5 meters square by 0.5 meters deep. This instrument provides a useful radiator area of 206 cm square and a total geometry factor for entry through the top of the detector of 8.7 m^2sr. It is essential that each of the detector systems have a dynamic range that covers the charge range $10\leq Z\leq96$ so that we can achieve the ZIM goals for measurement of heavy elements. Additionally, a charge resolution of ≤0.25 cu must be obtained over the full charge and energy range.

Silicon Detector

The ZIM module uses arrays of silicon detectors to measure dE/dx. There are four planes of silicon; two on top and two near the bottom of the detector stack (See Figure 4). We have confirmed with tests at the Brookhaven accelerator that silicon detectors have the necessary charge resolution, and have excellent linearity, at least over the range $10\leq Z\leq79$ at energies 1 to 10 GeV/nucleon. In our baseline instrument, each plane of silicon detectors consists of an array of 10 cm square silicon wafers with thickness 380 μm. Each of the 10 cm wafers will be segmented into smaller pixels. This reduces the capacitive noise on the silicon detector, thus making it possible, in principle, to extend the ZIM dynamic range down to $Z=1$ which is expected to be useful in distinguishing the primary particle that initiated the slower from backsplash calorimeter events, thus reducing the probability of misidentification by the calorimeter of the primary particle. In addition to providing dE/dx measurements the silicon detectors also provide a coarse hodoscope which will be used for consistency checks on the trajectory determined by the fiber hodoscope described below.

To measure cosmic rays with charge 10 to 100 for all energies $\geq$300 MeV/nucleon at incident angles from 0 to 60 deg, requires a dynamic range of approximately 300. The charge-sensitive pre-amplifier must provide sufficient dynamic range while minimizing electronic noise. Application specific integrated circuit (ASIC) designs now in space on ACE and in the ATIC balloon payload have solved these dynamic range and noise problems. With these ASICs the power requirement will be $\leq$4 mW per channel.

Coded Fiber Hodoscope and Time-of-Flight Detector

Two planes of a coded scintillating-fiber hodoscope are located just inboard of the top silicon array and at the bottom of the detector stack. Each plane of fibers is composed of two layers of 0.5 mm fibers, one layer for x-and one for y-measurements. The four hodoscope layers each have fibers combined into 8 modules, each having a width of 25 cm. A 16 element multi-anode PMT (Hamamatsu R5900-M16) is used to detect the light at either end of the fibers. In Figure 4 the fiber outputs (triangular regions) and MAPMT readouts are shown only on the left half of the instrument for clarity. The fibers in a module are grouped in pairs (elements) and are coded differently at opposite ends such that the position of a particle traversing the 25 cm width of 500 fibers (250 pairs) can be unambiguously resolved to within 0.3mm (See Binns, *et al.*, 1995 for explanation of the method). The fiber MAPMTs can also provide time-of-flight (TOF) measurements to a precision sufficient to distinguish upward from downward trajectories since the MAPMT signals are very fast (~1-2ns). We have concluded that this is an important capability to have as part of ZIM since some nuclei will enter the bottom of the detector and exit through the top. These events can result in degraded resolution if they are not clearly identified.

Aerogel Cherenkov Counter

Immediately below the top fiber detector is an aerogel Cherenkov counter. The aerogel is 3 cm thick, has refractive index 1.04 and density 0.22 g/cm^3. In Figure 4 it is shown mounted in a light collection box. The light box is viewed by 48 five-inch PMTs. The threshold energy for this detector is ~2.4 GeV/nucleon. This will enable us to distinguish nuclei that have energy above that threshold from those that are on the low energy branch (Binns *et al.*, 1997).

Acrylic Cherenkov Counter

Immediately below the aerogel Cherenkov counter is a second Cherenkov counter, which uses an acrylic based radiator with a refractive index of about 1.5 in an essentially identical light collection box. This counter will also be viewed by 48 five-inch PMTs. The detector threshold energy is 0.3 GeV/nucleon. Signals from this counter will be used as the primary charge identification for nuclei on the high energy branch which have a saturated aerogel Cherenkov signal.

3 Numbers of events

In Figure 5, we show an estimate of the number of nuclei whose charge would be measured by the ZIM instrument as presently configured in a 4 year flight (assuming a 75% duty cycle) in the Space Station 51° orbit at solar minimum. Particles that undergo nuclear interactions in the detector have been excluded from these estimates. (For those elements that have not been previously observed we have made arbitrary assumptions, consistent with the group abundances observed by HEAO-3.)

4 Summary

We have described the scientific goals that can be achieved with the ACCESS-ZIM instrument. ZIM should be capable of obtaining precise measurements of the cosmic-ray elemental abundances over the interval 10≤Z≤83 which will enable us to make significant advances in our understanding of the cosmic ray sources, their acceleration, and propagation to earth.

5 Acknowledgements

This work was supported by NASA Grants NAG5-5154 at Washington University, NAG5-5166 at Caltech, 344-10-05 at the Goddard Space Flight Center, and S818-74Z at the Naval Research Laboratory.

6 References

Bibring, J.P., and Cezarsky, C.J., 1981,*17ᵗʰ ICRC Proc* .,**2**, 289.

Binns, W.R.,Garrard, T.L., Gibner, P.S., Israel, M.H., Kertzman, M.P., Klarmann, J., Newport, B.J., Stone, E.C., and Waddington , C.J., 1989; *Ap. J.,* **346**, 997.

Binns, W.R. 1995, *Adv. Space Res.,* 15, (6)29.

Binns, W.R., Hink, P.L., Israel, M.H., Mewaldt, R.A., Garrard, T.L., Leske, R.A., Schindler, S.M., Wiedenbeck, M.E., Streitmatter, R.E., Barbier, L.M., Christian, E.R., Mitchell, J.W., Ormes, J.F., Waddington , C.J., 1997, *25th Intl. Cosmic Ray Conf,* **5**, 65.

Epstein, R.I., 1980, M.N.R.A.S., **193**, 723.

Ferrando, P., 1993, *Proc. 23ʳᵈ Intl. Cosmic Ray Conf,, Rapporteur Paper,* Ed. Leahy, D.A., Hicks, R. B. and Venkatesan, C., Pub World Scientific 279.

Fowler, P.H.,Walker, R.N. F., Masheder, M.R.W., Moses, R.T., Worley, A., and Gay, A.M. , 1987; *Ap. J.,* **314**, 739.

Keane, A.J., Thompson, A, O'Sullivan, D., Drury, L.O'C., and Wenzel, K-P., 1997; *25th Intl. Cosmic Ray Conf,* **3**, 361.

Meyer, J.P., L.O'C.Drury, D.C.Ellison, 1997; *Ap. J.,* **487**, 182.

Ptuskin, V.S & A. Soutoul, 1990; *Astonomy and Astrophysics,* 237, 445.

Ramaty, R., Kozlofsky, B., Lingenfelter, R.E., and Reeves, H., 1997; Light Elements and Cosmic Rays in the Early Galaxy, 1997, *Ap.J.*, 488, 730.

Sakuri, K., 1995, *Advances in Space Research* , **15**, (1) 35.

Shapiro, M.M., 1997; *25th Intl. Cosmic Ray Conf,* 4, 349.

Shapiro, M.M., 1997; *25th Intl. Cosmic Ray Conf,* 4, 353.

Stone, E.C., 1988, in *AIP Conference Proceedings 183, Cosmic Abundances of Matter,*
p. 72, Ed. C.J. Wadington.

Stone, E.C. et al., 1998, Space Sci. Rev., to be published.

Westphal, J.W., P.B. Price, B.A. Weaver, V.G. Afanasiev, 1998, *Nature*, **396**, 50.

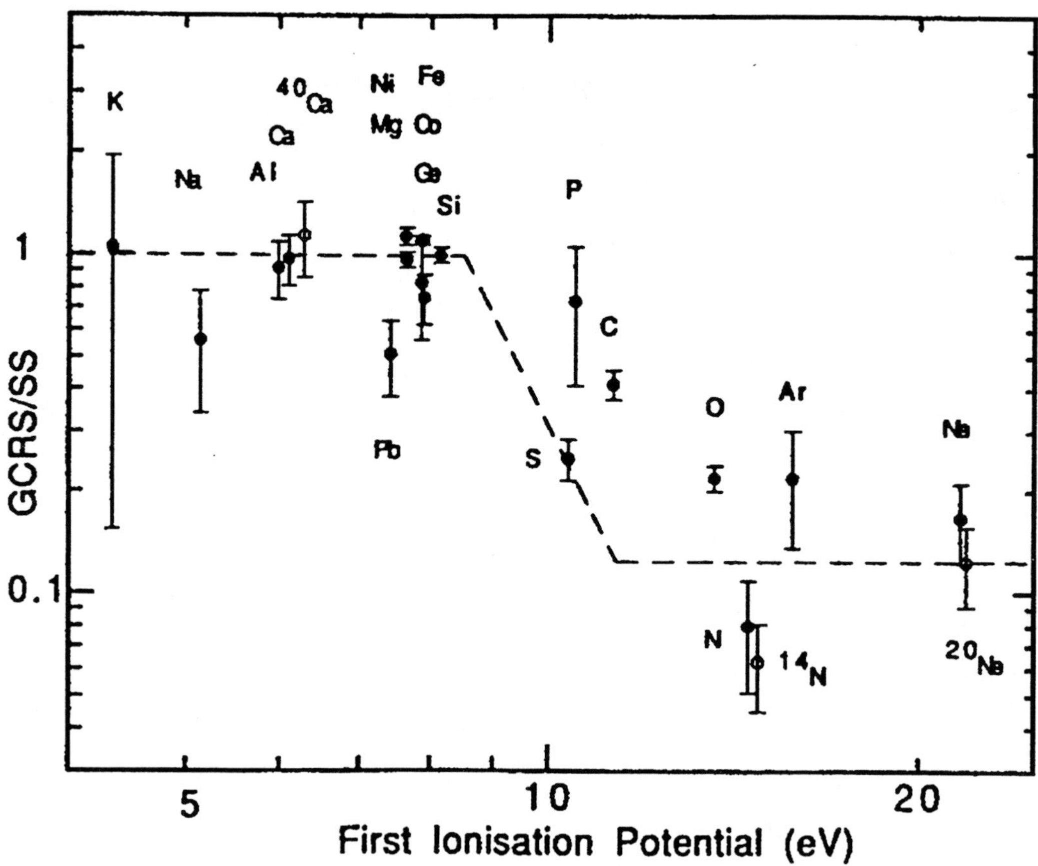

Figure 1. Ratio of cosmic-ray source abundance to solar-system abundance (GCRS/SS) vs. first ionization potential (FIP).

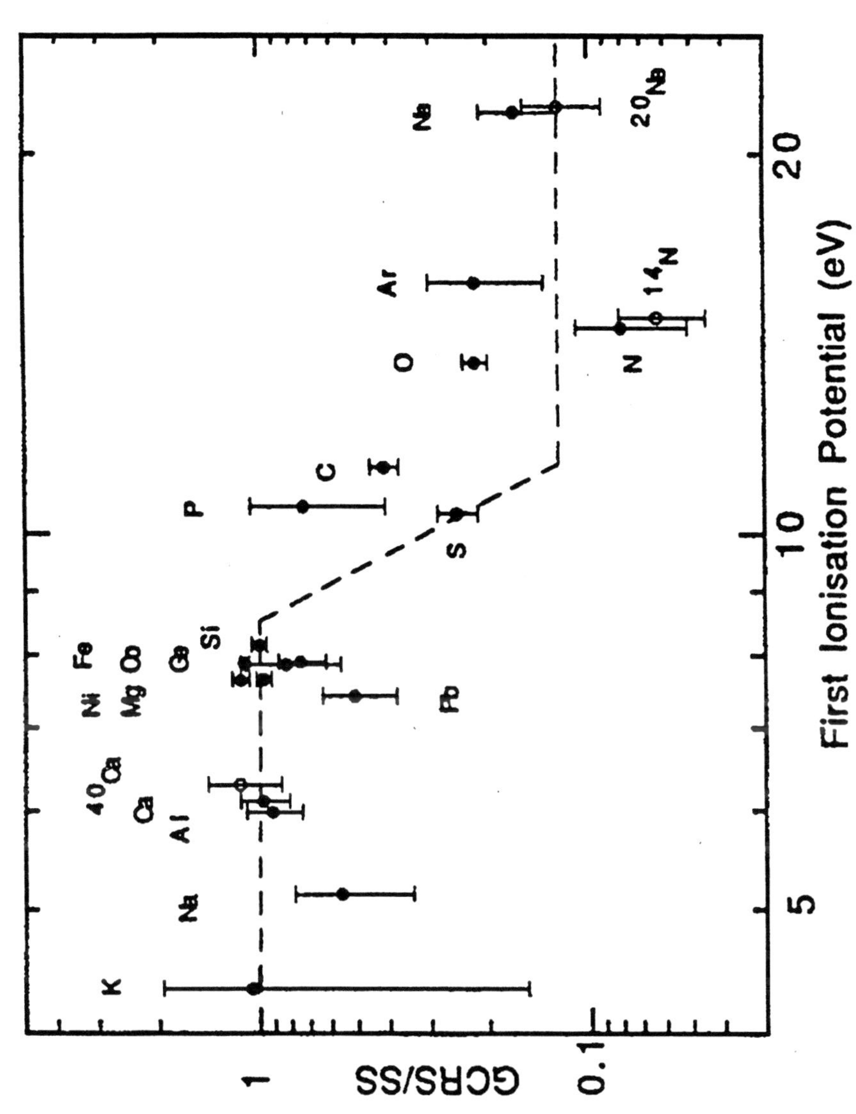

Figure 2. Ratio of observed cosmic-ray abundance to abundance expected if the cosmic-ray source has solar-system abundance vs. atomic number (Z) (a) with FIP fractionation at the cosmic-ray source (b) without FIP fractionation.

     W.R. Binns, J.H. Adams, Jr., L.M. Barbier, E.R. Christian, et al.

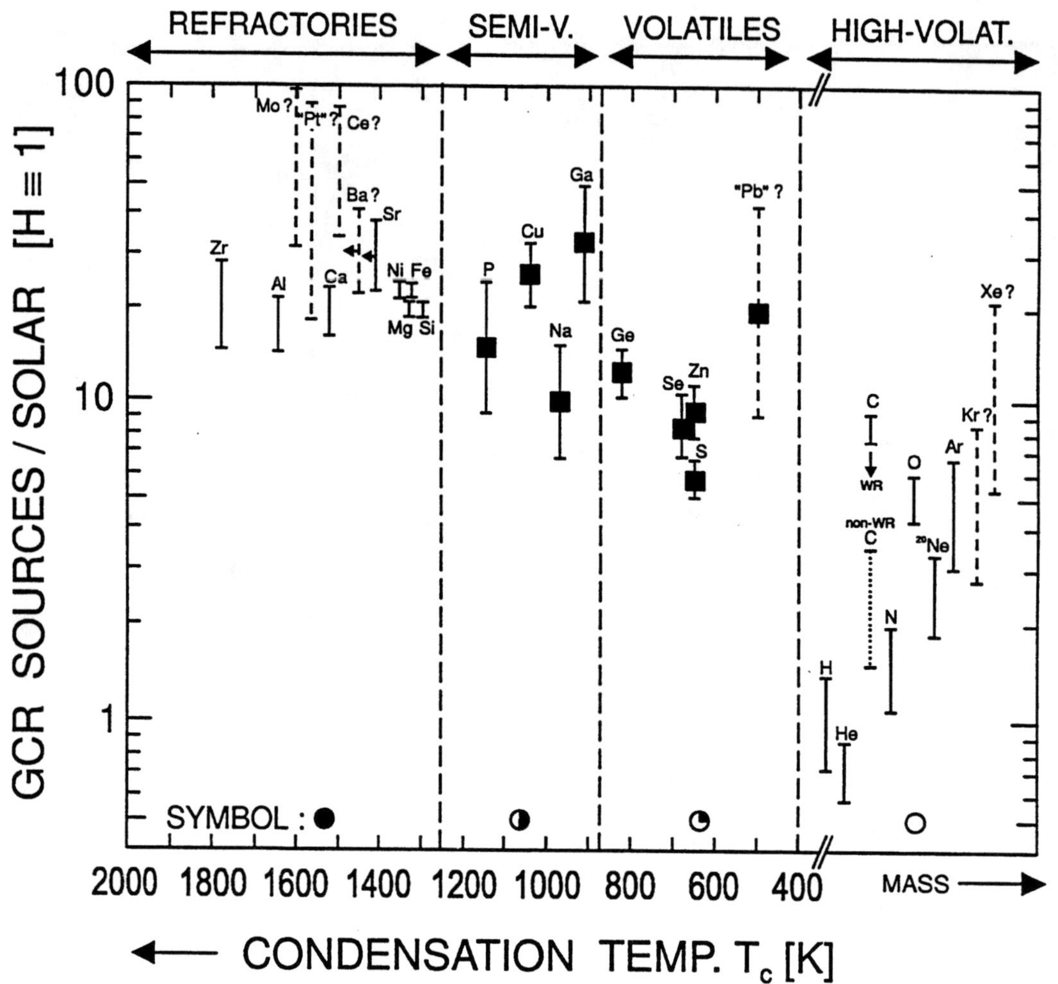

Figure 3. Ratio of cosmic-ray source abundance to solar-system abundance vs. condensation temperature.

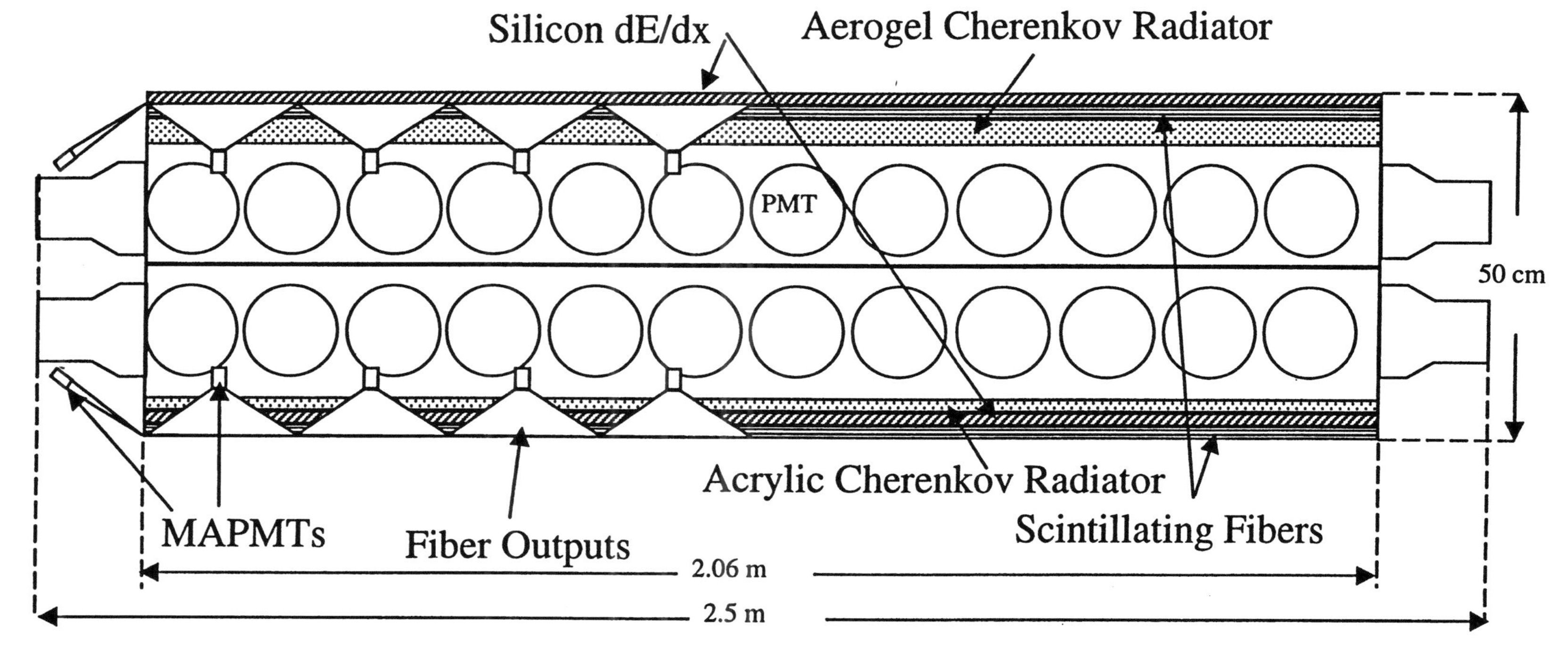

Figure 4. Schematic cross-section of charge identification module (ZIM) for ACCESS.

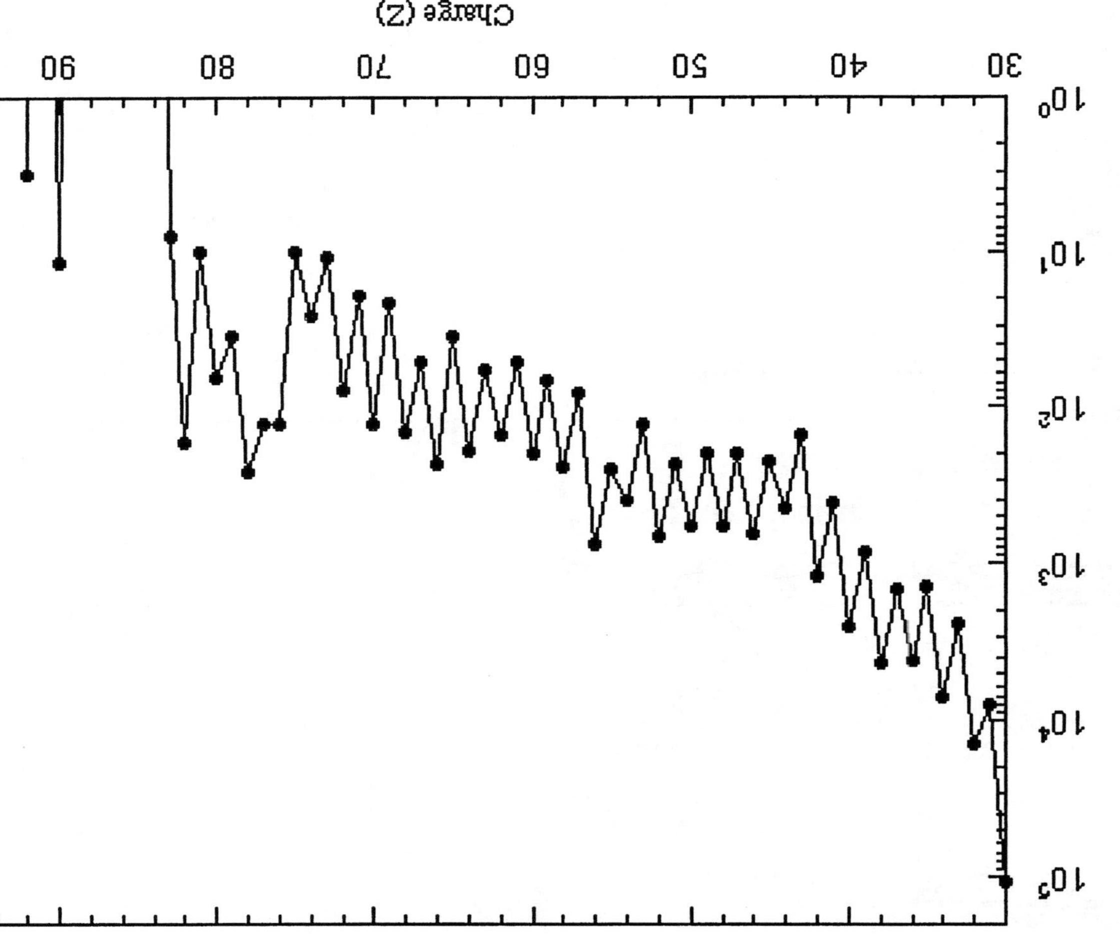

Figure 5. Expected numbers of nuclei seen by ZIM in 4 years on the Space Station.

V. Ultra-high energy cosmic rays

Acceleration and Interaction of Ultra High Energy Cosmic Rays

R.J. Protheroe

Department of Physics and Mathematical Physics
The University of Adelaide, Adelaide, Australia 5005

ABSTRACT

In this chapter I give an overview of shock acceleration, including
a discussion of the maximum energies possible and the shape of the
spectrum near cut-off, interactions of high energy cosmic rays with,
and propagation through, the background radiation, and the resulting
electron-photon cascade. Possible sources of the highest energy cosmic
rays are discussed including active galaxies, gamma ray bursts and
topological defects. I argue that while the origin of the highest energy
cosmic rays is still uncertain, it is not necessary to invoke exotic models
such as emission by topological defects to explain the existing data. It
seems likely that shock acceleration at Fanaroff-Riley Class II radio
galaxies can account for the existing data. However, new cosmic ray
data, as well as better estimates of the extragalactic radiation fields and
magnetic fields will be necessary before we will be certain of the origin
of the highest energy particles occurring in nature.

1.   Introduction

Cosmic rays with energies up to 100 TeV are thought to arise predominantly
through shock acceleration by supernova remnants (SNR) in our Galaxy (Lagage
& Cesarsky 1983). A fraction of the cosmic rays accelerated should interact within
the supernova remnant and produce γ–rays (Drury et al. 1994, Gaisser et al. 1998,
Baring et al. 1999) , and recent observations above 100 MeV by the EGRET
instrument on the Compton Gamma Ray Observatory have found γ-ray signals
associated with at least two supernova remnants – IC 443 and γ Cygni (Esposito
et al. 1996). However, Brazier et al. (1996) have suggested that the γ-ray emission
from IC 443 may be associated with a pulsar within the remnant rather than
the remnant itself. Further evidence for acceleration in SNR comes from the
Rossi X-ray Timing Explorer observations of Cassiopeia A showing a non-thermal
component in the spectrum (Allen et al. 1997), and the ASCA observation of
non-thermal X–ray emission from SN 1006 (Koyama et al. 1995). Reynolds (1996)
and Mastichiadis (1996) interpret the latter as synchrotron emission by electrons
accelerated in the remnant up to energies as high as 100 TeV, although Donea
and Biermann (1998) suggest it may be bremsstrahlung from much lower energy
electrons. The CANGAROO telescope appears to have detected TeV γ-rays from

SN1006 (Tanimori et al. 1998), while there has been a disappointing lack of detections of TeV γ-rays from IC443 and γ Cygni. This may be related to the matter density in which the SNR shocks propagate (Baring et al. 1999), higher densities yielding lower cut-off energies for IC443 and γ Cygni.

Acceleration to somewhat higher energies than 100 TeV may be possible (Markiewicz et al. 1990), but probably not to high enough energies to explain the smooth extension of the spectrum to 1 EeV. Several explanations for the origin of the cosmic rays in this energy range have been suggested: reacceleration of the supernova component while still inside the remnant (Axford 1991); by several supernovae exploding into a region evacuated by a pre-supernova star (Ip and Axford 1991); or acceleration in shocks inside the strong winds from hot stars or groups of hot stars (Biermann and Cassellini 1993). Very recently, an analysis of arrival direction data from AGASA (Hayashida et al. 1998) shows excesses in the directions of the Galactic Centre and the Cygnus region which could not easily be explained by charged particle propagation from these sources. If indeed the excess is due to these sources it may provide some evidence for a component of the cosmic rays at 1 EeV being neutrons. At 5 EeV the spectral slope changes, and there is evidence for a lightening in composition (Bird et al. 1994, Dawson et al. 1998) and it is likely this marks a change from galactic cosmic rays to extragalactic cosmic rays being dominant.

An alternative, although less popular, scenario is that almost all cosmic ray nuclei are of extragalactic origin (e.g. Brecher and Burbidge 1972, and references therein). In any case, whether or not the lower energy cosmic rays are indeed galactic, at the highest energies (above $\sim 10^{19}$ eV) it is very probably extragalactic.

The cosmic ray air shower events with the highest energies so far detected have energies of 2×10^{11} GeV (Hayashida et al. 1994) and 3×10^{11} GeV (Bird et al. 1995), and recent results from AGASA have shown there to be a continuous spectrum between 10^{11} GeV and 3×10^{11} GeV (Takeda et al. 1998)). The question of the origin of these cosmic rays having energy significantly above 10^{11} GeV is complicated by propagation of such energetic particles through the Universe. Nucleons interact with the cosmic background radiation fields, losing energy by Bethe-Heitler pair production, or interacting by pion photoproduction, and in the latter case may emerge as either protons or neutrons with reduced energy. The threshold for pion photoproduction on the microwave background is $\sim 2 \times 10^{10}$ GeV, and at 3×10^{11} GeV the energy-loss distance is about 20 Mpc. Propagation of cosmic rays over substantially larger distances gives rise to a cut-off in the spectrum at $\sim 10^{11}$ GeV as was first shown by Greisen (1966), and Zatsepin and Kuz'min (1966), the "GZK cut-off", and a corresponding pile-up at slightly lower energy (Hill and Schramm 1985, Berezinsky and Grigor'eva 1988). These processes occur not only during propagation, but also during acceleration, and may actually limit the maximum energies particles can achieve.

In this chapter I give an overview of shock acceleration, describe interactions of high energy protons and nuclei with radiation, discuss maximum energies obtainable during acceleration, and the shape of the spectrum near maximum energy, outline propagation of cosmic rays through the background radiation and the consequent electron-photon cascading, and finally discuss conventional and exotic models of the highest energy cosmic rays.

2. Interactions of High Energy Cosmic Rays

Interactions of cosmic rays with radiation are important both during acceleration when the resulting energy losses compete with energy gains by, for example, shock acceleration, and during propagation from the acceleration region to the observer. For ultra-high energy (UHE) cosmic rays, the most important processes are pion photoproduction and Bethe-Heitler pair production both on the microwave background, and synchrotron radiation. In the case of nuclei, photodisintegration on the microwave background is important. In this section, I shall describe how to calculate the mean free path for such interactions, and briefly discuss how to simulate the interactions using the Monte Carlo method.

2.1. Nucleons

The mean interaction length, $x_{p\gamma}$, of a proton of energy E is given by,

$$\frac{1}{x_{p\gamma}(E)} = \frac{1}{8\beta E^2} \int_{\varepsilon_{\min}(E)}^{\infty} \frac{n(\varepsilon)}{\varepsilon^2} \int_{s_{\min}}^{s_{\max}(\varepsilon,E)} \sigma(s)(s - m_p^2 c^4) ds d\varepsilon, \tag{1}$$

where $n(\varepsilon)$ is the differential photon number density of photons of energy ε, and $\sigma(s)$ is the appropriate total cross section for the process in question for a centre of momentum (CM) frame energy squared, s, given by

$$s = m_p^2 c^4 + 2\varepsilon E(1 - \beta \cos \theta) \tag{2}$$

where θ is the angle between the directions of the proton and photon, and βc is the proton's velocity.

For pion photoproduction

$$s_{\min} = (m_p c^2 + m_\pi c^2)^2 \approx 1.16 \text{ GeV}^2, \tag{3}$$

and

$$\varepsilon_{\min} = \frac{m_\pi c^2 (m_\pi c^2 + 2m_p c^2)}{2E(1 + \beta)} \approx \frac{m_\pi c^2 (m_\pi c^2 + 2m_p c^2)}{4E}. \tag{4}$$

For photon-proton pair-production the threshold is somewhat lower,

$$s_{\min} = (m_p c^2 + 2m_e c^2)^2 \approx 0.882 \text{ GeV}^2, \tag{5}$$

and

$$\varepsilon_{\min} \approx m_e c^2 (m_e c^2 + m_p c^2)/E. \tag{6}$$

For both processes,

$$s_{\max}(\varepsilon, E) = m_p^2 c^4 + 2\varepsilon E(1 + \beta) \approx m_p^2 c^4 + 4\varepsilon E, \tag{7}$$

and $s_{\max}(\varepsilon, E)$ corresponds to a head-on collision of a proton of energy E and a photon of energy ε.

Examination of the integrand in Equation 1 shows that the energy of the soft photon interacting with a proton of energy E is distributed as

$$p(\varepsilon) = \frac{x_{p\gamma}(E)n(\varepsilon)}{8\beta E^2 \varepsilon^2}\Phi(s_{\max}(\varepsilon, E)) \tag{8}$$

in the range $\varepsilon_{\min} \leq \varepsilon \leq \infty$ where

$$\Phi(s_{\max}) = \int_{s_{\min}}^{s_{\max}} \sigma(s)(s - m_p^2 c^4)ds. \tag{9}$$

Similarly, examination of the integrand in Equation 1 shows that the square of the total CM frame energy is distributed as

$$p(s) = \frac{\sigma(s)(s - m_p^2 c^4)}{\Phi(s_{\max})}, \tag{10}$$

in the range $s_{\min} \leq s \leq s_{\max}$.

The Monte Carlo rejection technique can be used to sample ε and s respectively from the two distributions, and Equation 2 is used to find θ. One then Lorentz transforms the interacting particles to the frame in which the interaction is treated (usually the proton rest frame), and samples momenta of particles produced in the interaction from the appropriate differential cross section by the rejection method. The energies of produced particles are then Lorentz transformed to the laboratory frame, and the final energy of the proton is obtained by requiring energy conservation. In this procedure, it is not always possible to achieve exact conservation of both momentum and energy while sampling particles from inclusive differential cross sections (e.g. multiple pion production well above threshold), and the momentum of the last particle sampled is therefore adjusted to minimize the error.

The mean interaction lengths for both processes, $x_{p\gamma}(E)$, are obtained from Equation 1 for interactions in the microwave background and are plotted as dashed lines in Fig. 1. Dividing by the inelasticity, $\kappa(E)$, one obtains the energy-loss distances for the two processes,

$$\frac{E}{dE/dx} = \frac{x_{p\gamma}(E)}{\kappa(E)}. \tag{11}$$

2.2.   Nuclei

In the case of nuclei the situation is a little more complicated. The threshold condition for Bethe-Heitler pair production can be expressed as

$$\gamma > \frac{m_e c^2}{\varepsilon}\left(1 + \frac{m_e}{Am_p}\right), \tag{12}$$

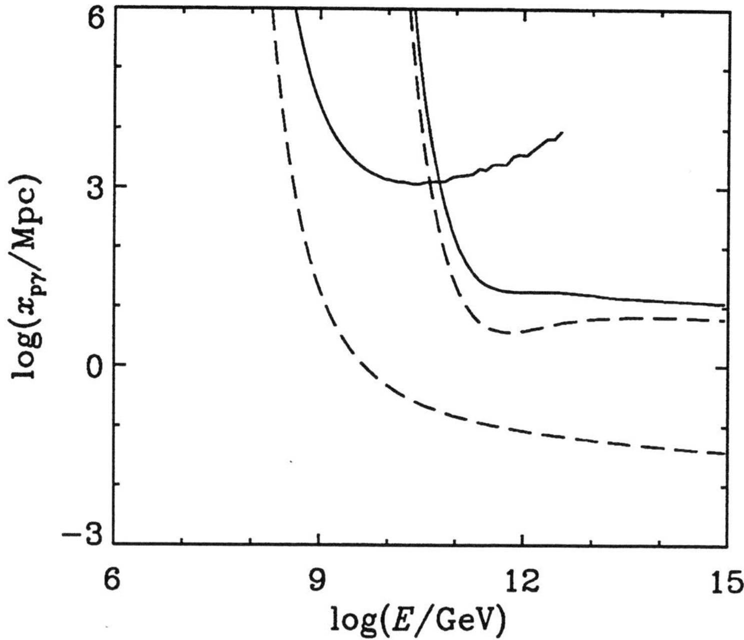

Fig. 1.— Mean interaction length (dashed lines) and energy-loss distance (solid lines), $E/(dE/dx)$, for proton-photon pair-production and pion-production in the microwave background (lower and higher energy curves respectively). (From Protheroe and Johnson 1995).

and the threshold condition for pion photoproduction can be expressed as

$$\gamma > \frac{m_\pi c^2}{2\varepsilon}\left(1 + \frac{m_\pi}{2Am_p}\right). \tag{13}$$

Since $\gamma = E/Am_p c^2$, where A is the mass number, we will need to shift both energy-loss distance curves in Fig. 1 to higher energies by a factor of A. We shall also need to shift the curves up or down as discussed below.

For Bethe-Heitler pair production the energy lost by a nucleus in each collision near threshold is approximately $\Delta E \approx \gamma 2 m_e c^2$. Hence the inelasticity is

$$K \equiv \frac{\Delta E}{E} \approx \frac{2m_e}{Am_p}, \tag{14}$$

and is a factor of A lower than for protons. On the other hand, the cross section goes like Z^2, so the overall shift is down (to lower energy-loss distance) by Z^2/A. For example, for iron nuclei the energy loss distance for pair production is reduced by a factor $26^2/56 \approx 12.1$.

For pion production the energy lost by a nucleus in each collision near threshold is approximately $\Delta E \approx \gamma m_\pi c^2$, and so, as for pair production, the inelasticity is

                                                                 R.J. Protheroe

factor A lower than for protons. The cross section increases approximately as $A^{0.9}$ giving an overall increase in the energy loss distance for pion production of a factor $\sim A^{0.1} \approx 1.5$ for iron nuclei. The energy loss distances for pair production and pion photoproduction are shown for iron nuclei in Fig. 2.

Photodisintegration is very important and has been considered in detail by Tkaczyk et al. (1975), Puget et al. (1976), Karakula and Tkaczyk (1993) , Epele and Roulet (1998) and Stecker and Salamon (1999). The photodisintegration distance defined by $A/(dA/dx)$ taken from Stecker and Salamon (1999) is shown in Fig. 2 together with an estimate made over a larger range of energy by Protheroe (unpublished) of the total loss distance based on photodisintegration cross sections of Karakula and Tkaczyk (1993). Since iron nuclei will be fragmented during pion photoproduction, the photodisintegration distance of Fe at high energies should be consistent with $\sim 56\times$ *the mean free path* for pion photoproduction by Fe at higher energies, and this is found to be the case (see Fig. 2).

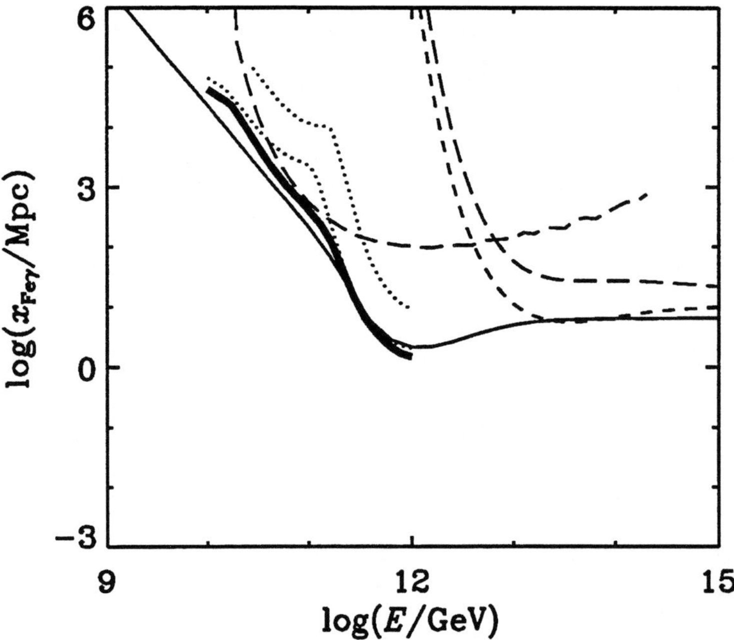

Fig. 2.— Energy-loss distance of Fe-nuclei in the CMBR for pair-production (leftmost long dashed line) and pion photoproduction (rightmost long dashed line), and mean interaction length for pion photoproduction multiplied by 56 (short dashed line) are obtained from curves in Fig. 1. The photodisintegration distances given by Stecker and Salamon (1999) for loss of one nucleon (lower dotted curve) and two nucleons (upper dotted line) are shown together with the total loss distance estimated by Stecker and Salamon (1999). The thin full curve shows an estimate over a larger range of energy (Protheroe, unpublished) of the total loss distance based on photodisintegration cross sections of Karakula and Tkaczyk (1993).

3. Cosmic Ray Acceleration

For stochastic particle acceleration by electric fields induced by motion of magnetic fields B, the rate of energy gain by relativistic particles of charge Ze can be written (in SI units)

$$\left.\frac{dE}{dt}\right|_{\text{acc}} = \xi Zec^2 B \tag{15}$$

where $\xi < 1$ and depends on the acceleration mechanism. I shall give a simple heuristic treatment of Fermi acceleration based on that given in Gaisser's excellent book (Gaisser 1990). I shall start with 2nd order Fermi acceleration (Fermi's original theory) and describe how this can be modified in the context of astrophysical shocks into the more efficient 1st order Fermi mechanism known as shock acceleration. More detailed and rigorous treatments are given in several review articles (Drury 1983a, Blandford and Eichler 1987, Berezhko and Krymsky 1988). See the review by Jones and Ellison (1991) on the plasma physics of shock acceleration which also includes a brief historical review and refers to early work.

3.1. Fermi's Original Theory

Gas clouds in the interstellar medium have random velocities of ~ 15 km/s superimposed on their regular motion around the galaxy. Cosmic rays gain energy on average when scattering off these magnetized clouds. A cosmic ray enters a cloud and scatters off irregularities in the magnetic field which is tied to the cloud because it is partly ionized.

In the frame of the cloud: (a) there is no change in energy because the scattering is collisionless, and so there is elastic scattering between the cosmic ray and the cloud as a whole which is much more massive than the cosmic ray; (b) the cosmic ray's direction is randomized by the scattering and it emerges from the cloud in a random direction.

Consider a cosmic ray entering a cloud with energy E_1 and momentum p_1 travelling in a direction making angle θ_1 with the cloud's direction. After scattering inside the cloud, it emerges with energy E_2 and momentum p_2 at angle θ_2 to the cloud's direction (Fig. 3). The energy change is obtained by applying the Lorentz transformations between the laboratory frame (unprimed) and the cloud frame (primed). Transforming to the cloud frame:

$$E_1' = \gamma E_1(1 - \beta \cos\theta_1) \tag{16}$$

where $\beta = V/c$ and $\gamma = 1/\sqrt{1 - \beta^2}$.

Transforming to the laboratory frame:

$$E_2 = \gamma E_2'(1 + \beta \cos\theta_2'). \tag{17}$$

The scattering is collisionless, being with the magnetic field. Since the magnetic field is tied to the cloud, and the cloud is very massive, in the cloud's rest

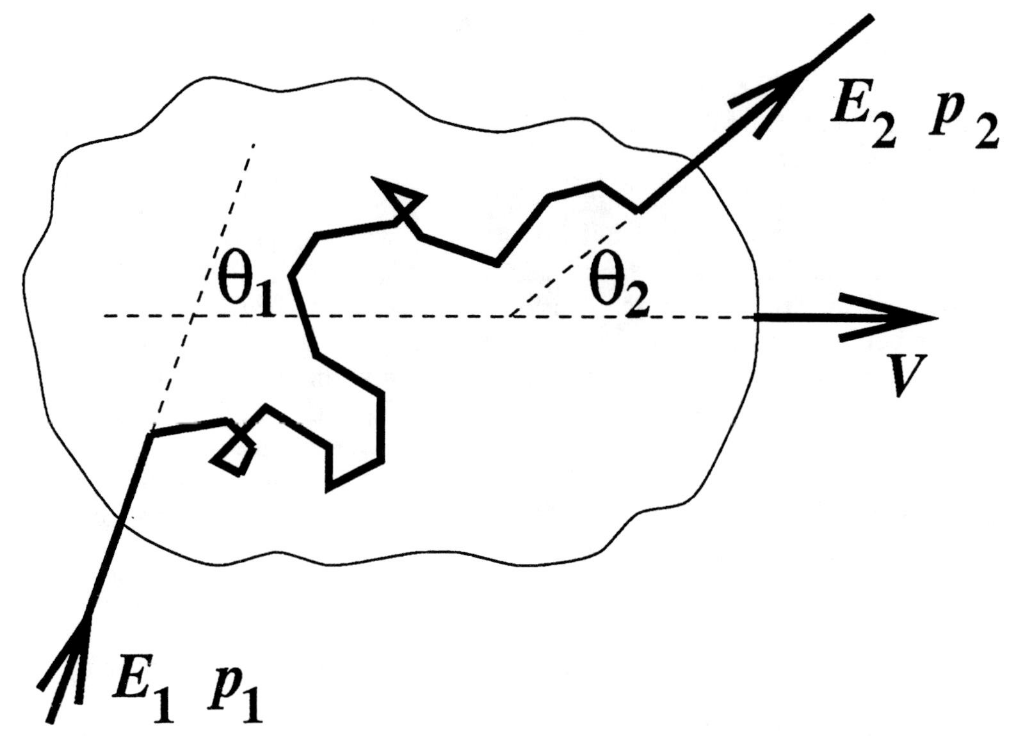

Fig. 3.— Interaction of cosmic ray of energy E_1 with "cloud" moving with speed V

frame there is no change in energy, $E'_2 = E'_1$, and hence we obtain the fractional change in LAB-frame energy $(E_2 - E_1)/E_1$,

$$\frac{\Delta E}{E} = \frac{1 - \beta\cos\theta_1 + \beta\cos\theta'_2 - \beta^2\cos\theta_1\cos\theta'_2}{1 - \beta^2} - 1. \tag{18}$$

We need to obtain average values of $\cos\theta_1$ and $\cos\theta'_2$. Inside the cloud, the cosmic ray scatters off magnetic irregularities many times so that its direction is randomized,

$$\langle\cos\theta'_2\rangle = 0. \tag{19}$$

The average value of $\cos\theta_1$ depends on the rate at which cosmic rays collide with clouds at different angles. The rate of collision is proportional to the relative velocity between the cloud and the particle so that the probability per unit solid angle of having a collision at angle θ_1 is proportional to $(v - V\cos\theta_1)$. Hence, for ultrarelativistic particles $(v = c)$

$$\frac{dP}{d\Omega_1} \propto (1 - \beta\cos\theta_1), \tag{20}$$

and we obtain

$$\langle\cos\theta_1\rangle = \int\cos\theta_1\frac{dP}{d\Omega_1}d\Omega_1 \Big/ \int\frac{dP}{d\Omega_1}d\Omega_1 = -\frac{\beta}{3}, \tag{21}$$

giving

$$\frac{\langle \Delta E \rangle}{E} = \frac{1 + \beta^2/3}{1 - \beta^2} - 1 \simeq \frac{4}{3}\beta^2 \tag{22}$$

since $\beta \ll 1$.

We see that $\langle \Delta E \rangle / E \propto \beta^2$ is positive (energy gain), but is 2nd order in β and because $\beta \ll 1$ the average energy gain is very small. This is because there are almost as many overtaking collisions (energy loss) as there are head-on collisions (energy gain).

3.2.   1st Order Fermi Acceleration at SN or Other Shocks

Fermi's original theory was modified in the 1970's (Axford, Lear and Skadron 1977, Krymsky 1977, Bell 1978, Blandford and Ostriker 1978) to describe more efficient acceleration (1st order in β) taking place at supernova shocks but is generally applicable to strong shocks in other astrophysical contexts. Our discussion of shock acceleration will be of necessity brief, and omit a number of subtleties.

Here, for simplicity, we adopt the test particle approach (neglecting effects of cosmic ray pressure on the shock profile), adopt a plane geometry and consider only non-relativistic shocks. Nevertheless, the basic concepts will be described in sufficient detail that we can consider acceleration and interactions of the highest energy cosmic rays, and to what energies they can be accelerated. We consider the classic example of a SN shock, although the discussion applies equally to other shocks. During a supernova explosion several solar masses of material are ejected at a speed of $\sim 10^4$ km/s which is much faster than the speed of sound in the interstellar medium (ISM) which is ~ 10 km/s. A strong shock wave propagates radially out as the ISM and its associated magnetic field piles up in front of the supernova ejecta. The velocity of the shock, V_S, depends on the velocity of the ejecta, V_P, and on the ratio of specific heats, γ through the compression ratio, R,

$$V_S/V_P \simeq R/(R-1). \tag{23}$$

For SN shocks the SN will have ionized the surrounding gas which will therefore be monatomic ($\gamma = 5/3$), and theory of shock hydrodynamics shows that for $\gamma = 5/3$ a strong shock will have $R = 4$.

In order to work out the energy gain per shock crossing, we can visualize magnetic irregularities on either side of the shock as clouds of magnetized plasma of Fermi's original theory (Fig. 4). By considering the rate at which cosmic rays cross the shock from downstream to upstream, and upstream to downstream, one finds $\langle \cos \theta_1 \rangle = -2/3$ and $\langle \cos \theta_2' \rangle = 2/3$, giving

$$\frac{\langle \Delta E \rangle}{E} \simeq \frac{4}{3}\beta \simeq \frac{4}{3}\frac{V_P}{c} \simeq \frac{4}{3}\frac{(R-1)}{R}\frac{V_S}{c}. \tag{24}$$

Note this is 1st order in $\beta = V_P/c$ and is therefore more efficient than Fermi's original theory. This is because of the converging flow – whichever side of the shock you are on, if you are moving with the plasma, the plasma on the other side of the shock is approaching you at speed V_p.

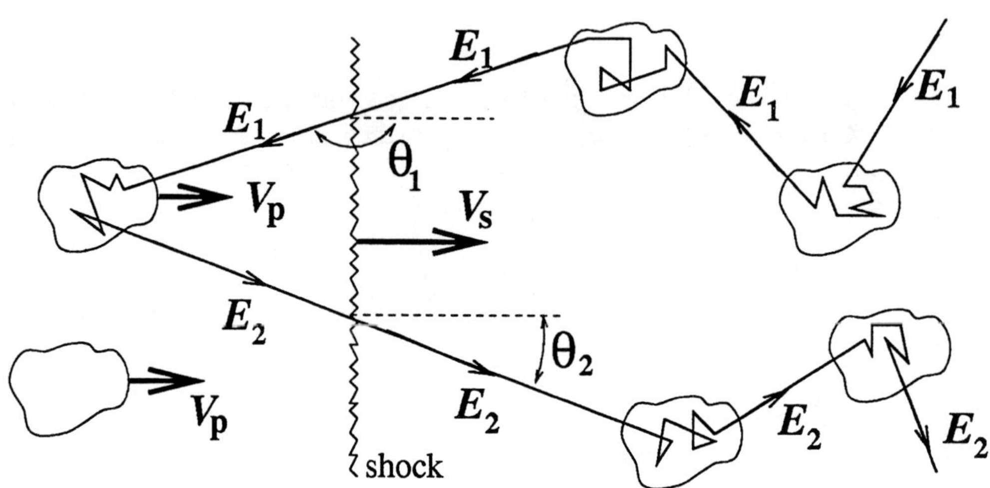

Fig. 4.— Interaction of cosmic ray of energy E_1 with a shock moving with speed V_s.

To obtain the energy spectrum we need to find the probability of a cosmic ray encountering the shock once, twice, three times, etc. If we look at the diffusion of a cosmic ray as seen in the rest frame of the shock (Fig. 5), there is clearly a net flow of the energetic particle population in the downstream direction.

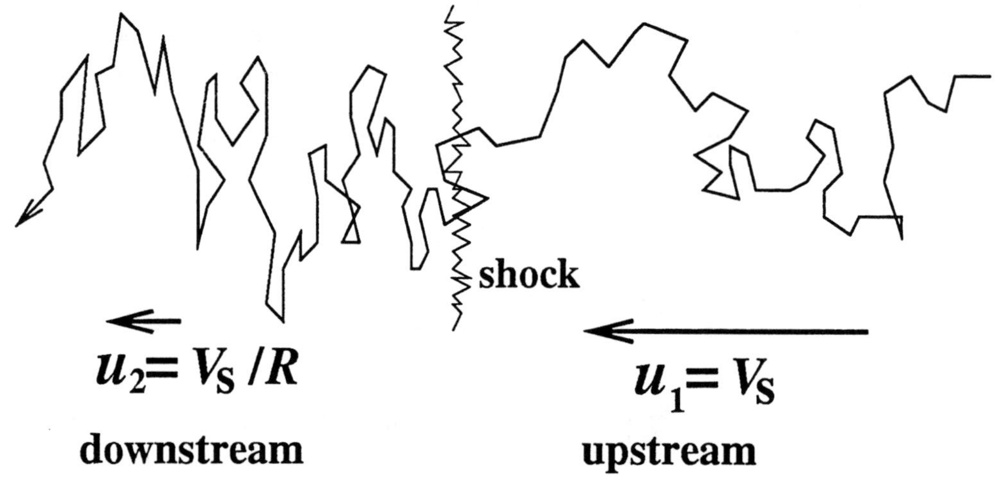

Fig. 5.— Diffusion of cosmic rays from upstream to downstream seen in the shock frame speed V_s.

The net flow rate downstream gives the rate at which cosmic rays are lost downstream

$$r_{\text{loss}} = n_{\text{CR}}V_S/R \qquad \text{m}^{-2}\text{s}^{-1} \tag{25}$$

since cosmic rays with number density n_{CR} at the shock are advected downstream with speed V_S/R (from right to left in Fig. 5) and we have neglected relativistic transformations of the rates because $V_S \ll c$.

Upstream of the shock, cosmic rays travelling at speed v at angle θ to the shock normal (as seen in the laboratory frame) approach the shock with speed $(V_S + v\cos\theta)$ as seen in the shock frame. Clearly, to cross the shock, $\cos\theta > -V_S/v$. Then, assuming cosmic rays upstream are isotropic, the rate at which they cross from upstream to downstream is

$$
\begin{aligned}
r_{\text{cross}} &= n_{\text{CR}}\frac{1}{4\pi}\int_{-V_S/v}^{1}(V_S + v\cos\theta)2\pi d(\cos\theta) \\
&\approx n_{\text{CR}}v/4 \quad \text{m}^{-2}\text{s}^{-1}.
\end{aligned}
\tag{26}
$$

The probability of crossing the shock once and then escaping from the shock (being lost downstream) is the ratio of these two rates:

$$
\text{Prob.(escape)} = r_{\text{loss}}/r_{\text{cross}} \approx 4V_S/Rv.
\tag{27}
$$

The probability of returning to the shock after crossing from upstream to downstream is

$$
\text{Prob.(return)} = 1 - \text{Prob.(escape)},
\tag{28}
$$

and so the probability of returning to the shock k times and also of crossing the shock at least k times is

$$
\text{Prob.(cross} \geq k) = [1 - \text{Prob.(escape)}]^k.
\tag{29}
$$

Hence, the energy after k shock crossings is

$$
E = E_0\left(1 + \frac{\Delta E}{E}\right)^k
\tag{30}
$$

where E_0 is the initial energy.

To derive the spectrum, we note that the integral energy spectrum (number of particles with energy greater than E) on acceleration must be

$$
Q(> E) \propto [1 - \text{Prob.(escape)}]^k
\tag{31}
$$

where

$$
k = \frac{\ln(E/E_0)}{\ln(1 + \Delta E/E)}.
\tag{32}
$$

Hence,

$$
\ln Q(> E) = A + \frac{\ln(E/E_0)}{\ln(1 + \Delta E/E)}\ln[1 - \text{Prob.(escape)}],
\tag{33}
$$

where A is a constant, and so

$$
\ln Q(> E) = B - (\Gamma - 1)\ln E
\tag{34}
$$

where B is a constant and

$$\Gamma = 1 - \frac{\ln[1 - \text{Prob.(escape)}]}{\ln(1 + \Delta E/E)} \approx \frac{R+2}{R-1}. \tag{35}$$

Hence we arrive at the spectrum of cosmic rays on acceleration

$$Q(> E) \propto E^{-(\Gamma-1)} \qquad \text{(integral form)} \tag{36}$$

$$Q(E) \propto E^{-\Gamma} \qquad \text{(differential form)}. \tag{37}$$

For $R = 4$ we have the well-known E^{-2} differential spectrum. The observed cosmic ray spectrum is steepened by energy-dependent escape of cosmic rays from the Galaxy.

3.3.   Effect of cosmic ray pressure

For simplicity, we have neglected the effect of cosmic ray pressure on the shock which can alter the shock profile. The original method of treating this non-linear effect is the two-fluid method, the two-fluids being plasma and cosmic rays, and this is reviewed by Drury (1983b). The shock profile, instead of being a step-function becomes smoothed, and this affects the acceleration of low and high energy particles differently. Lower energy particles with short diffusion mean free paths spanning only part of the shock profile (i.e. effectively seeing a lower R) will have a steeper spectrum, while high energy particles with longer diffusion mean free paths spanning most of the shock profile (i.e. effectively seeing a higher R) will have a flatter spectrum. The net result of this is a spectrum with upward curvature. In addition, as a result of the cosmic ray pressure far downstream providing additional slowing of the plasma, the overall compression ratio from far upstream to far downstream may be even higher than in the test particle case giving an even flatter spectrum for the highest energy particles (Ellison and Eichler 1984; see Baring 1997 for a brief review and additional references).

3.4.   Relativistic Shocks

Shocks in jets of active galactic nuclei (AGN) and gamma ray bursts (GRB) are likely to be relativistic, i.e. $u_1 > 0.1c$. This will affect the acceleration in two ways: (a) the adiabatic index of a relativistic gas is $\gamma = 4/3$ giving $R = 7$ for a strong shock, and so one would expect from Eq. 35 a flatter spectrum with $\Gamma = 1.5$. However, for relativistic plasma motion with bulk velocities comparable to those of the particles being accelerated, the approximations used to derive Eqns. 25 and 26 are no longer valid, the escape probability being greater and the particles being anisotropic, with the result that $\Gamma > 1.5$. Detailed studies have shown a trend in which shocks with larger u_1 generally have lower Γ (Kirk and Schneider 1987, Ellison and Jones 1990). However, the spectral index is very sensitive to the pitch angle (see Baring 1997 for additional references).

4. Shock Acceleration Rate

Here we again neglect effects of cosmic ray pressure and consider only a non-relativistic shock. The acceleration rate is defined by

$$r_{\text{acc}} \equiv \frac{1}{E}\frac{dE}{dt}\bigg|_{\text{acc}} = \frac{(\langle\Delta E\rangle/E)}{t_{\text{cycle}}} \approx \frac{4}{3}\frac{(R-1)}{R}\frac{V_S}{c}t_{\text{cycle}}^{-1} \tag{38}$$

where t_{cycle} is the time for one complete cycle, i.e. from crossing the shock from upstream to downstream, diffusing back towards the shock and crossing from downstream to upstream, and finally returning to the shock.

The rate of loss of accelerated particles downstream is the probability of escape per shock crossing divided by the cycle time

$$r_{\text{esc}} = \frac{\text{Prob.(escape)}}{t_{\text{cycle}}} \approx \frac{4}{R}\frac{V_S}{c}t_{\text{cycle}}^{-1} \tag{39}$$

We see immediately that the ratio of the escape rate to the acceleration rate depends on the compression ratio

$$\frac{r_{\text{esc}}}{r_{\text{acc}}} \approx \frac{3}{R-1} \tag{40}$$

and for a strong shock $(R = 4)$ the two rates are equal. As we shall see later, a consequence of this is that the asymptotic spectrum of particles accelerated by a strong shock is the well-known E^{-2} power-law.

We shall discuss these processes in the shock frame (see Fig. 5) and consider first particles crossing the shock from upstream to downstream and diffusing back to the shock, i.e. we shall work out the average time spent downstream. Since we are considering non-relativistic shocks, the time scales are approximately the same in the upstream and downstream plasma frames, and so in this section I shall drop the use of subscripts indicating the frame of reference.

Diffusion takes place in the presence of advection at speed u_2 in the downstream direction. The typical distance a particle diffuses in time t is $\sqrt{k_2 t}$ where k_2 is the diffusion coefficient in the downstream region. The distance advected in this time is simply $u_2 t$. If $\sqrt{k_2 t} \gg u_2 t$ the particle has a very high probability of returning to the shock, and if $\sqrt{k_2 t} \ll u_2 t$ the particle has a very high probability of never returning to the shock (i.e. it has effectively escaped downstream). So, we set $\sqrt{k_2 t} = u_2 t$ to define a distance k_2/u_2 downstream of the shock which is effectively a boundary between the region closer to the shock where the particles will usually return to the shock and the region farther from the shock in which the particles will usually be advected downstream never to return. There are $n_{\text{CR}}k_2/u_2$ particles per unit area of shock between the shock and this boundary. Dividing this by r_{cross} we obtain the average time spent downstream before returning to the shock

$$t_2 \approx \frac{4}{c}\frac{k_2}{u_2}. \tag{41}$$

Consider next the other half of the cycle after the particle has crossed the shock from downstream to upstream until it returns to the shock. In this case we can define a boundary at a distance k_1/u_1 upstream of the shock such that nearly all particles upstream of this boundary have never encountered the shock, and nearly all the particles between this boundary and the shock have diffused there from the shock. Then dividing the number of particles per unit area of shock between the shock and this boundary, $n_{\rm CR}k_1/u_1$, by $r_{\rm cross}$ we obtain the average time spent upstream before returning to the shock

$$t_1 \approx \frac{4}{c}\frac{k_1}{u_1},\qquad(42)$$

and hence the cycle time

$$t_{\rm cycle} \approx \frac{4}{c}\left(\frac{k_1}{u_1}+\frac{k_2}{u_2}\right).\qquad(43)$$

The acceleration rate is then given by

$$r_{\rm acc} \approx \frac{(R-1)u_1}{3R}\left(\frac{k_1}{u_1}+\frac{k_2}{u_2}\right)^{-1}.\qquad(44)$$

Assuming that the diffusion coefficients upstream and downstream have the same power-law dependence on energy, e.g.,

$$k_1 \propto k_2 \propto E^{\delta},\qquad(45)$$

then the acceleration rate also has a power-law dependence

$$r_{\rm acc} \propto E^{-\delta}.\qquad(46)$$

Note that more correctly, the diffusion coefficient will be a function of magnetic rigidity, ρ, which, for ultra-relativistic particles considered in this paper, is approximately equal to E/Ze where Ze is the charge. However, here we are mainly concerned with singly charged particles and shall work in terms of E rather than rigidity.

4.1. Maximum acceleration rate

We next consider the diffusion for the cases of parallel, oblique, and perpendicular shocks, and estimate the maximum acceleration rate for these cases. The diffusion coefficients required, k_1 and k_2, are the coefficients for diffusion parallel to the shock normal. The diffusion coefficient along the magnetic field direction is some factor η times the minimum diffusion coefficient, known as the Bohm diffusion coefficient,

$$k_{\|} = \eta\frac{1}{3}r_g c\qquad(47)$$

where r_g is the gyroradius, and $\eta > 1$.

Parallel shocks are defined such that the shock normal is parallel to the magnetic field ($\vec{B}\|\vec{u_1}$). In this case, making the approximation that $k_1 = k_2 = k_\|$ and $B_1 = B_2$ one obtains

$$t^\|_{\text{acc}} \approx \frac{20}{3}\frac{\eta E}{eB_1 u_1^2}. \tag{48}$$

For a shock speed of $u_1 = 0.1c$ and $\eta = 10$ one obtains an acceleration rate (in SI units) of

$$\left.\frac{dE}{dt}\right|_{\text{acc}} \approx 1.5 \times 10^{-4} ec^2 B. \tag{49}$$

For the oblique case, the angle between the magnetic field direction and the shock normal is different in the upstream and downstream regions, and the direction of the plasma flow also changes at the shock. The diffusion coefficient in the direction at angle θ to the magnetic field direction is given by

$$k = k_\| \cos^2\theta + k_\perp \sin^2\theta \tag{50}$$

where $k_\perp$ is the diffusion coefficient perpendicular to the magnetic field. Jokipii (1987) shows that

$$k_\perp \approx \frac{k_\|}{1 + \eta^2} \tag{51}$$

provided that η is not too large (values in the range up to 10 appear appropriate), and that acceleration at perpendicular shocks can be much faster than for the parallel case. For $k_{xx} = k_\perp$ and $B_2 \approx 4B_1$ and one obtains

$$t^\perp_{\text{acc}} \approx \frac{8}{3}\frac{E}{\eta eB_1 u_1^2}. \tag{52}$$

For a shock speed of $u_1 = 0.1c$ and $\eta = 10$ one obtains an acceleration rate (in SI units) of

$$\left.\frac{dE}{dt}\right|_{\text{acc}} \approx 0.04 ec^2 B. \tag{53}$$

Supernova shocks remain strong enough to continue accelerating cosmic rays for about 1000 years. The rate at which cosmic rays are accelerated is inversely proportional to the diffusion coefficient (faster diffusion means less time near the shock). For the maximum feasible acceleration rate, a typical interstellar magnetic field, and 1000 years for acceleration, energies of $10^{14} \times Z$ eV are possible (Z is atomic number) at parallel shocks and $10^{16} \times Z$ eV at perpendicular shocks.

In the case of acceleration by relativistic shocks (e.g. Bednarz & Ostrowski 1996, 1998, and Bednarz 1998, and Ostrowski 1998), the acceleration time depends strongly on u_1, $k_\perp/k_\|$, and θ and can be as low as $\sim (1\text{---}10)r_g/c$.

5. Maximum Energies

Protons and nuclei can be accelerated to much higher energies than electrons for a given magnetic environment. For stochastic particle acceleration by electric fields induced by motion of magnetic fields B, the rate of energy gain by relativistic particles of charge Ze can be written (in SI units) $dE/dt = \xi Zec^2 B$ as in Eq. 1, where $\xi < 1$ and depends on the acceleration mechanism; a value of $\xi = 0.04$ might be achieved by first order Fermi acceleration at a perpendicular shock with shock speed $\sim 0.1c$.

5.1.   Limits from Synchrotron Losses

The rate of energy loss by synchrotron radiation of a particle of mass Am_p, charge Ze, and energy γmc^2 is

$$-\left.\frac{dE}{dt}\right|_{\text{syn}} = \frac{4}{3}\sigma_T \left(\frac{Z^2 m_e}{Am_p}\right)^2 \frac{B^2}{2\mu_0}\gamma^2 c. \tag{54}$$

Equating the rate of energy gain with the rate of energy loss by synchrotron radiation places one limit on the maximum energy achievable by electrons, protons and nuclei:

$$E_e^{\text{max}} = 6.0 \times 10^2 \xi^{1/2} \left(\frac{B}{1\,\text{T}}\right)^{-1/2} \quad \text{GeV}, \tag{55}$$

$$E_p^{\text{max}} = 2.0 \times 10^9 \xi^{1/2} \left(\frac{B}{1\,\text{T}}\right)^{-1/2} \quad \text{GeV}, \tag{56}$$

$$E_{Z,A}^{\text{max}} = 2.0 \times 10^9 \xi^{1/2} \frac{A^2}{Z^{3/2}} \left(\frac{B}{1\,\text{T}}\right)^{-1/2} \text{GeV}. \tag{57}$$

The cut-off in proton spectrum may be caused by synchrotron radiation, and Totani (1998) has found that synchrotron radiation by highest energy protons could account for much of the observed cosmic γ-ray background. The maximum energies of protons and iron nuclei allowed by synchrotron radiation losses are shown in Figs. 6 and 7 respectively, and are plotted against magnetic field for three values of ξ. The straight lines are for: the maximum possible acceleration rate $\xi = 1$ (dashed), plausible acceleration at perpendicular shock $\xi = 0.04$ (solid), and plausible acceleration at parallel shock $\xi = 1.5 \times 10^{-4}$ (dot-dash).

Other limits on the maximum energy are placed by the dimensions of the acceleration region and the time available for acceleration. These limits were obtained and discussed in some detail by Biermann and Strittmatter (1987).

5.2.   Limits from Interactions with Radiation

Equating the total energy loss rate for proton–photon interactions (i.e. the sum of pion production and Bethe-Heitler pair production) in Fig. 1 to the rate of energy gain by acceleration gives the maximum proton energy in the absence of other loss processes. This is shown in Fig. 6 as a function of magnetic field which determines the rate of energy gain through Eq. 15. The result is shown by the curves for the maximum possible acceleration rate $\xi = 1$ (dashed), plausible acceleration at perpendicular shock $\xi = 0.04$ (solid), and plausible acceleration at parallel shock $\xi = 1.5 \times 10^{-4}$ (dot-dash). Also shown is the maximum energy determined by synchrotron losses (thick lines) for the three cases. As can be seen, for a perpendicular shock it is possible to accelerate protons to $\sim 10^{13}$ GeV in a $\sim 10^{-5}$ G field.

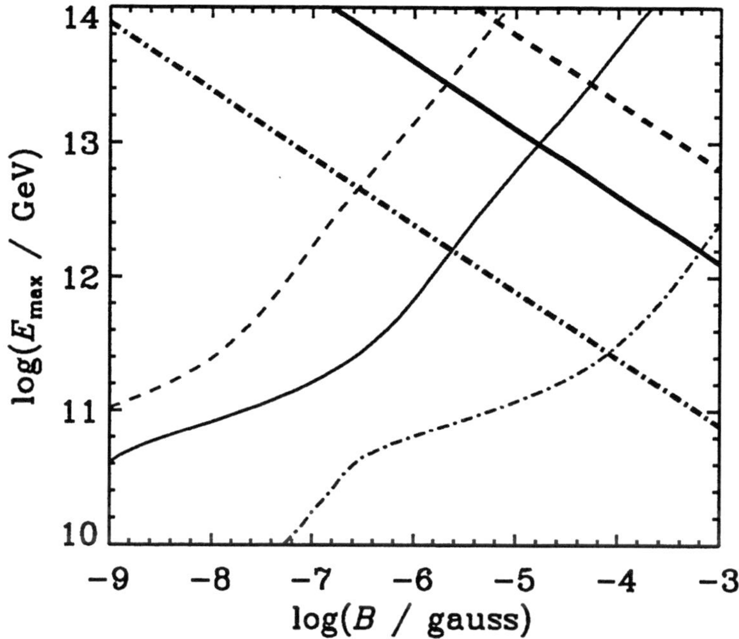

Fig. 6.— Maximum proton energy as a function of magnetic field. Straight lines give the limit from synchrotron loss, curved lines give the limit from pion photoproduction. E_{max} is given for $\xi = 1$ (dashed), $\xi = 0.04$ (solid), $\xi = 1.5 \times 10^{-4}$ (dot-dash).

The effective loss distance given in Fig. 2 is used together with the acceleration rate for iron nuclei to obtain the maximum energy as a function of magnetic field. This is shown in Fig. 7 which is analogous to Fig. 6 for protons. We see that for a perpendicular shock it is possible to accelerate iron nuclei to $\sim 2 \times 10^{14}$ GeV in a $\sim 3 \times 10^{-5}$ G field. While this is higher than for protons, iron nuclei are likely to get photodisintegrated into nucleons of maximum energy $\sim 4 \times 10^{12}$ GeV, and so there is not much to be gained unless the source is nearby. Of course, potential acceleration sites need to have the appropriate combination of size (much larger than the gyroradius at the maximum energy), magnetic field, and shock velocity (or other relevant velocity), and these criteria have been discussed in detail by Hillas (1984).

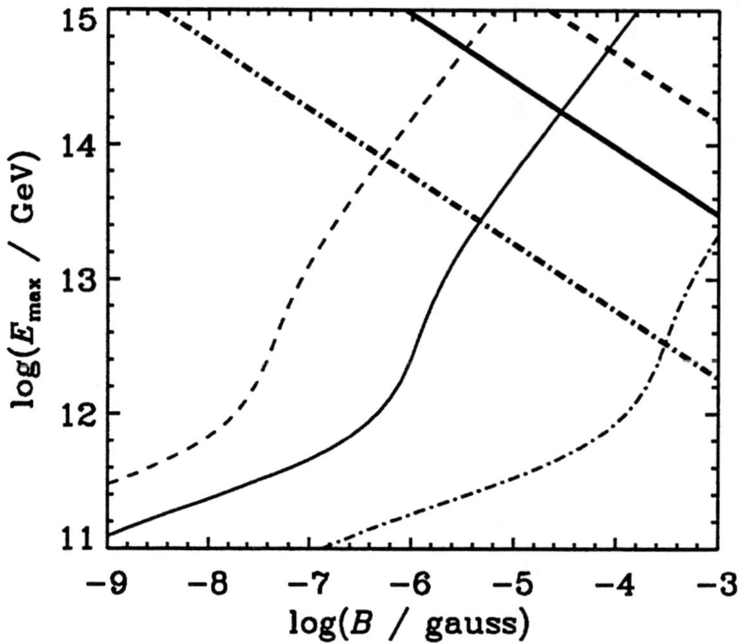

Fig. 7.— Maximum iron nucleus energy as a function of magnetic field. Straight lines give the limit from synchrotron loss, curved lines give the limit from photodisintegration. E_{max} is given for $\xi = 1$ (dashed), $\xi = 0.04$ (solid), $\xi = 1.5 \times 10^{-4}$ (dot-dash).

6.   Spectral Shape near Maximum Energy

To determine the spectral shape near maximum energy we use the leaky-box acceleration model (Szabo and Protheroe 1994) which may be considered as follows. A particle of energy E_0 is injected into the leaky box. While inside the box, the particle's energy changes at a rate $dE/dt = Er_{\mathrm{acc}}(E)$ and that in any short time interval Δt the particle has a probability of escaping from the box given by $\Delta t r_{\mathrm{esc}}(E)$. The energy spectrum of particles escaping from the box then approximates the spectrum of shock accelerated particles.

Let us consider first the case of no energy losses, interactions, or losses due to any other process. N_0 particles of energy E_0 are injected at time $t = 0$, and we assume the following acceleration and escape rates:

$$r_{\mathrm{acc}} = aE^{-\delta},\tag{58}$$

$$r_{\mathrm{esc}} = cE^{-\delta}.\tag{59}$$

The energy at time t is then obtained simply by integrating

$$dE/dt = aE^{(1-\delta)},\tag{60}$$

giving

$$E(t) = (E_0^\delta + \delta at)^{1/\delta}. \tag{61}$$

The number of particles remaining inside the accelerator at time t after injection is obtained by solving

$$dN/dt = -N(t)c[E(t)]^{-\delta}. \tag{62}$$

Using Eq. 61 and integrating, one has

$$\int_{N_0}^{N(t)} N^{-1}dN = -c \int_0^t (E_0^\delta + \delta at)^{-1}dt, \tag{63}$$

giving

$$N(t) = N_0[E(t)/E_0]^{-c/a}. \tag{64}$$

Since $N_0 - N(t)$ particles have escaped from the accelerator before time t, and therefore have energies between E_0 and $E(t)$, the differential energy spectrum of particles which have escaped from the accelerator is simply given by

$$dN/dE \;=\; N_0(\Gamma - 1)(E_0)^{-1}(E/E_0)^{-\Gamma}, \tag{65}$$

for $(E > E_0)$ where $\Gamma = (1 + c/a)$ is the differential spectral index. We note that for $r_{\rm esc}(E) = r_{\rm acc}(E)$ one obtains the standard result for acceleration at strong shocks $\Gamma = 2$.

6.1.   Cut-off due to finite acceleration volume, etc.

Even in the absence of energy losses, acceleration usually ceases at some energy due to the finite size of the acceleration volume (e.g. when the gyroradius becomes comparable to the characteristic size of the shock), or as a result of some other process. We approximate the effect of this by introducing a constant term to the expression for the escape rate:

$$r_{\rm esc} = cE^{-\delta} + cE_{\rm max}^{-\delta}. \tag{66}$$

where $E_{\rm max}$ is defined by the above equation and will be close to the energy at which the spectrum steepens due to the constant escape term. We shall refer to $E_{\rm max}$ as the "maximum energy" even though some particles will be accelerated to energies above this.

Following the same procedure as for the case of a purely power-law dependence of the escape rate, we obtain the differential energy spectrum of particles $(E > E_0)$ escaping from the accelerator,

$$\frac{dN}{dE} \;=\; N_0(\Gamma - 1)(E_0)^{-1}(E/E_0)^{-\Gamma}[1 + (E/E_{\rm max})^\delta]$$

$$\times \exp\left\{ -\frac{\Gamma - 1}{\delta}\left[\left(\frac{E}{E_{\rm max}}\right)^\delta - \left(\frac{E_0}{E_{\rm max}}\right)^\delta\right]\right\} \tag{67}$$

for $\delta > 0$. For $\delta = 0$ we note that from Eq. 66 $r_{\rm esc} = (2c)$ at all energies. Thus the situation is equivalent to the case of $E_{\rm max} \to \infty$ and the spectrum is given by Eq. 65 provided we replace c with $2c$, i.e. we must replace Γ with $\Gamma' = (2\Gamma - 1)$. We compare in Fig 8 the spectra for $\Gamma = 2$ and δ ranging from 1/3 to 1, and note that the energy dependence of the diffusion coefficient has a profound influence on the shape of the cut-off.

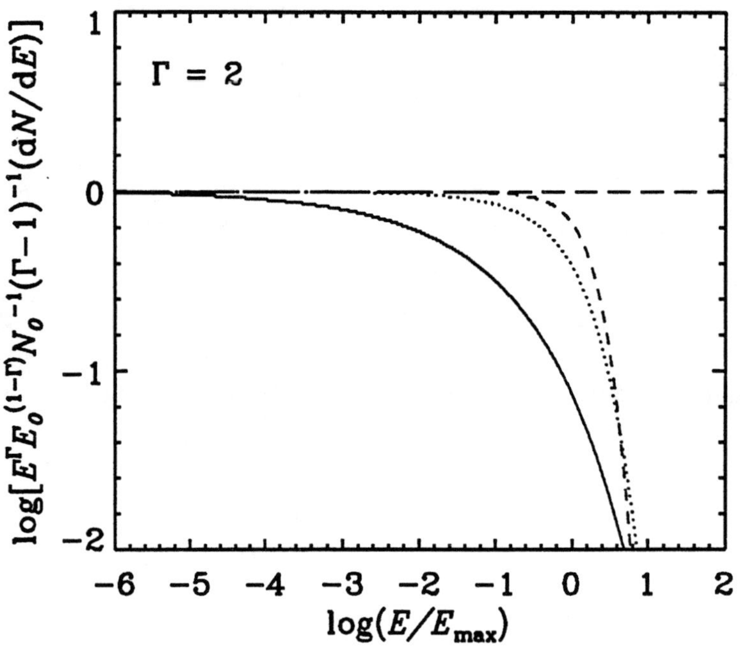

Fig. 8.— Differential energy spectrum for $\Gamma = 2$ and $\delta = 1/3$ (solid curve), 2/3 (dotted curve) and 1 (dashed curve). (From Protheroe & Stanev 1998).

Spectra such as those presented in Fig. 8 may be used to model the source spectra of high energy cosmic ray nuclei of various species if one replaces energy with magnetic rigidity, $\rho = pc/Ze \approx E/Ze$, where Z is the atomic number:

$$\frac{dN}{d\rho} = N_0(\Gamma - 1)(\rho_0)^{-1}(\rho/\rho_0)^{-\Gamma}[1 + (\rho/\rho_{\rm max})^\delta]$$

$$\times \exp\left\{-\frac{\Gamma - 1}{\delta}\left[\left(\frac{\rho}{\rho_{\rm max}}\right)^\delta - \left(\frac{\rho_0}{\rho_{\rm max}}\right)^\delta\right]\right\} \tag{68}$$

for $\rho > \rho_0$.

6.2. Cut-off due to E^2 energy losses

E^2 energy losses of protons or nuclei at extremely high energies result from synchrotron radiation. The introduction of continuous energy losses into the problem is, in principle, straightforward and accomplished simply by modifying Eq. 60,

$$dE/dt = aE^{(1-\delta)} - bE^2. \tag{69}$$

Setting $dE/dt = 0$ we obtain the cut-off energy

$$E_{\text{cut}} = (a/b)^{1/(1+\delta)}. \tag{70}$$

For this case, however, the problem is easier to solve using the Green's function approach adopted by Stecker (1971) when considering the ambient spectrum of cosmic ray electrons and galactic γ-rays. Using the appropriate Green's function one can obtain the steady-state spectrum of particles inside the leaky-box accelerator, and multiplying this by the escape rate one obtains the spectrum of particles leaving the accelerator,

$$\frac{dN}{dE} = \frac{c\,E^{-\delta} + E_{\text{max}}^{-\delta}}{a\,E^{1-\delta} - b\,E^2} \exp[-I(E)] \tag{71}$$

where

$$I(E) \equiv \int_{E_0}^{E} \frac{c\,E^{-\delta} + E_{\text{max}}^{-\delta}}{a\,E^{1-\delta} - b\,E^2}\,dE \tag{72}$$

$$= \left[\left(\frac{c}{a\,\delta}\right) \left(\frac{E}{E_{\text{max}}}\right)^{\delta} \right.$$

$$\times \,{}_2F_1\left(\frac{\delta}{1+\delta}, 1, 1 + \frac{\delta}{1+\delta}, \frac{b\,E^{1+\delta}}{a}\right)$$

$$\left. - \frac{c}{a(1+\delta)} \ln\left(b - \frac{a}{E^{1+\delta}}\right) \right]_{E_0}^{E}, \tag{73}$$

and ${}_2F_1$ is the hypergeometric function.

The result depends on the parameters δ, Γ, E_0, E_{cut}, E_{max}. As a result of the energy loss by particles near the maximum energy a pile-up in the spectrum may be produced just below E_{cut}. The size of the pile-up will be determined by the relative importance of r_{acc} and r_{esc} at energies just below E_{cut}, and so should depend only on Γ and δ provided $E_0 \ll E_{\text{cut}} \ll E_{\text{max}}$. In this case, the shape of the spectrum is given by

$$\frac{dN}{dE} = N_0(\Gamma - 1)(E_0)^{-1} \left(\frac{E}{E_0}\right)^{-\Gamma}$$

$$\times \left[1 - \left(\frac{E}{E_{\text{cut}}}\right)^{(1+\delta)}\right]^{(\Gamma-2-\delta)/(1+\delta)}. \tag{74}$$

We compare in Fig. 9 the spectra for this case for $\delta = 1/3$, $2/3$ and 1, and for $\Gamma = 1.5$, 2, and 2.5, and note that the pile-ups are higher for flatter spectra, and

that for steep spectra the pile-up may be absent or the spectrum may steepen
before the cut-off if δ is small. The effect of synchrotron losses on multiple diffusive
shock acceleration have been considered by Melrose and Crouch (1997) , and
similar effects were found.

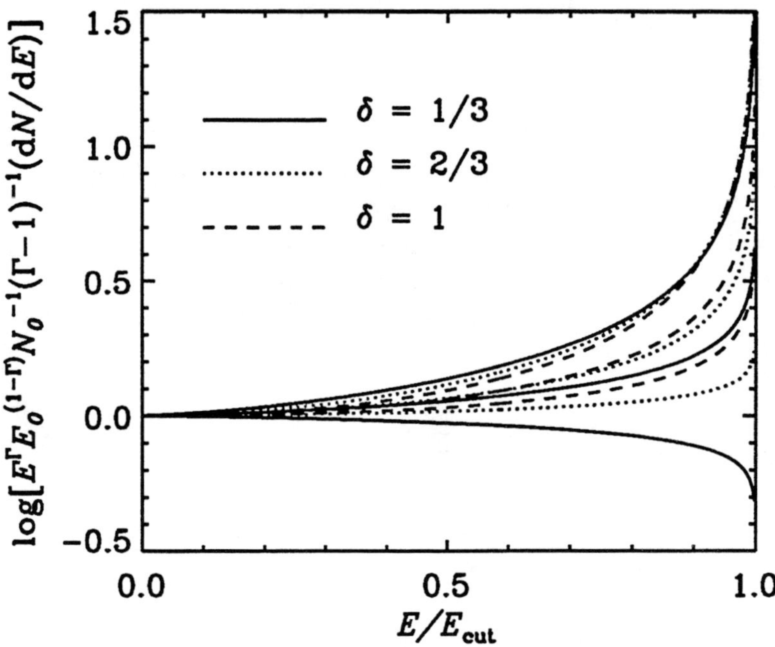

Fig. 9.— Differential energy spectra for $E_0 \ll E_{\rm cut} \ll E_{\rm max}$ for $\delta = 1/3$ (solid
curves), 2/3 (dotted curves) and 1 (dashed curves), and $\Gamma = 1.5$ (upper curves), 2.0
(middle curves) and 2.5 (lower curves). (From Protheroe & Stanev 1998).

6.3. Cut-off due to pion photoproduction

Here I discuss how one can use the Monte Carlo method to investigate the
shape of the cut-off or pile-up which results when the nominal cut-off energy is
determined by interactions rather than continuous energy losses. This technique
was used by Protheroe and Stanev (1998) to investigate cut-offs in electron
spectra due to inverse Compton scattering in the Klein-Nishina regime, and by
Szabo and Protheroe(1994) to investigate cut-offs in the proton spectrum due to
photoproduction in a radiation field.

The technique uses the leaky-box acceleration model described above. We
describe here the case considered by Szabo and Protheroe: $\Gamma = 2, \delta = 1$. Particles
of energy E_0 are injected and accelerated at a constant rate $a = dE/dt$. The
time scale for escape from the leaky box is equal to the acceleration time scale,
$t_{\rm esc} = t_{\rm acc} = E/a$, and so the probability of reaching an energy E without escaping

is $P_{\text{surv}}(E) = E_0/E$. Such a model produces an E^{-2} spectrum at energies above E_0 and may be easily adapted to calculations in which other processes take place in addition to acceleration, and are simulated by the Monte Carlo method.

To calculate the spectrum of particles produced during acceleration per injected particle, we use a method of weights. If a particle is injected with energy E_0, then the "weight" of the particle by the time it reaches an energy E_1 is set to the probability of it not having escaped, $W_1 = E_0/E_1$. If the particle interacts, and has energy E'_1 after the interaction, etc., then the weight of the particle at the nth interaction is given by

$$W_n = \frac{E_0}{E_1}\frac{E'_1}{E_2}\cdots\frac{E'_{n-1}}{E_n}. \tag{75}$$

For the case in which E_{max} is determined by proton-photon collisions, the following procedure was adopted. (i) Inject a particle with energy E_0 into the accelerator and after $(n-1)$ interactions (assuming it has survived catastrophic losses) it will have energy E'_{n-1} and weight W_{n-1}; (ii) Sample the nth path length for two competing interactions, the competing interaction with the shortest path length being assumed to be the one that occurs.

Before the interaction, the proton has energy E_n and the interaction is modelled by the Monte Carlo method. The energies of the produced particles are binned in energy with the appropriate weight, in this case W_n. The 'fraction' of the particle which does not interact, in this case $(1 - W_n)$, is assumed to escape from the accelerator and have an E^{-2} energy distribution over the energy range E'_{n-1} to E_n. This contribution is then added to the escaping proton spectrum. The process is repeated until the particle suffers a catastrophic loss in the form of a pion photoproduction reaction in which a neutron is produced.

In this way one can calculate the spectra of protons, neutrons, pions (neutral and charged) and electrons (including positrons) produced during acceleration. Results for $E_{\text{cut}} = 2 \times 10^{12}$ GeV due to photoproduction on the cosmic microwave background radiation are shown in Figure 10.

7. Cascading During Propagation

As well as particles being produced during the acceleration process as a result of interactions, during propagation to Earth cascading occurs and the accompanying fluxes of γ-rays and neutrinos must not exceed the observed flux or flux limits. Waxman and Bahcall (1998) estimate the maximum allowed neutrino flux to be very low and model independent in the important TeV to PeV range. However, Mannheim et al. (1998) obtain a limit in this energy range which is orders of magnitude higher. By measuring the accompanying fluxes, we may well provide additional clues to the nature and origin of the highest energy cosmic rays. Hence it is important to calculate these fluxes resulting from cascading.

There are several cascade processes which are important for UHE cosmic rays propagating over large distances through a radiation field: protons interact with

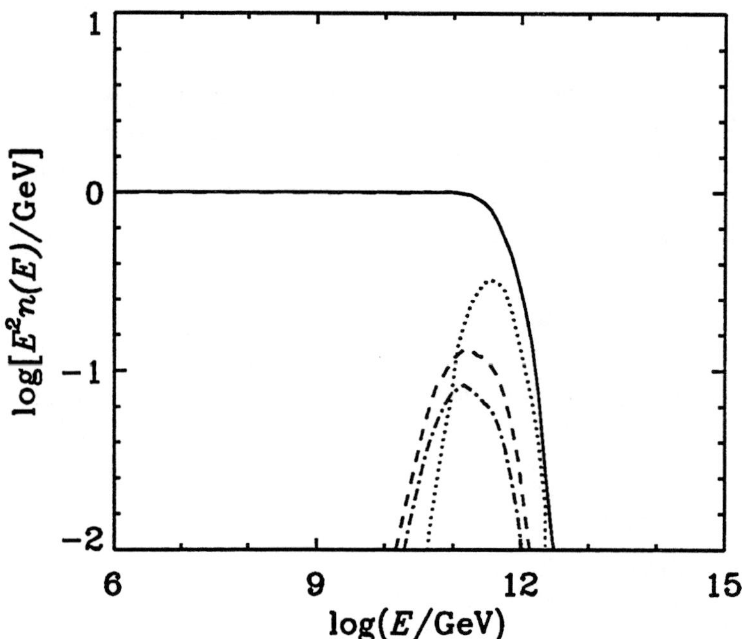

Fig. 10.— The spectrum of particles produced during acceleration (multiplied by E^2) per proton injected into the accelerator: protons (full curve), neutrons (dotted curve), charged pions (dashed curve) and neutral pions. Results are shown for $E_{cut} = 2 \times 10^{12}$ GeV due to photoproduction on the cosmic microwave background. (Adapted from Fig. 7 of Szabo and Protheroe 1994).

photons resulting in pion production and pair production; electrons interact via inverse-Compton scattering and triplet pair production, and emit synchrotron radiation in the intergalactic magnetic field; γ-rays interact by pair production. Energy losses due to cosmological redshifting of high energy particles and γ-rays can also be important, and the cosmological redshifting of the background radiation fields means that energy thresholds and interaction lengths for the above processes also change with epoch (see e.g. Protheroe et al. (1995)).

The energy density of the extragalactic background radiation is dominated by that from the cosmic microwave background at a temperature of 2.73 K. Other components of the extragalactic background radiation are discussed in the review of Ressel and Turner (1990). The extragalactic radiation fields important for cascades initiated by UHE cosmic rays include the cosmic microwave background, the radio background and the infrared–optical background. The radio background was measured over twenty years ago (Bridle 1967, Clarke et al. 1970), but the fraction of this radio background which is truly extragalactic, and not contamination from our own Galaxy, is still debatable. Berezinsky (1969) was first to calculate the mean free path on the radio background. More recently Protheroe and Biermann (1996) have made a new calculation of the extragalactic radio background radiation down to kHz frequencies. The main contribution to the background is from normal galaxies and is uncertain due to uncertainties in their evolution. The mean free path of photons in this radiation field as well as in the microwave and infrared backgrounds is shown in Fig. 11. Also shown is the mean interaction length for muon pair-production which is negligible in comparison with interactions with pair production on the radio background and double pair production on the microwave background.

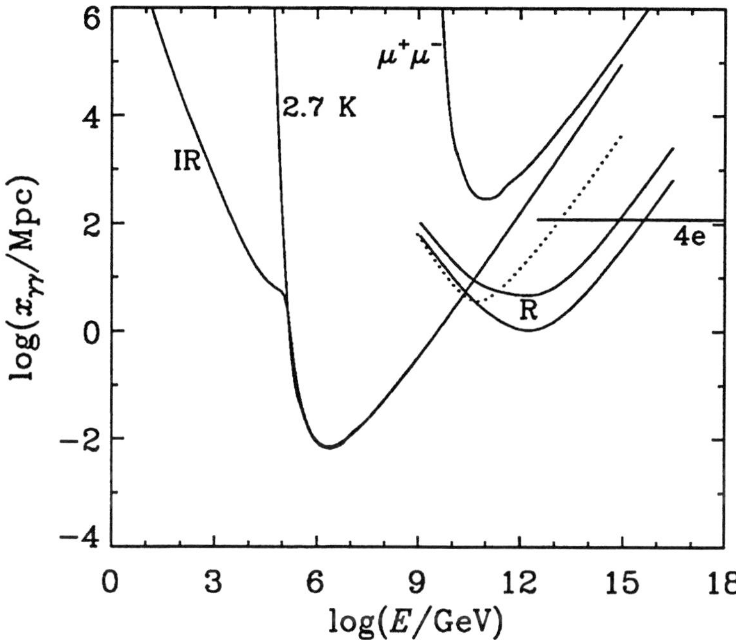

Fig. 11.— The mean interaction length for pair production for γ-rays in the Radio Background calculated in the present work (solid curves labelled R: upper curve – no evolution of normal galaxies; lower curve – pure luminosity evolution of normal galaxies) and in the radio background of Clark (1970) (dotted line). Also shown are the mean interaction length for pair production in the microwave background (2.7K), the infrared and optical background (IR), and muon pair production ($\mu^+\mu^-$) and double pair production (4e) in the microwave background (Protheroe and Johnson 1995). (From Protheroe and Biermann 1996).

Inverse Compton interactions of high energy electrons and triplet pair production can be modelled by the Monte Carlo technique (e.g. Protheroe 1986, Protheroe 1990, Protheroe et al. 1992, Mastichiadis et al. 1994), and the mean interaction lengths and energy-loss distances for these processes are given in Fig. 12. Synchrotron losses must also be included in calculations and the energy-loss distance has been added to Fig. 12 for various magnetic fields.

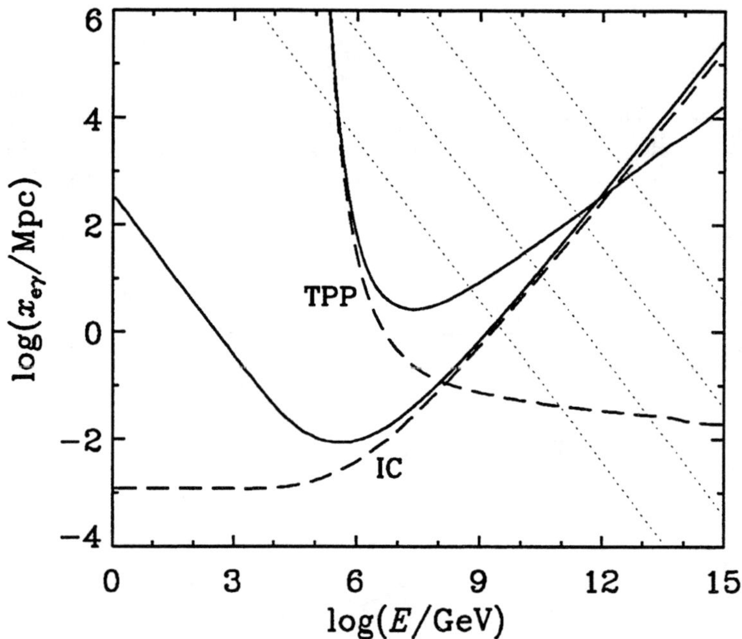

Fig. 12.— The mean interaction length (dashed line) and energy-loss distance (solid line), $E/(dE/dx)$, for electron-photon triplet pair production (TPP) and inverse-Compton scattering (IC) in the microwave background. The energy-loss distance for synchrotron radiation is also shown (dotted lines) for intergalactic magnetic fields of 10^{-9} (bottom), 10^{-10}, 10^{-11}, and 10^{-12} gauss (top). (From Protheroe and Johnson 1995.)

7.1. Practical Aspects of the Cascade

Where possible, to take account of the exact energy dependences of cross-sections, one can use the Monte Carlo method. However, direct application of Monte Carlo techniques to cascades dominated by the physical processes described above over cosmological distances takes excessive computing time. Another approach based on the matrix multiplication method has been described by Protheroe (1986) and developed in later papers (Protheroe and Stanev 1993, Protheroe and Johnson 1995). A Monte Carlo program is used to calculate the yields of secondary particles due to interactions with radiation, and spectra of produced pions are decayed (e.g. using routines in SIBYLL (Fletcher et al. 1994)) to give yields of γ-rays, electrons and neutrinos. For the pion photoproduction interactions a new program called SOPHIA is available (Mücke et al. 1998). The yields are then used to build up transfer matrices which describe the change in the spectra of particles produced after propagating through the radiation fields for a distance δx. Manipulation of the transfer matrices as described below enables one to calculate the spectra of particles resulting from propagation over arbitrarily large distances.

7.2. Matrix Method

In the work of Protheroe and Johnson (1995), fixed logarithmic energy bins were used, and the energy spectra of particles of type α ($\alpha = \gamma, e, p, n, \nu_e, \bar{\nu}_e, \nu_\mu, \bar{\nu}_\mu$) at distance x in the cascade are represented by vectors $F_j^\alpha(x)$ which give the total number of particles of type α in the jth energy bin at distance x. Transfer matrices, $\mathrm{T}_{ij}^{\alpha\beta}(\delta x)$, give the number of particles of type β in the bin j which result at a distance δx after a particle of type α and energy in bin i initiates a cascade. Then, given the spectra of particles at distance x one can obtain the spectra at distance $(x + \delta x)$

$$F_j^\beta(x + \delta x) = \sum_\alpha \sum_{i=j}^{180} \mathrm{T}_{ij}^{\alpha\beta}(\delta x) F_i^\alpha(x) \tag{76}$$

where $F_i^\alpha(x)$ are the input spectra (number in the ith energy bin) of species α.

We could also write this as

$$[\mathrm{F}(x + \delta x)] = [\mathrm{T}(\delta x)][\mathrm{F}(x)] \tag{77}$$

where

$$[\mathrm{F}] = \begin{bmatrix} F^\gamma \\ F^e \\ F^p \\ \vdots \end{bmatrix}, \quad [\mathrm{T}] = \begin{bmatrix} \mathrm{T}^{\gamma\gamma} & \mathrm{T}^{e\gamma} & \mathrm{T}^{p\gamma} & \cdots \\ \mathrm{T}^{\gamma e} & \mathrm{T}^{ee} & \mathrm{T}^{pe} & \cdots \\ \mathrm{T}^{\gamma p} & \mathrm{T}^{ep} & \mathrm{T}^{pp} & \cdots \\ \vdots & \vdots & \vdots & \ddots \end{bmatrix}. \tag{78}$$

The transfer matrices depend on particle yields, $\mathrm{Y}_{ij}^{\alpha\beta}$, which are defined as the probability of producing a particle of type β in the energy bin j when a primary particle of type α with energy in bin i undergoes an interaction. To calculate $\mathrm{Y}_{ij}^{\alpha\beta}$ a Monte Carlo simulation can be used (see Protheroe and Johnson (1995) for details).

7.3. Matrix Doubling

From Fig. 12 we see that the smallest effective interaction length is that for synchrotron losses by electrons at high energies. We require δx be much smaller than this distance which is of the order of parsecs for the highest magnetic field considered. To follow the cascade for a distance corresponding to a redshift of $z \sim 9$, and to complete the calculation of the cascade using repeated application of the transfer matrices would require $\sim 10^{12}$ steps. This is clearly impractical, and one must use the more sophisticated approach described below.

The matrix method and matrix doubling technique have been used for many years in radiative transfer problems (van de Hulst 1963, Hovenier 1971). The method described by Protheroe and Stanev (1993) is summarized below. Once the transfer matrices have been calculated for a distance δx, the transfer matrix for a distance $2\delta x$ is simply given by applying the transfer matrices twice, i.e.

$$[\mathrm{T}(2\delta x)] = [\mathrm{T}(\delta x)]^2. \tag{79}$$

In practice, it is necessary to use high-precision during computation (e.g. double-precision in FORTRAN), and to ensure that energy conservation is preserved after each doubling. The new matrices may then be used to calculate the transfer matrices for distance $4\delta x$, $8\delta x$, and so on. A distance $2^n\delta x$ only requires the application of this 'matrix doubling' n times. The spectrum of electrons and photons after a large distance Δx is then given by

$$[F(x + \Delta x)] = [T(\Delta x)][F(x)] \tag{80}$$

where $[F(x)]$ represents the input spectra, and $\Delta x = 2^n\delta x$. In this way, cascades over long distances can be modelled quickly and efficiently.

8.　The Origin of Cosmic Rays between 100 TeV and 300 EeV

The highest energy cosmic rays show no major differences in their air shower characteristics to cosmic rays at lower energies. One would therefore expect the highest energy cosmic rays to be protons particularly, if as is most likely, they are extragalactic in origin. However, it is possible that they are not single nucleons. Obvious candidates are heavier nuclei (e.g. Fe), γ-rays and neutrinos. In general it is even more difficult to propagate nuclei than protons, because of the additional photonuclear disintegration which occurs (Tkaczyk et al. 1975, Puget et al. 1976, Karakula and Tkaczyk 1993, Elbert and Sommers 1995, Anchordoqui et al. 1997, Stecker and Salamon 1999). The possibility that the 300 EeV event is a γ-ray has been discussed recently (Halzen et al. 1995) and, although not completely ruled out, the air shower development profile seems inconsistent with a γ-ray primary. Weakly interacting particles such as neutrinos will have no difficulty in propagating over extragalactic distances, of course. This possibility has been considered, and generally discounted (Halzen et al. 1995, Elbert and Sommers 1995), mainly because of the relative unlikelihood of a neutrino interacting in the atmosphere, and the necessarily great increase in the luminosity required of cosmic sources. Magnetic monopoles accelerated by magnetic fields in our Galaxy have also been suggested (Kephart and Weiler 1996) and can not be ruled out as the highest energy events until the expected air shower development of a monopole-induced shower is worked out.

The subject of possible acceleration sites of cosmic rays at these energies has been reviewed by Hillas (1984) and Axford (1994), and one of the very few plausible acceleration sites may be associated with the radio lobes of powerful radio galaxies, either in the hot spots (Rachen and Biermann 1993) or possibly the cocoon or jet (Norman et al. 1995). One-shot processes such as magnetic reconnection (e.g. in jets or accretion disks) comprise another possible class of sources (Haswell et al. 1992, Sorrell 1987).

Acceleration at the termination shock of the galactic wind from our Galaxy has also been suggested by Jokipii and Morfill (1985), but due to the lack of any statistically significant anisotropy associated with the Galaxy it is unlikely to be the explanation. However, a recent re-evaluation of the world data set of cosmic rays has shown that there is a correlation of the arrival directions of cosmic rays above 40 EeV with the supergalactic plane (Stanev et al. 1995), lending support to an extragalactic origin above this energy, and in particular to models where "local" sources (< 100 Mpc) would appear to cluster near the supergalactic plane. Such a correlation would also be consistent with a Gamma Ray Burst (GRB) origin as several GRB have now been identified with galaxies.

8.1. Active Galactic Nuclei

Rachen and Biermann (1993) have demonstrated that cosmic ray acceleration in Fanaroff-Riley Class II radio galaxies can fit the observed spectral shape and the normalization at $10 - 100$ EeV to within a factor of less than 10. The predicted spectrum below this energy also fits the proton spectrum inferred from Fly's Eye data (Rachen et al. 1993). Protheroe and Johnson (1995) have repeated Rachen and Biermann's calculation to calculate the flux of diffuse neutrinos and γ-rays which would accompany the UHE cosmic rays, and their result is shown in Fig. 13. The flux of extremely high energy neutrinos may give important clues to the origin of the UHE cosmic rays. (For a review of high energy neutrino astrophysics see Protheroe 1998)

Very recently, Farrar and Biermann (1998) have found a remarkable correlation between the arrival directions of the five highest energy cosmic rays having well measured arrival directions and radio-loud flat spectrum radio quasars with redshifts ranging from 0.3 to 2.2. The probability of obtaining the observed correlation by chance is estimated to be 0.5%. Although the statistical significance is not overwhelming, if further evidence is provided for this correlation the consequences would be far reaching. The distances to these AGN are far in excess of the energy-loss distance for pion photoproduction by protons. Furthermore, given the existence of intergalactic magnetic fields, any charged particle would be significantly deflected and there should be no arrival direction correlation with objects at such distances. Hence, the particles responsible must be stable, neutral, and have a very low cross section for interaction with radiation. Of currently known particles, only the neutrino fits this description, however supersymmetric particles are another possibility. The difficulty of having high energy neutrinos producing the highest energy cosmic rays directly is circumvented if the neutrinos interact well before reaching Earth and produce a particle or particles which will produce a normal looking air shower. Farrar and Biermann (1998) note that, as suggested by Weiler (1998), this may occur due to interactions with the 1.9 K cosmic background neutrinos. Very recent calculations by Yoshida et al (1998) have shown that as a result of such interactions, and subsequent cascading, a flux of ultra high energy γ-rays would result. The clustering of massive relic light neutrinos in hot dark matter galactic halos would give an even denser nearby target for ultra high energy neutrinos as suggested by Fargion et al. (1997).

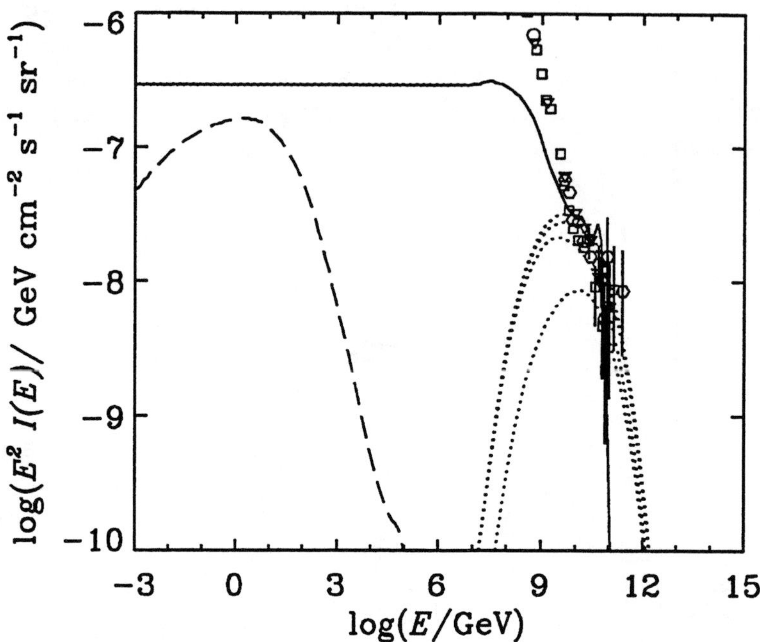

Fig. 13.— Cosmic ray proton intensity multiplied by E^2 in the model of Rachen and Biermann for $H_0 = 75$ km s^{-1} Mpc^{-1} with proton injection up to 3×10^{11} GeV (solid line). Also shown are intensities of neutrinos (dotted lines, $\nu_\mu, \bar{\nu}_\mu, \nu_e, \bar{\nu}_e$ from top to bottom), and photons (long dashed lines). Data are from Stanev (1992); large crosses at EeV energies are an estimate of the proton contribution to the total intensity based on Fly's Eye observations. (From Protheroe and Johnson 1995).

8.2. Gamma ray bursts

Gamma ray bursts (GRB) provide an alternative scenario for producing the highest energy cosmic rays (Waxman 1995, Vietri 1995) , and this possibility has received renewed attention following the discovery that they are cosmological. GRB are observed to have non-thermal spectra with photon energies extending to MeV energies, and in some GRB to much higher energies (GRB 940217 was observed by EGRET up to 18 GeV, Hurley et al. 1994). Recent identification of GRB with galaxies at large redshifts (e.g. GRB 971214 at $z = 3.42$, Kulkarni et al 1998) show that the energy output in γ-rays alone from these objects can be as high as 3×10^{53} erg if the emission is isotropic, making these the most energetic events in the Universe. GRB 980425 has been tentatively identified by Galama et al (1998) with an unusual supernova in ESO 184-G82 at a redshift of $z = 0.0085$ implying an energy output of 10^{52} erg. These high energy outputs, combined with the short duration and rapid variability on time-scales of milliseconds, require highly relativistic motion to allow the MeV photons to escape without severe photon-photon pair production losses. The energy sources of GRB may be neutron star mergers with neutron stars or with black holes, collapsars associated with supernova explosions of very massive stars, hyper-accreting black holes, hypernovae, etc. (see Popham et al 1998, Iwamoto et al 1998 for references to these models).

In the relativistic fireball model of GRB (Meszaros and Rees 1994) a relativistic fireball sweeps up mass and magnetic field, and electrons are energized by shock acceleration and produce the MeV γ-rays by synchrotron radiation. Protons will also be accelerated, and may interact with the MeV γ-rays producing neutrinos via pion photoproduction and subsequent decay at energies above $\sim 10^{14}$ eV (Waxman and Bahcall 1997, Rachen and Meszaros 1998). Acceleration of protons may also take place to energies above 10^{19} eV, producing a burst of neutrinos at these energies by the same process (Vietri 1998). These energetic protons may escape from the host galaxy to become the highest energy cosmic rays (Waxman 1995, Vietri 1995). Additional neutrinos due to interactions of the highest energy cosmic rays with the CMBR will be produced as discussed in the previous section.

8.3. Topological defects

Finally, I discuss perhaps the most uncertain of the components of the diffuse high energy neutrino background, that due to topological defects (TD). In a series of papers (Hill 1983, Aharonian et al. 1992, Bhattacharjee et al. 1992, Gill and Kibble 1994) , TD have been suggested as an alternative explanation of the highest energy cosmic rays. In this scenario, the observed cosmic rays are a result of top-down cascading, from somewhat below (or even much below, depending on theory) the GUT scale energy of $\sim 10^{16}$ GeV (Amaldi 1991), down to 10^{11} GeV and lower energies. These models put out much of the energy in a very flat spectrum of neutrinos, photons and electrons extending up to the mass of the "X–particles" emitted.

Protheroe and Stanev (1996) argued that these models appear to be ruled out by the GeV γ-ray intensity produced in cascades initiated by X-particle decay for GUT scale X-particle masses. The γ-rays result primarily from synchrotron radiation of cascade electrons in the extragalactic magnetic field and were found to just exceed the observed diffuse γ-ray background for a magnetic field of 10^{-9} G and X-particle mass of 1.3×10^{14} GeV if the observable particle intensity was normalized to the UHE cosmic ray data. Thus , for such magnetic fields and higher X-particle masses (e.g. GUT scale), TD cannot explain the highest energy cosmic rays. Indeed there is evidence to suggest that magnetic fields between galaxies in clusters could be as high as 10^{-6} G (Kronberg 1994). However, for lower magnetic fields and/or lower X-particle masses the TD models might explain the highest energy cosmic rays without exceeding the GeV γ-ray limit. For example, Sigl et al. (1997) show that a TD origin is not ruled out if the extragalactic field is as low as 10^{-12} G, and Birkel & Sarkar (1998) adopt an X-particle mass of 10^{12} GeV.

I emphasize that the TD model predictions are *not* absolute predictions, but the intensity of γ-rays and nucleons in the resulting cascade is normalized in some way to the highest energy cosmic ray data. It is my opinion that GUT scale TD models are neither necessary nor able to explain the highest energy cosmic rays without violating the GeV γ-ray flux observed. The predicted neutrino intensities are therefore *extremely* uncertain. Nevertheless, it is important to search for such emission because, if it is found, it would overturn our current thinking on the origin of the highest energy cosmic rays and, perhaps more importantly, our understanding of the Universe itself.

9. Observability of Ultra High Energy Gamma Rays

In at least one of the origin models discussed above, a significant fraction of the "observable particles" at 300 EeV are γ-rays, and so it is appropriate to consider the detectability of such energetic photons. Above 100 EeV the interaction properties of γ-rays in the terrestrial environment are very uncertain. Two effects may play a significant role: interaction with the geomagnetic field, and the Landau-Pomeranchuk-Migdal (LPM) effect (Landau and Pomeranchuk 1953, Migdal 1956).

Energetic γ-rays entering the atmosphere will be subject to the LPM effect (the suppression of electromagnetic cross-sections at high energies) which becomes very important at ultra-high energies. The radiation length changes as $(E/E_{\mathrm{LPM}})^{1/2}$, where $E_{\mathrm{LPM}} = 6.15 \times 10^4 \ell_{\mathrm{cm}}$ GeV, and ℓ_{cm} is the standard Bethe-Heitler radiation length in cm (Stanev et al. 1982). Protheroe and Stanev (1996) found that average shower maximum will be reached below sea level for energies 5×10^{11} GeV, 8×10^{11} GeV, and 1.3×10^{12} GeV for γ-rays entering the atmosphere at $\cos\theta = 1$, 0.75, and 0.5 respectively. Such showers would be very difficult to reconstruct by experiments such as Fly's Eye and at best would be assigned a lower energy.

Before entering the Earth's atmosphere γ-rays and electrons are likely to interact on the geomagnetic field (see Erber (1968) for a review of the theoretical and experimental understanding of the interactions). In such a case the γ-rays propagating perpendicular to the geomagnetic field lines would cascade in the geomagnetic field, i.e. pair production followed by synchrotron radiation. The cascade process would degrade the γ-ray energies to some extent (depending on pitch angle), and the atmospheric cascade would then be generated by a bunch of γ-rays of lower energy. Aharonian et al. (1991) have considered this possibility and conclude that this bunch would appear as one air shower, having the energy of the initial gamma-ray outside the geomagnetic field, being made up of the superposition of many air showers of lower energy where the LPM effect is negligible. If this is the case, then γ-rays above 300 EeV would be observable by Fly's Eye, etc. There is however, some uncertainty as to whether pair production will take place in the geomagnetic field. This depends on whether the geomagnetic field spatial dimension is larger than the formation length of the electron pair, i.e. the length required to achieve a separation between the two electrons that is greater than the classical radius of the electron (see also Stanev and Vankov 1996).

10. Affect of large-scale magnetic structure

At low energies, propagation of cosmic rays through the Galaxy smears out their arrival directions, giving rise to an almost isotropic distribution. However, simulations of cosmic rays through a model of the galactic magnetic field including a turbulent component shows that for energies greater than $\sim Z$ EeV significant anisotropies may be expected from sources in our Galaxy (see, e.g., Lee and Clay 1995). Clearly, at much higher energies the magnetic field of our galaxy plays an insignificant role in smearing directions of extragalactic cosmic ray nuclei. Ultra-high energy cosmic rays can be subject, however, to significant deflection and energy dependent time delays in large scale extragalactic or halo magnetic fields. Lemoine et al. (1997) have performed Monte Carlo simulations of propagation which show how the duration of cosmic ray emission in distant sources has a dramatic effect on the observed spectrum. For example, in one model, a bursting source at 30 Mpc with an E^{-2} spectrum gave a spectrum very strongly peaked at 110 EeV.

Ryu et al. (1998) consider the possibility that the cosmic magnetic field, instead of being uniformly distributed, is strongly correlated with the large scale structure of the Universe. They derive an upper limit to the magnetic field in filaments and sheets of $1\mu G$ which is $\sim 10^3$ times higher than the previously quoted values. Clearly, if such cosmic structures have high magnetic fields this will have important consequences for cosmic ray acceleration and propagation. For example, Tanco (1998b) has considered propagation through well ordered compressed magnetic fields ($\sim 0.1\mu G$) inside cosmological walls and found the cosmic ray flux leaving a wall to be highly anisotropic (by 2 or 3 orders of magnitude), being greatest near the central perpendicular of the wall. Thus, we might expect an anisotropy in the direction perpendicular to the nearest cosmological wall (Centaurus). He also found that arrival directions of CR are smeared, and correlations between γ-rays and CR in bursting events are lost.

Sigl, Lemoine and Biermann (1998) have considered propagation in the local supercluster with $B_{\mathrm{rms}} \sim 0.1\mu G$, with a maximum eddy length of 10 Mpc and a distance to the nearest source of 10 Mpc. For a soft injection spectrum, $E^{-2.4}$, and with both the source and the observer in the local supercluster (sheet with thickness 10 Mpc) they found excellent agreement with the spectrum observed above 10 EeV. Below 100 EeV cosmic rays diffuse, with arrival directions at Earth being spread out over $\sim 80°$ while above 200 EeV they are just deflected but still have a significant spread in arrival directions of $\sim 10°$ explaining the lack of correlation of the highest energy cosmic ray with any known nearby source. Tanco (1998a) noted that the local distribution of luminous matter is far from uniform, and using a redshift distribution based on the CfA redshift catalog they found agreement with the observed super-GZK spectrum. A similar conclusion was reached by Blasi and Olinto (1998).

11. Conclusion

While the origin of the highest energy cosmic rays remains uncertain, there appears to be no necessity to invoke exotic models. Shock acceleration, which is believed to be responsible for the cosmic rays up to at least 100 GeV, is a well-understood mechanism and there is evidence of shock acceleration taking place in radio galaxies (Biermann and Strittmatter 1987), and the conditions there are favourable for acceleration to at least 300 EeV. Such a model, in which cosmic rays are accelerated in Fanaroff-Riley Class II radio galaxies, can readily account for the flat component of cosmic rays which dominates the spectrum above ~ 10 EeV (Rachen et al. 1993). Indeed, one of the brightest FR II galaxies, 3C 134, is a candidate source for the 300 EeV Fly's Eye event (Biermann, 1997).

Whatever the source of the highest energy cosmic rays, because of their interactions with the radiation and magnetic fields in the Universe, the cosmic rays reaching Earth will have spectra, composition and arrival directions affected by propagation. In particular the composition of the arriving particles may differ drastically from that on acceleration. For example, the accompanying fluxes of neutrinos produced in the sources and by cascade processes during propagation to Earth may dominate the spectrum at highest energies. The astrophysics ingredients to the propagation (e.g., magnetic field intensities and configurations) are still somewhat uncertain, and we need better statistics on the arrival directions, energy spectra and composition, as well as the intensity of the diffuse backgrounds of very high energy neutrinos and γ-rays (partly produced in cascades initiated by cosmic ray interactions).

The Auger Project (Auger Collaboration 1995), an international collaboration to build two UHE cosmic ray detectors, one in the United States and one in Argentina, each having a collecting area of about 3000 km^2 , and will measure the energy spectrum and anisotropy of the highest energy cosmic rays with the required precision, and will also be sensitive to extremely high energy neutrinos (Capelle et al. 1998). The next generation neutrino telescopes, such as the planned extension of AMANDA, ICECUBE (Shi et al 1998), and ANTARES (Blanc et al. 1997), may have effective areas of 0.1 km^3, or larger, and be sufficiently sensitive to detect bursts of neutrinos from extragalactic objects and to map out the spectrum of the diffuse high energy neutrino background.

Once we have better statistics of particle fluxes we will be better placed to understand the origin of the highest energy particles occurring in nature. Furthermore, once the astrophysics is worked out, cosmic rays will be the tool for exploring particle physics well above terrestrial accelerator energies.

Acknowledgments

This chapter is based in part on a lecture given at Erice (Protheroe 1996). I thank Qinghuan Luo, Anita Mücke and Peter Biermann for reading the manuscript. My research is supported by a grant from the Australian Research Council.

REFERENCES

Aharonian, F.A., Kanevsky, B.L., and Sahakian, J., *J. Phys. G,* **17** (1991) 1909.

Aharonian, F.A., Bhattacharjee, P., and Schramm, D., *Phys. Rev. D* **46** (1992) 4188.

Allen, G.E., et al., *Astrophys. J. Lett.* **487** (1997) 97

Amaldi, U., de Boer, W., and Fürstenau, H., *Phys. Lett.* **B260** (1991) 447.

Anchordoqui, L.A., Dova, M.T., Epele, L.N., Swain, J.D., *Phys. Rev. D* **57** (1998) 7103.

Auger Collaboration, "The Pierre Auger Project Design Report" (Fermilab 1995).

Axford, W.I., *Ap. J. Suppl.* **90** (1994) 937.

Axford, W.I., Lear, E., and Skadron, G., *Proc. 15th Int. Cosmic Ray Conf., Plovdiv,* **11** (1977) 132.

Axford W.I. *Astrophysical Aspects of Cosmic Rays* ed. M. Nagano and F. Takahara (World Scientific, Singapore, 1991) p. 46.

Baring, M.G., Proc. of XXXIInd Rencontres de Moriond, "Very High Energy Phenomena in the Universe", eds. Giraud-Heraud, Y. & Tran Thanh Van, J., (Editions Frontieres, Paris, 1997), p. 97.

Baring, M.G., Ellison, D.C., Reynolds, S.P., Grenier, I., and Goret, P., *Ap. J.* **513** (1999) in press. astro-ph/9810158

Bednarz, J., Ostrowski, M., *Mon. Not. R. Astr. Soc.,* **283** (1996) 447.

Bednarz, J., Ostrowski, M., *Phys. Rev. Lett.* **80** (1998) 3911.

Bednarz, J., "Proceedings of 16th Europ. Cosmic Ray Symp.", in press (1998).

Bell, A.R., *Mon. Not. R. Astr. Soc.,* **182** (1978) 443.

Berezhko, E.G., and Krymski, G.F., *Usp. Fiz. Nauk* **154** (1988) 49.

Berezinsky, V.S. and Grigor'eva, S.I., *Astron. Astrophys.* **199** (1988) 1.

Berezinsky, V.S., *Yad. Fiz.* **11** (1970) 339; English translation in *Sov. J. Nucl. Phys.,* **11** (1970) 222.

Bhattacharjee, P., Hill, C.T., and Schramm, D.N., *Phys. Rev. Lett.* **69** (1992) 567.

Biermann, P.L., and Strittmatter, P.A., *Ap. J.* **322** (1987) 643.

Biermann, P.L., and Cassellini, J.P., *Astron. Astrophys.* **277** (1993) 691.

Biermann, P.L., *J. Phys. G: Nucl. Part. Phys.* **23** (1997) 1.

Bird, D.J., *et al., Ap. J.* **424** (1994) 491.

Bird, D.J., *et al., Ap. J.* **441** (1995) 144.

Birkel, M., and Sarkar, S., preprint (1998) hep-ph/9804285

Blanc, F., et al., ANTARES proposal (1997) astro-ph/9707136

Blandford, R.D., and Ostriker, J.P., *Astrophys. J. Lett.* **221** (1978) L29.

Blandford, R, and Eichler, D., *Phys. Rep.* **154** (1987) 1.

Blasi, P., Olinto, A.V., submitted to *Phys. Rev. D* (1998) astro-ph/9806264

Brazier, K.T.S., Kanbach, G., Carraminana, A., Guichard, J., and Merck, M., *Mon. Not. R. Astr. Soc.*, **281** (1996) 1033.

Brecher, K., and Burbidge, G.R., *Ap. J.* **174** (1972) 253.

Bridle, A.H., *Mon. Not. R. Astron. Soc.* **136** (1967) 14.

Capelle, K.S., Cronin, J.W., Parente, G., Zas, E., *Astropart. Phys.* **8** (1998) 321.

Clarke, T.A., Brown, L.W. and Alexander, J.K., *Nature* **228** (1970) 847.

Dawson, B.R., Meyhandan, R., Simpson, K.M., *Astropart. Phys.* in press (1998) astro-ph/9801260 .

Donea, A.C., and Biermann, P.L., (1996) personal communication.

Drury, L.O'C., *Space Sci. Rev.* **36** (1983) 57.

Drury, L.O'C., *Rep. Prog. Phys.* **46** (1983) 973.

Drury, L.O'C., Aharonian, F.A., and Völk, H.J., *Astron. Astrophys.* **287** (1994) 959.

Elbert, J.W., and Sommers, P., *Ap. J.* **441** (1995) 151.

Ellison, D.C, Jones, F.C., Reynolds, S.P., *Ap. J.* **360** (1990) 702.

Ellison, D.C, Eichler, D., *Ap. J.* **286** (1984) 691.

Epele, L.N., Roulet, E., *J. High Energy Phys.* **9810** (1998) 9.

Erber, T., *Phys. Rev.* **38** (1968) 626.

Esposito, J.A., Hunter, S.D., Kanbach, G., and Sreekumar, P., *Astrophys. J.* **461** (1996) 820.

Fargion, D., Mele, B., Salis, A., (1997) astro-ph/9710029

Farrar, G.R., and Biermann P.L., *Phys. Rev.* **81** (1998) 3579.

Fletcher, R.S., Gaisser, T.K., Lipari, P., Stanev, T., *Phys. Rev. D* **50** (1994) 5710.

Gaisser, T.K., *Cosmic Rays and Particle Physics*, (Cambridge University Press, Cambridge, 1990).

Gaisser, T.K., Protheroe, R.J., and Stanev, T., *Astrophys. J.* **492** (1998) 219.

Galama, T.J., et al., *Nature*, **395** (1998) 670 astro-ph/9806175

Gill, A.J., and Kibble, T.W.B., *Phys. Rev. D* **50** (1994) 3660.

Greisen, K., *Phys. Rev. Lett.* **16** (1966) 748.

Hayashida, N., et al., *Phys. Rev. Lett.* **73** (1994) 3491.

Hayashida, N., et al., preprint (1998) astro-ph/9807045

Halzen F., Vazquez R.A., Stanev T. and Vankov H.P., *Astroparticle Phys.*, **3** (1995) 151.

Halzen, F., 18th Int. Conf. on Neutrino Phys. and Astrophys. (Neutrino 98), Takayama, Japan (1998) hep-ex/9809025

Haswell, C.A., Tajima, T., and Sakai, J.-L., *Ap. J.* **401** (1992) 495.

Hill, C.T., *Nucl. Phys. B* **224** 469 (1983)

Hill, C.T., and Schramm, D.N., *Phys. Rev. D* **31** (1985) 564.

Hillas, A.M, *Ann. Rev. Astron. Astrophys.* **22** (1984) 425.

Hovenier, J.V., *Astron. Astrophys.* **13** (1971) 7.

van de Hulst, H.C., *A New Look at Multiple Scattering*, Report, NASA Institute for Space Studies, New York, 1963).

Hurley, K., et al., *Nature,* **372** (1994) 652.

Ip, W.-H., and Axford, W.I., in *Particle Acceleration in Cosmic Plasmas*, eds. T.K. Gaisser and G.P. Zank (AIP Conference Proceedings No. 264, 1991) p. 400.

Iwamoto, K., et al., *Nature,* **395** (1998) 672

Jokipii, J.R., and Morfill G.E., *Ap. J. Lett.* **290** (1985) L1.

Jokipii, J.R., *Astrophys. J.* **313** (1987) 842.

Jones, F.C., and Ellison, D.C., *Space Sci. Rev.* **58** (1991) 259.

Karakula, S., and Tkaczyk, W., *Astroparticle Phys.* **1** (1993) 229.

Kephart, T.W., and Weiler, T.J., *Astroparticle Phys.,* **4** (1996) 271.

Kirk, J.G., Schneider, P., *Ap. J.* **315** (1987) 425.

Koyama, K., Petre, R., Gotthelf, E.V., Hwang, U., Matsura, M., Ozaki, M., and Holt, S.S., *Nature* **378** (1995) 255.

Kronberg, P.P., *Rep. Prog. Phys.* **57** (1994) 325.

Krymsky, G.F., *Dokl. Akad. Nauk. SSSR,* **243** (1977) 1306.

Kulkarni, S.R., et al., *Nature,* **393** (1998) 35

Lagage, P.O., and Cesarsky, C.J., *Astron. Astrophys.* **118** (1983) 223

Landau, L.D., and Pomeranchuk, I., *Dok. Akad. Nauk. SSSR* **92** (1953) 535.

Lee, A.A., and Clay, R.W., *J. Phys. G: Nucl. Part. Phys.* **21** (1995) 1743.

Lemoine, M., Sigl, G., Olinto, A.V.,Schramm, D.N., *Astrophys. J. Lett.* **486** (1997) L115.

Mannheim, K., Rachen, J.P., Protheroe, R.J., in preparation (1998)

Markiewicz W.J., Drury L. O'C., and Volk H.J., *Astron. Astrophys.* **236** (1990) 487.

Mastichiadis, A., Protheroe, R.J. and Szabo, A.P. *Mon. Not. R. Astron. Soc.* **266** (1994) 910.

Mastichiadis, A., *Astron. Astrophys.,* **305** (1996) 53.

Melrose, D.B., and Crouch, A.D., *Publ. Astron. Soc. Aust.* **14** (1997) 251.

Meszaros, P., and Rees, M., *Mon. Not. R. Astr. Soc.* **269** (1994) 41P

Mücke, A., Rachen, J.P., Engel, R., Protheroe, R.J., Stanev, T., *Publ. of the Astron. Soc. of Australia*, submitted (1998) astro-ph/9808279

Migdal, A.B., *Phys. Rev.* **103** (1956) 1811.

Norman, C.A., Melrose, D.B., and Achterberg, A., *Astrophys. J.* **454** (1995) 60.

Ostrowski, M., Proc. Vulcano Workshop 1998: "Frontier Objects in Astrophysics and Particle Physics", in press (1998).

Popham, R., Woosley, S.E., and Fryer, C., *Ap. J.*, submitted (1998) astro-ph/9807028

Protheroe, R.J., and Stanev, T.S., *Mon. Not. R. Astron. Soc.* **264** (1993) 191.

Protheroe, R.J., Stanev, T., and Berezinsky, V.S., *Phys. Rev. D15*, **151** (1995) 4134.

Protheroe, R.J., and Biermann, P.L., *Astroparticle Phys.*, **6** (1996) 45.

Protheroe, R.J., *Mon. Not. R. Astron. Soc.* **221** (1986) 769.

Protheroe, R.J., *Mon. Not. R. Astr. Soc.* **246** (1990) 628.

Protheroe, R.J., Mastichiadis A. and Dermer C.D. *Astroparticle Phys.* **1** (1992) 113.

Protheroe, R.J., and Johnson, P.A., *Astroparticle Phys.* **4** (1995) 253 and erratum **5** (1996) 215.

Protheroe, R.J., and Stanev, T., *Phys. Rev. Lett.* **77** (1996) 3708, and Erratum **78**, 3420 (1997)

Protheroe, R.J., in Towards the Millennium in Astrophysics: Problems and Prospects, Erice 1996, eds. M.M. Shapiro and J.P. Wefel (World Scientific, Singapore), in press. astro-ph/9612212

Protheroe, R.J., and Stanev, T., *Astroparticle Phys.*, in press (1998) astro-ph/9808129

Protheroe, R.J., 18th Int. Conf. on Neutrino Phys. and Astrophys. (Neutrino 98), Takayama, Japan (1998) astro-ph/9809144

Puget, J.L., Stecker, F.W., Bredekamp, J.H., *Ap. J.* **205** (1976) 638.

Rachen, J.P., Stanev, T., and Biermann, P.L., *Astron. Astrophys.* **273** (1993) 377.

Rachen, J.P., and Biermann, P.L., *Astron. Astrophys.* **272** (1993) 161.

Rachen, J.P., and Meszaros, P., *Phys. Rev. D*, **58** (1998) 12-30-05 astro-ph/9802280

Reynolds, S.P., *Astrophys. J. Lett.* **459** (1996) L13.

Ressell, M.T., and Turner, M.S., *Comm. Astrophys.* **14** (1990) 323.

Ryu, D., Kang, H., and Biermann, P.L., *Astron. Astrophys.* **335** (1998) 19.

Shi, X., Fuller, G.M., Halzen, F., submitted to *Phys. Rev. Lett.* (1998) astro-ph/9805242

Sigl, G., Lee, S., Schramm, D., and Coppi, P., *Phys. Lett. B* **392** (1997) 129.

Sigl, G., Lemoine, M., Biermann, P.L., *Astropart. Phys.* in press (1998) astro-ph/9806283

Sorrell, W.H., *Ap. J.* **323** (1987) 647.

Stanev, T., Biermann, P.L., Lloyd-Evans, J., Rachen, J.P., and Watson, A.A., *Phys. Rev. Lett.* **75** (1995) 3056.

Stanev, T., in *Particle Acceleration in Cosmic Plasmas*, edited by G.P. Zank and T.K. Gaisser (American Institute of Physics, New York, 1992) p. 379.

Stanev, T., et al., *Phys. Rev. D* **25** (1982) 1291.

Stanev, T., and Vankov, H.P., *Phys. Rev. D.* **55** (1996) (1997) 1365.

Stecker, F.W., "Cosmic Gamma Rays" (Baltimore: Mono Book Co., 1971) p. 133.

Stecker, F.W., and Salamon, M., *Ap. J.* **512** (1999) in press.

Szabo, A.P., and Protheroe, R.J., *Astropart. Phys.* **2** (1994) 375.

Takeda, M., et al., *Phys. Rev. Lett.* **81** (1998) 1163.

Tanco, G.M., *Astrophys. J. Lett.* (1998a) astro-ph/9810366

Tanco, G.M., *Astrophys. J. Lett.* **505** (1998b) 79

Tanimori, T., et al. *Ap. J. Lett.* **497** (1998) 25

Tkaczyk, W., Wdowczyk, J., and Wolfendale, A.W., *J. Phys. A* **8** (1975) 1518.

Totani, T., submitted to *Phys. Rev. Lett.* astro-ph/9810207

Vietri, M., *Ap. J.*, *453* (1995) 883.

Vietri, M., *Phys. Rev. Lett.* **80** (1998) 3690

Waxman, E., *Phys. Rev. Lett.* **75** (1995) 386

Waxman, E., and Bahcall, J., *Phys. Rev. Lett.* **78** (1997) 2292.

Waxman, E., and Bahcall, J., *Phys. Rev. D* submitted (1998) hep-ph/9807282

Weiler, T., *Astroparticle Phys.*, in press (1998) hep-ph/9710431

Yoshida, S., Sigl, G., Lee, S., *Phys. Rev. Lett.* submitted (1998) hep-ph/9808324

Zatsepin, G.T., and Kuz'min, V.A., *JETP Lett.* **4** (1966) 78.

Air shower Cherenkov light measurements using balloon–borne detector

R.A. Antonov[1], D.V. Chernov[1], E.A. Petrova[1], W. Tkaczyk[2]

1 - Institute of Nuclear Physics, Moscow State University, Moscow, Russia e-mail:
antr@dec1.npi.msu.su
2 - Department of Experimental Physics, University of Lodz, Lodz, Poland

Abstract

A new method of making the high energy cosmic ray energy spectrum measurements is presented. The wide-angle balloon-borne small detector is lifted above a snow field during cloudless, moonless nights to make "photos" of the air shower Cherenkov light spots on the snow field. This method makes it possible to have a sensitive area of up to some hundred km^2. The integral flux of the Cherenkov light is a good measure of the primary particle energy. At the current stage of the experiment, the detector is lifted by the tethered balloon to 1 km above the snow field.

1 Method

The main difficulty of the measurements in energy range $10^{15} - 10^{20}$ eV is the very small flow of particles. The detection of Cherenkov light radiated in the atmosphere by the particles of so called extensive air shower (EAS) generated by the primary particle and the detection of the fluorescence light generated in the process of the air atoms and molecules ionization by shower particles are the best methods of particle energy measurement. The Cherenkov and fluorescence light detection is possible only in clear cloudless and moonless nights.

The total flux of Cherenkov light as well as the fluorescence light is proportional to the total ionization energy losses of the shower charged particles in the atmosphere. On the other hand, the ionization losses of the shower constitute about 80% of the primary particle energy of the energy range $10^{15} - 10^{20}$ eV. So the total flux of the Cherenkov light of EAS is proportional to the primary particle energy and the method of energy measurements by detection of the Cherenkov light integral flux is the "calorimetric" one.

The ground based experimental arrays are used for Cherenkov light detection. Such an array consists of many detector points and covers the area up to 100 km^2 in real arrays and up to more then 1000 km^2 in project arrays. The total Cherenkov light flux is not measured by the ground-based arrays but only estimated by light flux density detected at some distance from the EAS core, for example at 400 m as at Yakutsk array. The primary energy evaluation is based on the calculations.

A.E. Chudakov suggested the other way (Chudakov 1972) to detect total Cherenkov light flux. Wide-angle "photo camera" lifted by the balloon or aircraft above snow field will make photos of the air shower Cherenkov light spots on the snow field. Cherenkov light is strongly collimated and radius of the spot is about 0.3–0.5 km. Light is reflected from the snow according to the Lambert reflection law. The observed snow area depends on the detector altitude H as $\sim H^2$.

The correlation between the number of detected Cherenkov photons and the energy of primary particle is very simple if you know the reflection coefficient of the snow and the transparency of the atmosphere that may be measured experimentally. The accuracy of energy measurements is possible to be reached some percent independently of primary particle type. If the altitude of detector H is lower then 3–5 km the calculation must be used to take into account that some part of the Cherenkov light spot of air shower may fall outside the area observed by the detector.

The inclination angles of the shower axis can be determined by the detection of the time structure of the light detector pulses.

The first unaccomplished attempt of such an experiment was undertaken by C. Castagnoli, G. Navarra et al. (Castagnoli 1981)

2   SPHERE experiments

Experiment SPHERE is an attempt to realise the Chudakov's method of very high energy cosmic ray measurements.

SPHERE detector array (Figure 1) was elaborated for balloon–borne experiments at the Institute of Nuclear Physics of the Moscow State University (Antonov 1975, 1986, 1997). Figure 2 shows the scheme of the optical part of the array. The light spots are detected by 19 photomultipliers FEU–110 located on the focal surface of the 1.2-m diameter spherical mirror. The diaphragm is used to diminish the picture distortion caused by the spherical aberration. Dark violet filters and shifters are used with photomultipliers to decrease the influence of the starlight background. The angular aperture of the detector is about $50° \times 50°$. Detector lifted to the altitude H make it possible to have a sensitive area $\sim H^2$. The energy threshold is several times 10^{15} eV at 1 km altitude and rise to 10^{18} eV at $H = 30$ km.

The first measurements with SPHERE array were carried out in the Thien–Shan mountains in winter 1993 (Antonov 1997). Detector was situated on the 160-m high mountain ledge nearby the B. Alma–Ata lake (2500 m above sea level) to detect Cherenkov light reflected from the snow surface of the lake. The area of the lake was about 0.7 km^2. The average inclination angle of the detector optical axis to the horizon was $10°$. Primary cosmic ray flux at the energy 10^{17} eV obtained in this experiment is in agreement with results of other experiments. This experiment showed that the method may be applied for the cosmic ray energy measurements.

During 1995–98 the improved balloon–borne detector was created. The detector electronics measures the integral of light pulse in each of 19 channels during 2.0 μs. Onboard computer controls the electronics and accumulates the data. The parameters of electronics, the temperature inside the box with electronics, the gain and the current of photomultipliers are controlled periodically. In winter 1997–98 the first liftings of the detector by fastened balloon to 1 km level were carried out (Antonov 1998). The primary cosmic ray spectrum in the energy region $10^{16} - 10^{17}$ eV was obtaned in experiment, it is in agreement with other experimental data.

One of the difficulties of the measurements with SPHERE detector is the presence of background of Cherenkov light emitted by charged particles in the glass of the filter and photomultiplier tubes. Another problem is the presence of the starlight background. In the experiments this backgrounds were rejected by the comparison of the experimental data with the laboratory one.

Now SPHERE detector is improved to have the time information to measure the angles of shower axis by detecting the PMT pulse shape. Some steps are taken to reject particle and light background by apparatus but not by analysis.

3   Development of the method

The main aim of the SPHERE detector is to measure cosmic ray energy spectrum in the energy range $10^{15} - 10^{20}$ eV but it may be used for another problem (Antonov 1998). SPHERE detector may serve as a base of prototype of the spacecraft–borne AIRWATCH detector to measure cosmic ray energy spectrum beyond 10^{20} eV by detecting of shower fluorescent tracks of EAS in the atmosphere (Linsley 1997).

Spacecraft–borne detector ($H_{space} = 500\text{--}800$ km) with $S \sim 1\text{--}3$ m^2 mirror and $\sim 10^4$ photodetector pixels will able to have the sensitive area of several times 10 km^2 and to measure the energy of primary particle, its arrival direction and the shape of the EAS cascade curve in the atmosphere. Estimated threshold energy of EAS fluorescent track detection for such a spacecraft detector is $E_{thr} \sim 10^{19}$ eV.

In the balloon-borne experiment the detector with the same mirror area may ensure threshold energy $E_{thr} \sim 10^{18}$ eV and the same angular resolution with the pixel number ~ 100 due to the lower detector altitude ($H_{balloon} = 40$ km vs H_{space}). SPHERE may be used as such a detector for 40-km level balloon–borne experiment. All is need is to increase the light detector pixel number up to ~ 100. Electronics for space detector may be tested in that balloon experiment. Array will rather simple and cheap enough.

About 150 events of energy $> 10^{19}$ eV may be detected during 20-day polar night balloon flight around the Pole at the 40 km altitude above the surface. This event number is compatible with one detected by all experimental arrays by now. The integral flux of fluorescent light as well as the integral flux of Cherenkov light is a good measure of primary particle energy. But due to the anisotropy of Cherenkov radiation the Cherenkov image of EAS is easier to detect than the fluorescent one. Simultaneous use of these two methods (see Figure 3) might increase the methodical accuracy of primary particle energy determination. In this experiment the EAS will look as the fluorescent track that ends with the bright point of the Cherenkov light spot.

The same array elevated to $H = 1\text{--}3$ km makes it possible to study the primary cosmic ray energy spectrum structure in the region $10^{15} - 10^{17}$ eV and the shape of Cherenkov light lateral distribution function that is sensitive to the cosmic ray composition by detection of the Cherenkov light reflected from the snow.

4 Summary

To sum up, there are the main advantages of this method.

1. The method is the calorimetric one. It makes it possible to carry out measurements in wide energy range $10^{15} - 10^{20}$ eV using one balloon–borne detector at different balloon elevations. One or two longtime polar flights of the balloon with detector allow to obtain the same number of events with energy $10^{19} - 10^{20}$ eV as detected at all ground–based arrays by now.

2. Sensitive area in such an experiment is continuous but not discrete as for large ground-based EAS arrays. The results of the measurements do not depend on the shape of Cherenkov light lateral distribution and its fluctuations. It is important especially for extremely high energy EAS ($10^{18} - 10^{20}$ eV).

3. The detector is very small and is not expensive particularly in comparison with ground–based arrays.

4. SPHERE detector may serve as a base of the prototype of the spacecraft-borne detector to measure cosmic ray energy spectrum beyond 10^{20} eV.

References

[1] Antonov, R.A., et al. 1975, *Proc. 14th ICRC, Munich*, **9**, 3360.

[2] Antonov, R.A., et al. 1986, *Izvestiya Academii Nauk*, (Russian), **50**, 2217-2220.

[3] Antonov, R.A., et al. 1997, *Proc. 25th ICRC, Durban*, **4**, 149.

[4] Antonov, R.A., et al. 1998, *Proc. of the NASA workshop "Observing giant cosmic ray air showers from > 10^{20} eV particles from space", Maryland*, 367.

[5] Antonov, R.A., et al. 1998, *Proc. of 16th European Cosmic Ray Symposium*, will be published.

[6] Castagnoli C., et al. 1981, *Proc. 17th ICRC, Paris*, **6**, 103–107.

[7] Chudakov, A.E, 1972, in *Trudy conf. po cosm. lutcham, Yakutsk* (Russian) 69.

[8] Linsley, J., et al. 1997, *Proc. 25th ICRC, Durban*, **5**, 385.

Figure 1: View of the SPHERE array

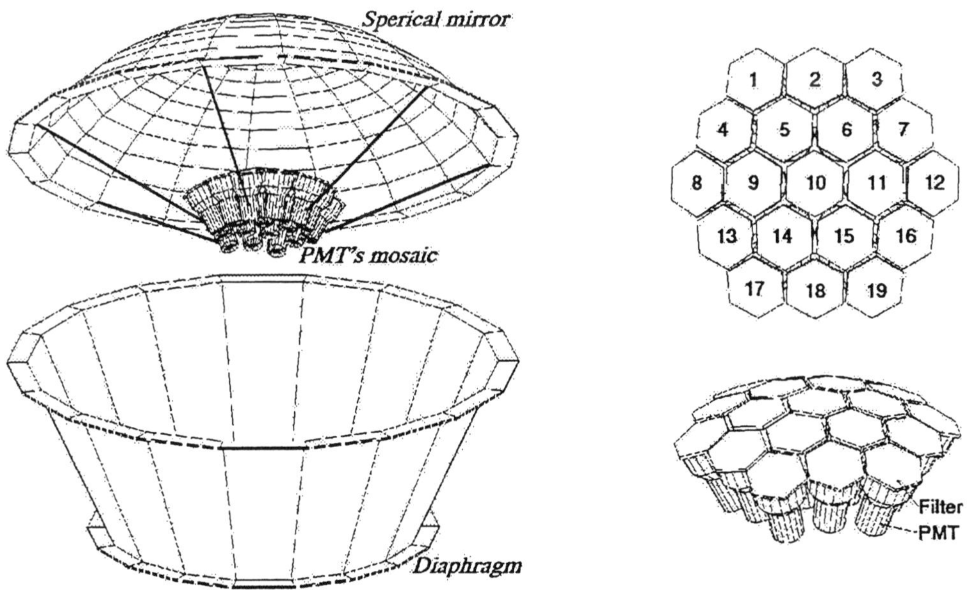

Figure 2: 3-D scheme of the optical part of the SPHERE detector

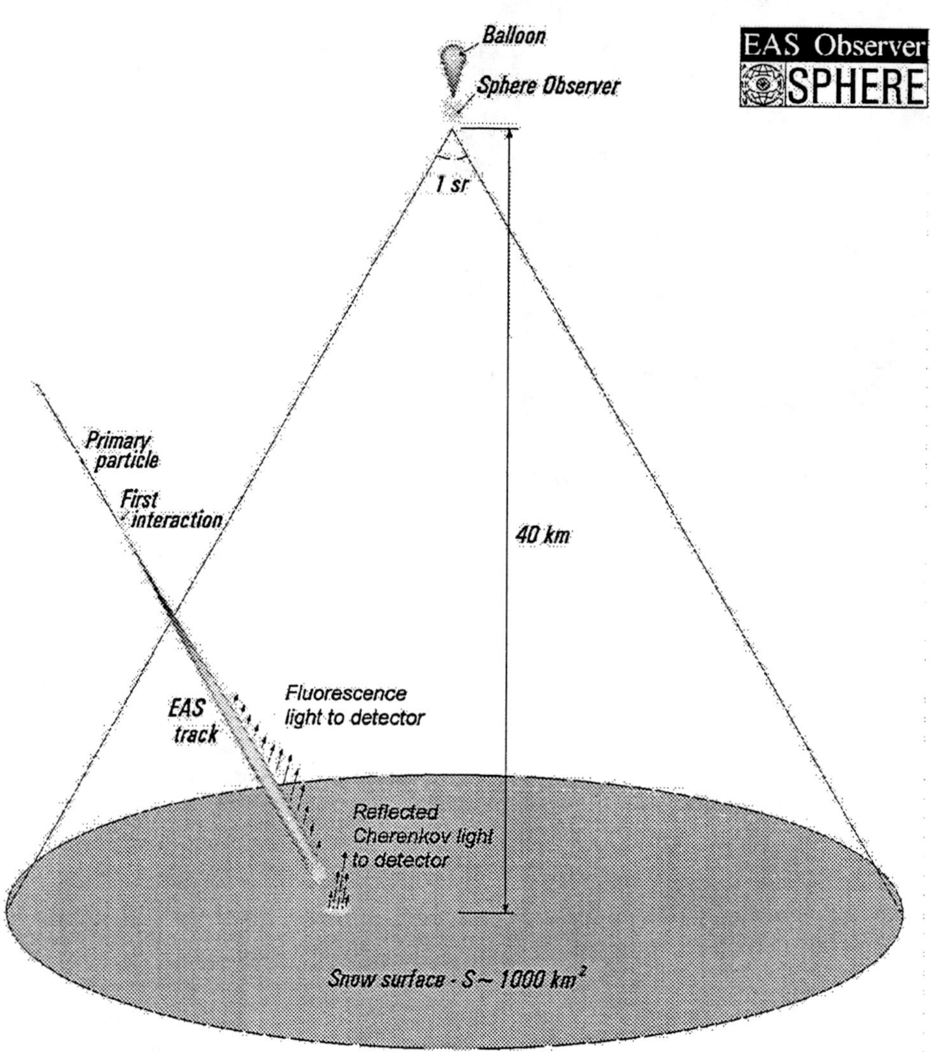

Figure 3: Geometry of the free balloon flight around the Pole.

Ultra high-energy comic rays: probing the local Universe

Gustavo Medina Tanco

Instituto Astronômico e Geofísico - Universidade de São Paulo, Brasil
Department of Physics and Astronomy, University of Leeds, United Kingdom
gustavo@iagusp.usp.br

Abstract

A general view is presented on the problem of propagation of ultra high-energy cosmic rays through the intergalactic and galactic magnetic fields. Especial emphasis is given to the possibility of correlating the present events with potential sources face to the uncertainties in the intervening magnetic fields. Given high enough statistics, the latter problem can be transformed into a powerful tool for the study of cosmic magnetic fields. Finally, the three pairs of events pointed out by the AGASA collaboration as a possible evidence for clustering are analyzed.

Introduction

Ultra high energy cosmic rays (UHECR), have challenged our imagination for the last four decades. They pose in some respects the same kind of questions that gamma ray bursts (GRB) do. They are relatively rare events of, very likely, extragalactic origin with no certain optical counterpart, poor angular determination, undetermined distance scale to the sources and unknown powering mechanism. To complicate things further, depending on the energy threshold we use to define UHECR, the detection rate can be as low as 1 event per square kilometer per century; this means 1 event per year at the largest experiment running at present (AGASA). Even the nature of the primary is uncertain.

So, what do we know? Although other particles cannot be disregarded, the ratio of muon density to charge particle density in the observed showers seems to point to a hadronic primary; protons, in particular, seem to be favored (Gaisser 1993). This information alone immediately sets an upper limit to the distance-scale to the sources: photo-pion production via cosmic microwave background radiation (CMBR) interactions will hinder proton propagation farther than some few tens of Mpc (e.g., Berezinsky and Grigor'eva 1988, Aharonian and Cronin 1994). In fact, one way of defining UHECR is by using the energy threshold for this interaction ($\approx 3 \times 10^{19}$ eV) as a lower limit. The photo-pion interaction with the CMBR should produce a bump followed by a cut-off at the highest energy end of the cosmic ray spectrum. This feature is known as the GZK cut-off (Greisen 1966; Zatsepin and Kuz'min 1966), and its exact position depends on the characteristic distance and distribution of the sources (e.g., Berezinsky and Grigor'eva 1988, Yoshida and Teshima 1993, Medina Tanco 1998c). (The GZK cut-off is a very solid prediction, and should hold as long as Lorentz symmetry is not violated at relativistic factors $\gamma \geq 10^{11}$ Gonzalez-Mestres 1997, 1998.)

Furthermore, charged particles interact with the magnetic field that permeates the propagation region, hampering the location of the sources. Unfortunately, relatively few is known about the galactic magnetic field (GMF) structure in the Milky Way, specially as we go into the Halo (Vallée 1997). Our knowledge of the intergalactic magnetic field (IGMF) is even more speculative and observationally poor, limited mostly to upper limits and few punctual determinations (Arp 1988, Kim et al. 1989, Kronberg 1994, 1996, Vallée 1997). However, commonly adopted fiducial values for the GMF and IGMF are $B_{gal} \approx 10^{-6}$ G and $B_{IGM} \approx 10^{-9}$ G respectively. Therefore, at 10^{19} eV protons have a gyroradius of ≈ 10 kpc in the GMF, which is large enough compared to the thickness of the galactic disk to forbid confinement inside the galaxy, but still large enough as to produce a considerable deflection of the incoming particle. The particles are probably extragalactic, but any information on the location of individual sources is lost. However, at 10^{20} eV protons have a gyroradius of $\approx 10^{2}$ kpc in the GMF, i.e., $\approx 10^{3}$ times the thickness of the galactic disk. Furthermore, the gyroradius of a 10^{20} eV proton in the IGMF is $\approx 10^{2}$ Mpc, i.e., larger than the probable distance to the sources. The latter means that at the highest energies observed, the charged particles are not only very likely extragalactic, but they also point in principle to their sources (Medina Tanco et al. 1997). This opens the remarkable possibility of doing an "UHECR astronomy", with charged particles in a similar way as it is done with photons. A peculiar kind of astronomy, though, as the images of the sources should be strongly coupled with the intervening magnetic field. UHECR represent, in this sense, a potentially powerful tool for the mapping of the GMF in the obscured central regions of our galaxy and the weak and elusive IGMF.

In the following sections we show what should be expected if the sources of the particles are distributed in the same way as the luminous matter in the nearby universe (Medina Tanco 1998c), and analyze the effects of the GMF (Medina Tanco 1997a, Medina Tanco et al. 1998) and different configurations of the IGMF on UHECR propagation (Medina Tanco 1997b, 1998b). Finally, we also study what constrains, if any, can be imposed on the source location problem by the detection of possibly correlated cluster (pairs) of events, suggested by AGASA data (Hayashida et al. 1996, Medina Tanco 1997c, 1998a).

Propagation through the galactic magnetic field

While propagating from the source to the detector, UHECR particles have to cross several different regions from the point of view of the magnetic field: the immediate environment of the source, the intergalactic medium, the galactic halo, the galactic disk, the heliosphere and the magnetosphere. The first region is completely unknown, but mainly irrelevant to us here — one can always redefine a source large enough to enclose any "immediate environment". The last two regions have much larger magnetic fields than the GMF or IGMF, but their size is relatively small and so the deflection they produce is negligible. They can be important, however, as sources of differential deflection which may have relevance for composition determination (e.g., Medina Tanco and Watson 1998, A. Cillis, S. J. Sciutto 1997). Therefore, from the astronomical point of view, only the GMF and IGMF matter.

The magnetic field inside the galactic disk, as far as Faraday rotation measures indicate (see Kronberg 1998 and references therein), appears organized on a grand scale with, possibly, field reversals. There is, however, a random component comparable in intensity to ordered field. Few can be said about the symmetry of the GMF by looking to our own galaxy. Nevertheless, for the more than 20 spiral galaxies with measured large scale magnetic fields (Beck et al 1996), the field usually follows the spiral arms and there is also evidence for either an axisymmetric or a bisymmetric pattern, and even for a combination of both. The GMF seems to be compatible with these topologies but, at the present stage, choosing between one model and another is just a working hypothesis.

Measurements of the galactic magnetic halo are also very difficult, but they suggest a scale height of $\approx$ 4 kpc and an intensity of $\approx$ 0.1 μG at and outside the solar circle (Kronberg 1998). Comparable or larger scale heights are observed in a few nearby edge-on spiral galaxies (Hummel, Beck and Dahlem 1991).

Therefore, to described the regular large scale component of the GMF, we adopt below the axisymmetric model as Stanev (1997), including a component perpendicular to the galactic plane, plus a superimposed random component of variable correlation length L_c, and amplitude $\delta B/B \approx 1$ Abramenkov and Krymkin 1990):

$$\vec{B}_{GMF}(\vec{r}) = \vec{B}_{reg}(\vec{r}) + B_z \times \hat{e}_z + \vec{B}_{RND}(\vec{r})$$

where the correlation length is taken as L_c = 100 pc in the Sun's vicinity, and scales throughout the galaxy according to $L_c(r) \propto [B(r)]^{-2}$.

The principle of reversibility is applied to the propagation of protons through the GMF. Approximately 10^4 antiprotons are uniformly injected at Earth with momentum '-p' over 4π sr, and their trajectories are integrated until the interface between halo and IGM, represented by a spherical galactocentric surface of radius R_H = 20 kpc, is reached. This schematically represented in figure 1. In this way, the arrival directions of the UHECR onto the border of the galactic halo (which, if the IGMF is neglected, point to the true location of the sources) can be mapped onto the celestial sphere seen by the observer. Two maps can be constructed in this way, one with the actual position of the sources on the sky and another with the apparent position of their images in the sky seen by the detector due to GMF deflection (for further details and several examples see Medina Tanco 1997a, Medina Tanco et al. 1998).

In figure 2a-b we show the resulting source location error boxes as a function of galactic coordinates for incoming protons, with energies 4×10^{19} and 10^{20} eV respectively, due to the GMF. It can be seen that the effects are strongly dependent on the direction on the sky and the particle's arrival energy, ranging from almost negligible at the galactic polar regions to severe towards the galactic center. Consequently, identification of extragalactic sources near the inner galactic plane and bulge should be much hampered unless protons of $> 10^{20}$ eV are used. On the other hand, this relatively large deflections mean that, once a source or distribution of sources is identified, UHECR could be used map the presently poorly known GMF in the central regions of our galaxy. Probably, even an unexpectedly high magnetic field in the halo (Hillas 1998) should produce a recognizable signature on top of a distribution of sources related with an independently known spatial distribution (like the distribution of luminous matter in the nearby Universe).

The effects are substantially larger if UHECR comprise heavier nuclei (Medina Tanco 1997a).

Propagation through the intergalactic magnetic field

The main problem lies in our lack of knowledge of the IGMF, with the exception of some few observational determinations and upper limits (e.g., Arp 1988, Kim et al. 1989, Kronberg 1994) and numerical simulations of cosmological structure formation (Biermann 1996 and references there in). It seems, however, that fields ≈ 0.1 μG are generally associated with the presence of intergalactic hot gas and galaxies (Kronberg, 1998), whereas field in the general intergalactic medium and in the interior of voids remains basically undetected. On the other hand, evidence from rotation measure in galaxy clusters suggests that the largest reversal scale is $L_c \approx 1$ Mpc. This, together with upper limits in rotation measure out to $z \approx 2.5$ indicates an upper limit for the IGMF, $B_{IGMF} \leq 10^{-9}$ G (Kronberg, 1994). The latter result is also compatible with Vallée's (1990) determination of $B_{IGMF} < 10^{-9}$ G for a regular cosmic magnetic field (outside clusters of galaxies) and mean particle density 10^{-7} cm^{-3}.

Furthermore, for those spatial scales where measurements are available, the intensity of astrophysical magnetic fields seems to correlate remarkably well with the density of thermal gas in the medium. Figure 3 (an adaptation of figure 1 of Vallée 1997) shows that we have probably two regimes: one at small ($\leq$ galactic) scales, and other at large (extragalactic) scales. It is apparent that B can be reasonably well fitted by a single power law over $\approx$ 14 orders of magnitude in thermal gas density at sub-galactic scales. Besides, a correlation is also suggested at very large scales from, galactic halos to the environments outside galactic clusters, over $\approx$ 4 orders of magnitude in thermal gas density.

Thus, the observational clues/constrains given heretofore point to a possible picture in which the IGMF correlates with the distribution of matter as traced, for example, by the distribution of galaxies. A high degree of non-homogeneity should be expected, with relatively high values of B_{IGMF} over small regions ($\approx$ 1 Mpc) of high matter density (c.f., Arp's 1988 determination of $B_{IGMF} \approx 3 \times 10^{-7}$ G for the Virgo cluster or Kim's et al. 1989 $B_{IGMF} \approx 10^{-6}$ G for the Coma cluster). These systems should be immersed in vast low density/low B_{IGMF} regions with $B_{IGMF} < 10^{-9}$ G. Furthermore, from rotation measure, the topology of the field should be such that it is structured coherently on scales of the order of the correlation length L_c which, in turn, scales with IGMF intensity: $L_c \propto B_{IGMF}^{-2}(\mathbf{r})$. B_{IGMF} vectors should be independently oriented at distances $> L_c$. Therefore, a scenario can be built in which the IGMF presents a cell-like spatial structure, with cell size given by the correlation length, L_c; and such that: $L_c \propto B_{IGMF}^{-2}(\mathbf{r})$ and $B_{IGMF} \propto \rho_{gal}^{0.35}(\mathbf{r})$, where ρ_{gal} is the galaxy density, and the IGMF is uniform inside cells of size L_c and randomly oriented with respect to adjacent cells (Medina Tanco 1997b, Medina Tanco et al 1997). The observed IGMF value at a given point, like the Virgo cluster ($\approx 10^{-7}$ G, Arp 1988), can be used as the normalization condition for the magnetic field intensity.

The density of galaxies, ρ_{gal}, is estimated using the latest release (version of Jul 27, 1998) of the CfA Redshift Catalogue (Huchra et al 1992). Figure 4 shows an Aitoff (equal area) projection of the galaxy distribution up to a depth of 5100 km/sec (equivalent to $\approx$ 100 Mpc for H=50 km/sec/Mpc).

The spatial distribution of the sources of UHECR is tightly linked to the nature of the main acceleration mechanism involved.

Two kinds of acceleration mechanisms can be envisaged for the generation of UHECR: bottom-up and top-down mechanisms. Bottom-up mechanisms consist in the acceleration particles injected at lower energies. Some examples are: particle acceleration in the accretion flows of cosmological structures (e.g., Norman, Melrose and Atcherberg 1995, Kang, Ryu and Jones 1996), galaxy collisions (Cesarsky and Ptuskin 1993, Al-Dargazelli et al. 1997, but see Jones 1998), galactic wind shocks (Jokipii and Morfill 1987), pulsars (Hillas 1984, Shemi 1995), active galactic nuclei (Biermann and Streitmatter 1987), powerful radio galaxies (Rawlings and Saunders 1991, Biermann 1998), gamma ray bursts (Vietri 1995, 1998, Waxmann 1995, but see Stanev, Schaefer and Watson 1996), etc.. Top-down mechanisms, already form the particles at high energies and simply cascade down their energy up to the observed values, avoiding in that way severe problem like energy losses during acceleration. Some examples are: the decay of topological defects into super-heavy gauge and Higgs bosons, which then decay into high energy neutrinos, gamma rays and nucleons with energies up to the GUT scale ($\approx 10^{25}$ eV) (e.g., Bhattacharge, Hill and Schramm 1992, Sigl, Schramm and Bhattacharge 1994, Berezinsky, Kachelrie and Vilenkin 1997, Berezinsky 1998), high energy neutrino annihilation on relic neutrinos (Waxmann 1998), etc.

In the case of bottom-up mechanisms the sources of the particles should be related to the distribution of luminous matter in the Universe. For top-down mechanisms, on the other hand, the distribution of the sources may or may not be related to the distribution of luminous matter. However, an isotropic distribution of sources is expected in most of the models.

In what follows, we consider a distribution of sources that closely resembles the distribution of luminous matter in the nearby universe, as represented by the observed ρ_{gal}, and assume that the scenario described above is an acceptable representation of the IGMF up to a distance 100 Mpc from our galaxy. Protons are injected at the sources (one source at the location of each galaxy) with a power law injection spectrum, $dN_{inj}/dE \propto E^{-\nu}$, with $\nu = 3$ above a threshold of 4×10^{19} eV. Each source is a standard candle, i.e., the luminosity in UHECR, L_{UHECR} = const. The 3D trajectories of the particles are calculated from the sources to the detector, taking into account adiabatic energy losses due to redshift, pair production and photo-pion production due to interactions with the CMBR. The interaction mean free times $[(1/E)dE/dt]^{-1}$ as a function of energy calculated by Berezinsky and Grigor'eva (1988), are used to determine, randomly, the interaction of protons with CMBR photons during the propagation and the subsequent evolution of UHECR particle energy.

Figure 5 shows an all sky Aitoff projection, in galactic coordinates with the galactic anticenter at the coordinate origin, of the arrival probability distribution of UHECR protons with energy $E_{arrival} > 4 \times 10^{19}$ eV. The available data on UHECR for the same energy interval is also displayed in the same figure for the experiments of AGASA, Haverah Park, Yakutsk and Volcano Ranch. The thick strip at zero galactic latitude covers the position of the galactic plane in the figure. Obscuration by dust in this region severely hinders the observation of galaxies, and so our approximation does not apply.

In figure 6, the median of the deflection (defined as the angle between the arrival direction of a particle and the true direction to its source) as a function of energy is shown, together with 63% and 95% confidence levels. It can be seen that, despite de fact that the median decreases strongly with increasing energy, there is considerable dispersion and an appreciable fraction of particles can arrive with large angular deflections ($> 20°$) at 10^{20} eV.

At the present low level of statistics, it is difficult to say something conclusive about any possible correlation; however, the angular distribution of observed events seems very much isotropic, in contrast to the inferred arrival probability. It is premature to say whether this is telling something about the distribution of the sources or the structure of the IGMF (c.f., Hillas 1998).

Table I. Possible Clusters of UHECRs observed by AGASA experiment (adapted from Hayashida et al, 1996).

Pair No.	Date	Δt_{arr} [yr]	Energy [EeV]	*Type*	l^g	b^g
1	93/12/03	1.90	210	A	131.2	-41.1
	95/10/29		51		130.2	-42.3
2	92/08/01	2.49	55	B	143.5	56.9
	95/01/26		78		145.8	55.3
3	91/04/20	3.21	43	B	77.9	18.6
	94/07/06		110		77.6	21.1

The first identification of UHECR sources?

Undoubtedly, if a point-like UHECR source exist (i.e., spatially confined to less than the angular resolution of the current experiments, $\approx 1°$) at a distance of no more than a few photo-interaction mean-free paths, it will eventually emerge observationally as the accumulation of events in a relatively small solid angle. How small this solid angle should be to be statistically significant is still open to debate. However, it is very exciting that some few "hot spot" may be already appearing in the data, as reported some time ago by the AGASA collaboration (Hayashida et al., 1996). Table I reproduce the three published pairs of events. Actually, there may be as much as 8 pairs of events and at least two triples if data from other experiments are combined. Nevertheless, based on past experiences of a long history of possible GRB repeaters (e.g., Fishman 1996, Meegan et al. 1995, Hartmann et al. 1995, Brainerd et al., 1995), caution must be exercised when associating two events only by their angular proximity on the sky, specially when the data is so scarce.

If UHECR are charged particles, protons as it is more likely, and the components of the pairs have a common origin, then the observed clustering impose severe constraints on the characteristics of the propagation region and/or their sources (e.g., Cronin 1996, Sigl et al 1996). Catastrophic extragalactic events, like GRB or the decay of topological defects (TD), which are able to produce the particles over a very short period of time, should only be consistent with the data for a suitable combination of low intergalactic magnetic field (IGMF) and distance to the source. Nevertheless, the stirring of the intergalactic medium (IGM) by large agglomerates of galaxies, shocks excited in binary collisions of galaxies or the bow shocks preceding fast moving galaxies in dense environments are examples of quiescent sources that could produce chance pairings of UHECR events on the sky. If these quiescent sources are traced by the distribution of luminous matter in the nearby universe, then the probability of the corresponding chance pairing can be estimated and compared with the observations.

In order to analyze the pairs, the propagation problem can be divided into two different calculations: one involving the IGMF, from the source to the border of the galactic halo and the other the GMF, through the galactic halo and galactic plane to detector (see Medina Tanco 1998a for details).

First, the trajectories of the individual particles through the GMF are calculated using the model described above (Medina Tanco et al. 1998, Medina Tanco 1997a). If the particles are simultaneously released at the source, this constrains the amount of time delay due to intergalactic propagation alone and, consequently, the range of IGMF values and source distances allowed. The separation angle between the momenta of the particles at their arrival at the border of the halo, θ_{HALO} ,can also be estimated. The latter can be seen as a matching condition to be met by the particle trajectories at the border of the halo.

Second, the same numerical scheme of Medina Tanco et al (1997) is used to estimate the arrival relative-deflection distribution function for some allowed combinations of IGMF and distance to the source. The comparison of this distribution function with the previously calculated θ_{HALO} , gives a quantitative idea of the likelihood of the observed events being the result of point-like sources.

Pair 1 is the only one in which the high energy event arrives first. Therefore, it could be produced, in principle, in a burst. Figure 7.a shows the distribution function of arrival time delays, Δt_{arr}, at the border of the halo for protons of the observed energies. It is assumed that the particles originated with a power law energy spectrum at the source, $dN_{inj}/dE \propto E^{-2}$. Two possible combinations of IGM and source distance are shown which would be able to reproduce the observed $\Delta t_{arr} \approx 1.3$ yr at the border of the halo.

In figure 7.b, the distribution functions of the angle between the momenta of the pair of paricles at the border of the halo, θ_{pp}, are shown for the same cases of figures 7.a. It can be seen that if both, Δt_{arr} and θ_{pp} are to be satisfied simultaneously, then a very rear event was observed.

On the other hand, in pairs 2 and 3 the lower energy particle arrives first. This means that a point source cannot have emitted both UHECR simultaneously. Therefore, if the point source hypothesis is to be maintained, we must assume either that the source is quiescent or, if bursting, that a finite acceleration time is involved which delays the emission of the high energy component. In this case, the sum of the arrival time delay, the time delay due to the propagation through the galactic disc and halo, and the time delay due to intergalactic portion of the trajectories, is a lower limit to the lifetime of the source. Again, the galactic and intergalactic trajectories must verify the matching of θ_{HALO}

at the border of the galactic halo. Applying the same procedure as for pair 1, it is found that θ_{HALO}(pair 2) $\approx 2°$ and θ_{HALO}(pair 3) $\approx 2°\text{-}5.5°$, depending on the model adopted for the galactic magnetic field. Numerical simulations for the IGM propagation of the proton components of pairs 2 and 3 are also shown in figures 7.a and 7.b. A distance of 30 Mpc to the source and two different values of the IGMF, $B_{IGM} = 10^{-12}$ and 10^{-9} Gauss, were used. The lower value of the IGMF is the one imposed by a bursting pair 1, and the second is the current upper limit for the IGMF. It can be seen from figures 7.a and 7.b that, as for pair 1, the assumption of a 10^{-12} Gauss IGMF leads to a very low probability for an event with θ_{HALO} on the order of a few degrees. Taking into account the galactic propagation, the lower limits for the lifetime of single sources for pairs 2 and 3, with $B_{IGM} \approx 10^{-12}$ Gauss, are ≈ 10 and ≈ 100 yr respectively. On the other hand, from the point of view of θ_{HALO}, a consistent picture can be obtained for a higher value of the IGMF, say near 10^{-9} Gauss. However, the lower limit for the lifetime of a single source scales up to few times 10^5 yr and so the source should be quiescent. Furthermore, the sources should probably extend over a large volume of space, perhaps enclosing more than one galaxy, in order to be able to confine $\approx 10^{20}$ eV particles.

The previous results seem to point to a chance clustering of the three proposed pairs of events. However, the problem remains that the chance probability for the pairs quoted by Hayashida et al (1996) is only 2.9%. We note, however, that this chance probability was derived under the assumption that the arrival direction distribution is uniform over the sky, which is arguable.

In fact, under the assumption that the sources are distributed following ρ_{gal} , the arrival probability is by no means isotropic. A clearer picture can be obtained if maps similar to figure 5 are produced for sources at different distance intervals (Medina Tanco 1998a). It is possible to see then that pair 2 is on top of a maximum of the arrival probability for sources located between 20 and 50 Mpc, while pair 1 is also located on a high arrival probability region for sources at more than 50 Mpc. This is in contrast with the chance probability estimated by Hayashida et al. (1996), and points to either different non-correlated sources of the components of each pair, or to very extended quiescent sources involving several galaxies or clusters.

The third pair, on the other hand, comes from a region of space where no large clustering of galaxies exist up to the depths considered. As the components cannot have originated simultaneously at the same extragalactic source because of galactic propagation constraints, they must have come from isolated sources. This could be interpreted as an indicative that very large agglomerates of galaxies working co-operatively in the acceleration process are not needed in order to accelerate UHECR.

Conclusions

UHECR are obviously an extremely interesting object of research in itself. They are also, undoubtedly, a promising tool for probing the GMF and the IGMF. They can be especially powerful in helping to understand the large scale of the magnetic field in the central regions of our own galaxy and in the galactic halo. Furthermore, they can be the only way to measure the average intensity and to study the spatial structure of the IGMF inside walls and voids at scales of tens of Mpc. Unfortunately, the available data at present seems insufficient to reach any definite conclusion about the UHECR origin. More data is badly needed, and just some few more years at the present detection rate will not do. The more attractive applications will require a change of orders of magnitude in our capacity of gathering data. Hopefully, the Southern site of the Auger experiment should be operative in few years time, and OWL is on the run for the next century (Streitmatter, 1998).

Finally, it must be stressed that all the previous analysis must be severely compromised if the IGMF were coherently structured on large scales ($\approx$ 10-20 Mpc) with an IGMF on the order of $\approx \mu G$ inside cosmological walls and almost negligible in the surrounding voids (see Medina Tanco 1998b) as recently proposed by Ryu, Kang and Biermann (1998).

ACKNOWLEDGMENTS:

I wish to thank the kind hospitality of the cosmic rays group at the University of Leeds (UK), and to Prof. A. A. Watson so many discussions and valuable comments.

References

Abramenkov and Krymkin 1990, in Galactic and Intergalactic Magnetic Fields, R. Beck, Kronberg P. and Wielebinski (eds), Kluwer Acad. Publ., Dodrecht, p 49.

Aharonian F. A. and Cronin J. W., 1994, Phys. Rev. **D50**, 1892.

Al-Dargazelli S. S., Lipiski M., Smialkowski A., Wdowdczyk J, Wolfendale A., 1997, 25th ICRC (Durban), 4, 465.

Arp H., 1988, Phys. Lett. A, **129**, 135.

Beck R., Brandemburg A., Moss D., Shukurov A and Sokoloff D., 1996, Ann. Rev. Astron. Astrophys., **34**, 155.

Berezinsky V., 1998, hep-ph/9802351.

Berezinsky V., Kachelrie M. and Vilenkin A., 1997, astro-ph/9708217.

Bhattacharjee P., Hill C. T. and Schramm D. N., 1992, Phys. Rev. Letters, **69**, 567.

Biermann P. L., 1998, Workshop on Observing Giant Cosmic Ray Air Showers, J. F. Krizmanic, J. F. Ormes and R. E. Streitmatter (eds.), AIP Conf. Proc., 433, 22.

Biermann P. L., Kang H., Ryu D., 1996, Proc. International Symp. on EHECR: Astrophys. And Future Obs., M. Nagano (ed.), Institute for CR Research, University of Tokyo, p.79.

Biermann P. L. and Streitmatter P. A. 1987, Ap. J., 322, 643.

Berezinsky V. S. and Grigor'eva S. I., 1988, Astron. Astrophys., **199**, 1.

Brainerd J. J., et al., 1995, Ap.J.

Cesarsky C. and Ptuskin V., 1993, 23rd ICRC (Calgary), 2, 341.

Cillis A. and Sciutto S. J. ,1997, astro-ph/9712345.

Cronin J. W., 1996, Proc. International Symp. on EHECR: Astrophys. And Future Obs., M. Nagano (ed.), Institute for CR Research, University of Tokyo, p.2.

Fishman G. J., 1996, in Compact Stars in Binaries, J. van Paradijs et al. (eds.), IAU, Netherlads, p. 467.

Gaisser T. K. et al., 1993, Phys. Rev. **D47**, 1919.

Gonzalez-Mestres L., 1997, Proc. 25 ICRC, **6**, 113.

Gonzalez-Mestres L., 1998, Proc. Workshop on Observing Giant Air Showers from Space, J. F. Krizmanic (ed.), p. 148.

Greisen K., 1966, Phys. Rev. Lett., **16**, 748.

Hartmann D. et al., 1995, Ap. J.

Hayashida et al 1996, Phys. Rev. Lett, **77**, 1000.

Hillas A. M., 1984, Ann. Rev. Astron. Astrophys., **22**, 425.

Hillas A. M., 1998, Nature, **395**, 15.

Huchra, J. Geller, M., Clemens, C., Tokarz, S and Michel, A. 1992, Bull. C.D.S., **41**, 31.

Hummel E., Beck R. and Dahlem M., 1991, A&A, **248**, 23.

Jokipii J. R. and Morfill G., 1987, Ap. J., 312, 170.

Jones F. C., 1998, Workshop on Observing Giant Cosmic Ray Air Showers, J. F. Krizmanic, J. F. Ormes and R. E. Streitmatter (eds.), AIP Conf. Proc., 433, 37.

Kang H., Ryu D. and Jones T. W., 1996, Ap.J., 456, 422.

Kim K.-T., et al., 1989, Nature, **341**, 720.

Kronberg P.P., 1994, Rep. Prog. Phys., **57**, 325.

Kronberg P. P., 1996, Proc. International Symp. on EHECR: Astrophys. And Future Obs., M. Nagano (ed.), Institute for CR Research, University of Tokyo, p.89.

Kronberg P. P., 1998, Proc. Workshop on Observing Giant Air Showers from Space, J. F. Krizmanic (ed.), p. 199.

Medina Tanco G. A. 1997a, Proc. 25th ICRC, South Africa, **4**, 485.

Medina Tanco G. A. 1997b, Proc. 25th ICRC, **4**, 477.

Medina Tanco G. A. 1997c, Proc. 25th ICRC, **4**, 481.

Medina Tanco G. A., 1998a, Ap. J. Lett., **495**, L71-L74.

Medina Tanco G. A., 1998b, Ap. J. Lett., in press.

Medina Tanco G. A., 1998c, Phys. Rev. Lett., submitted.

Medina Tanco G. A., Gouveia Dal Pino E. M., Horvath J. E., 1997, Astroparticle Phys., **6**, 337.

Medina Tanco G. A., Gouveia Dal Pino E. M., Horvath J. E., 1998, Ap. J., **492**, 200

Medina Tanco G. A. and Watson A. A., 1998, Astropart. Phys., submitted.

Meegan C. et al., 1995, ApJ.

Norman C. A., Melrose D. B. and Atcherberg A., 1995, Ap.J., 454, 60.

Rawlings S. and Saunders R., 1991, Nature, 349, 138.

Ryu D., Kang H. and Biermann P. L., 1998, A&A, **335**, 19.

Stanev T., 1997, Ap J, **479**, 290.

Stanev T., Schaefer R. and Watson A. A., 1996, Astroparticle Phys., 5, 75.

Shemi A., 1995, MNRAS, 275, 115.

Sigl G., Schramm D. N. and Bhattacharge P., 1994, 2, 401.

Sigl G., Schramm D. N., Lee S., Coppi P., and Hill C. T., 1996, FERMILAB-Pub-96/121-A

Streitmatter R. E., 1996, Proc. International Symp. on EHECR: Astrophys. And Future Obs., M. Nagano (ed.), Institute for CR Research, University of Tokyo, p.95.

Vallée J. P., 1990, Ap. J., **360**, 1.

Vallée J. P., 1997, Fundamental of Cosmic Phys., {\bf 19}, 1.

Vietri M., 1995, Ap. J., 453, 883.

Vietri M., 1998, Workshop on Observing Giant Cosmic Ray Air Showers, J. F. Krizmanic, J. F. Ormes and R. E. Streitmatter (eds.), AIP Conf. Proc., 433, 50.

Waxmann E., 1995, Ap.J., 452, L1.

Waxmann E., 1998, astro-ph/9804023.

Yoshida S. and Dai H., 1998, J. Phys. G: Nucl. Part. Phys., **24**, 905.

Zatsepin G. T. and Kuz'min V. A., 1966, JETP Letters, **4**, 78.

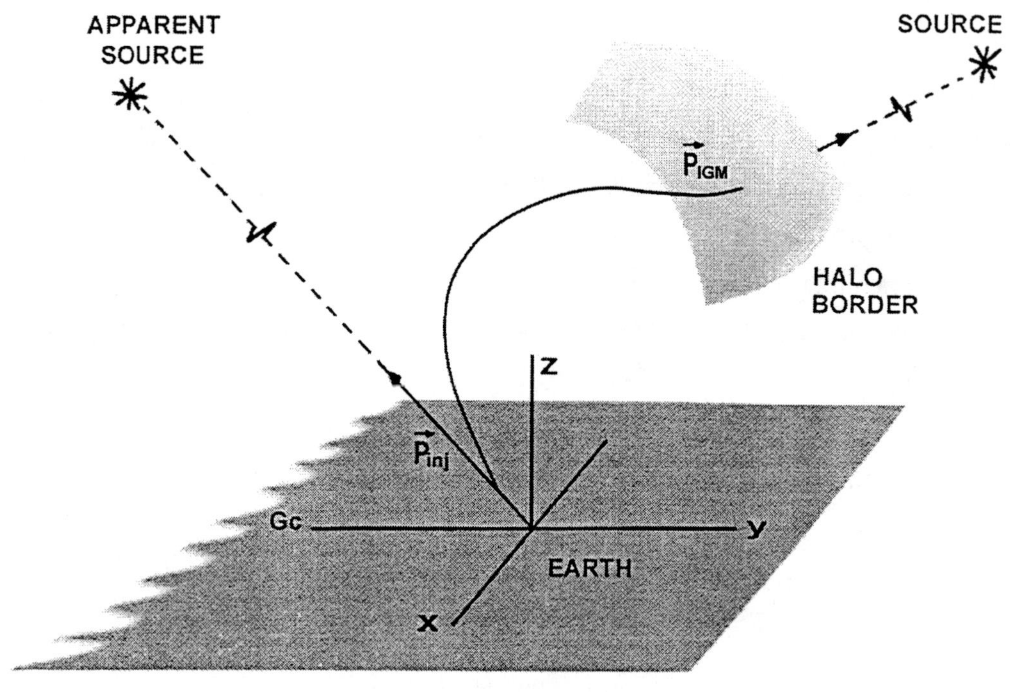

Figure 1:   Propagation through the GMF. Reversibility of the trajectories is applied. particles of momentum $\mathbf{p}_{inj} = - \mathbf{p}_{arrival}$ are injected at Earth and follow through the system until they reach the border of the galactic halo (a spherical surface of R=20 Mpc centered on the galactic center)

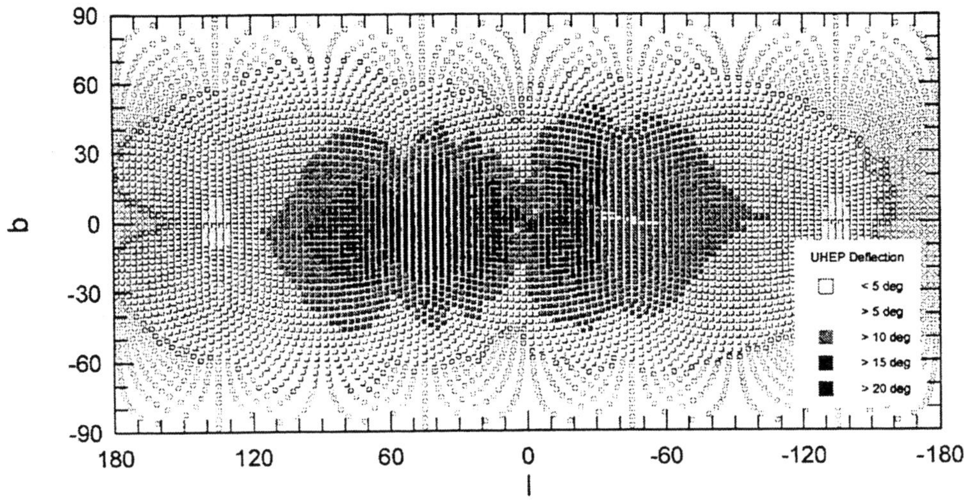

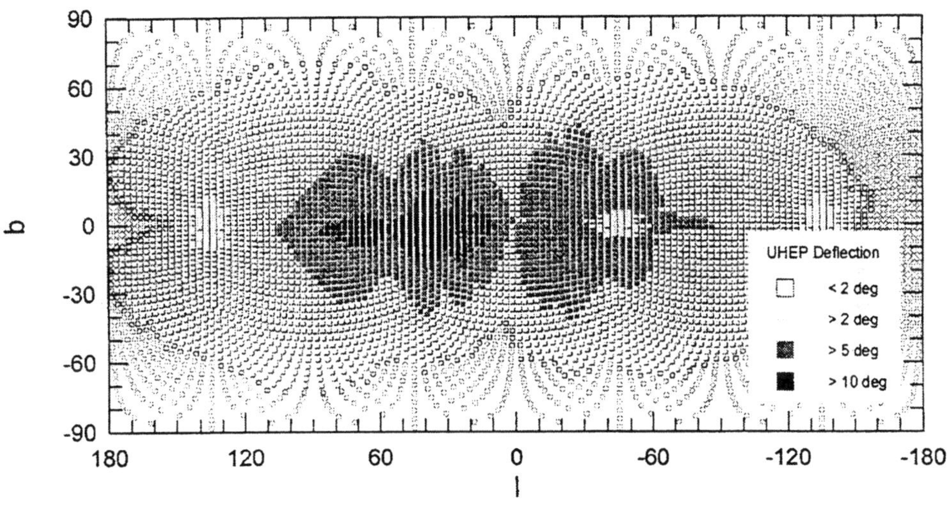

figure 2.b

Figure 2: Error box due to GMF propagation for protons of (a) $E=4 \times 10^{19}$ eV and (b) $E=10^{20}$ eV.

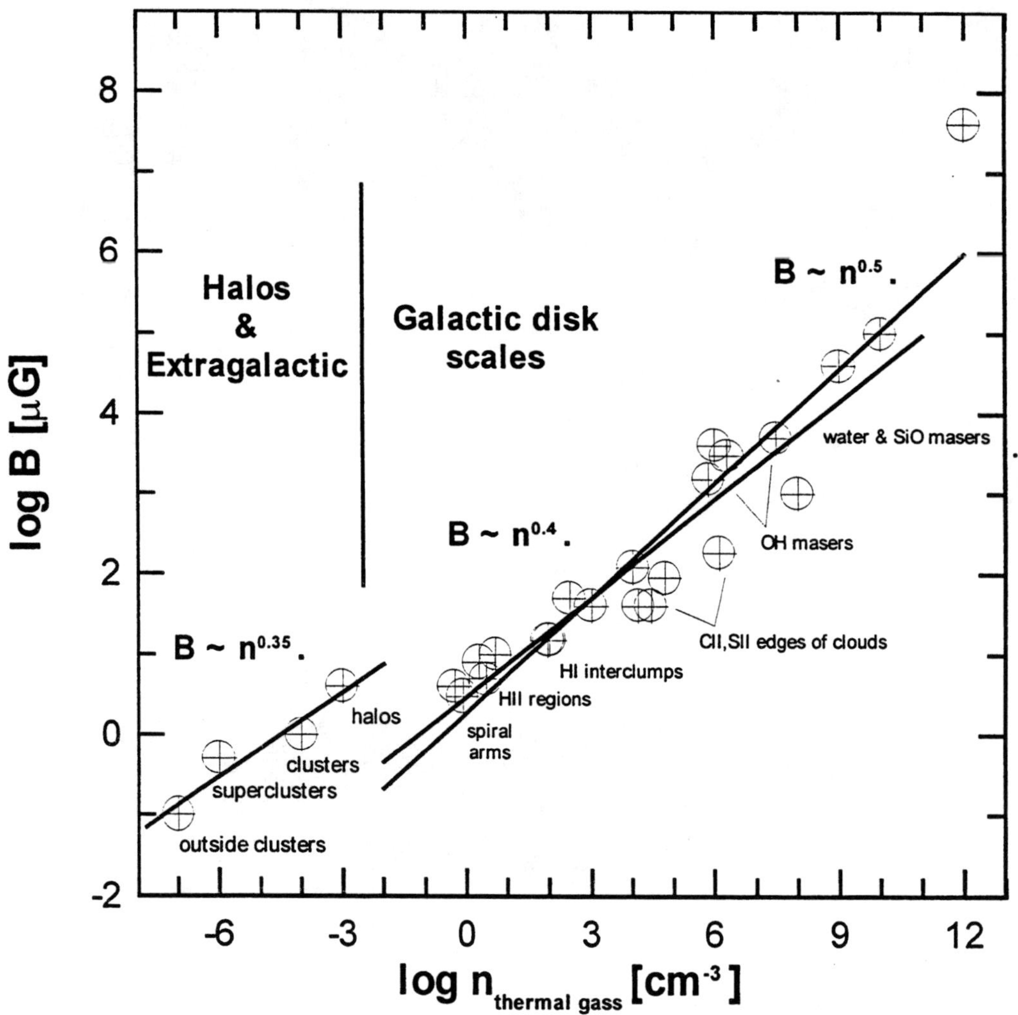

Figure 3: Scaling of magnetic field intensity with thermal gas density for different astrophysical environments. Note that the scale of the systems grows, roughly, to the left. (Adapted from Vallée 1997)

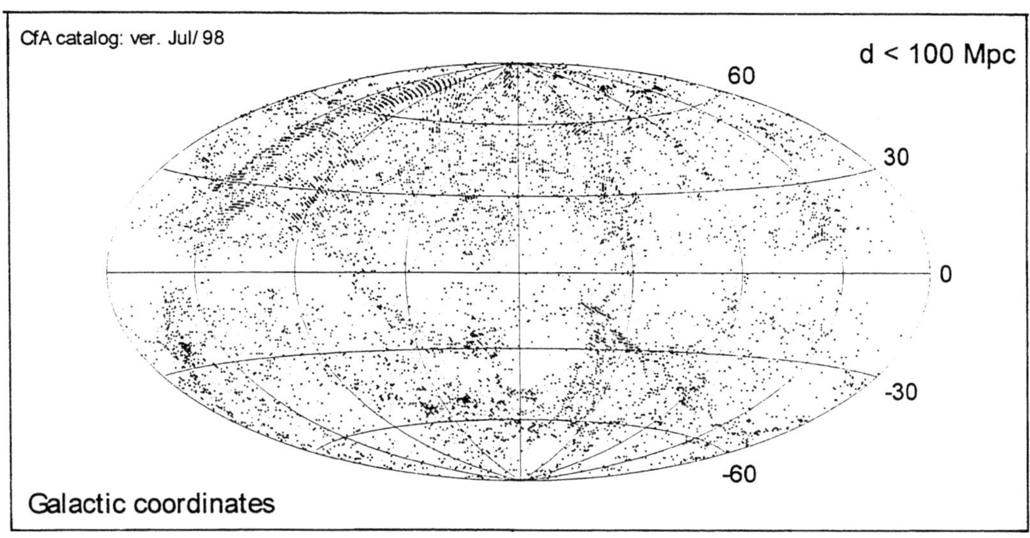

Figure 4: All sky Aitoff (equal area) projection of the galaxies with known redshift up to distances of $\approx$ 100 Mpc (H_0=50 km/sec/Mpc). Data from the July 1998 version of the CfA catalog (Huchra et al, 1992).

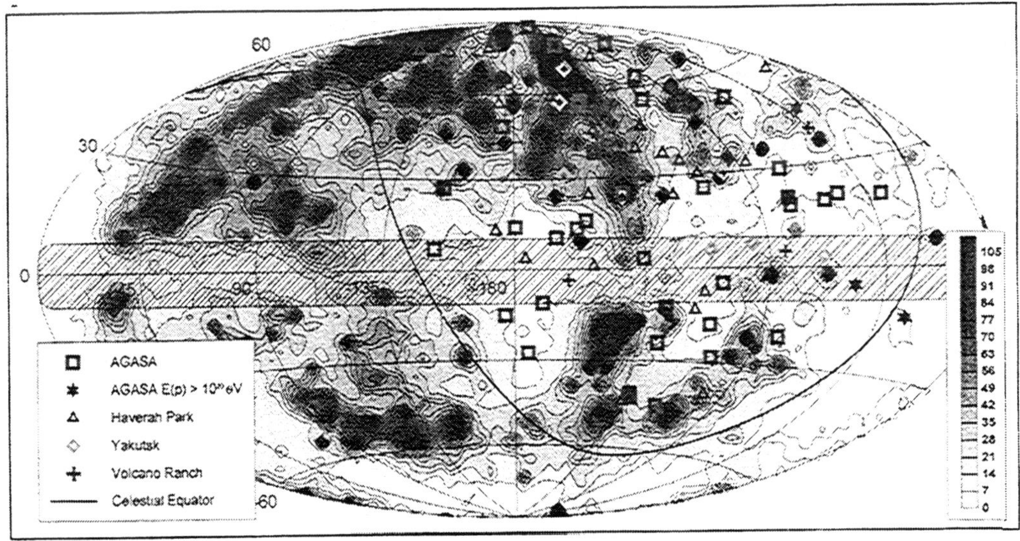

Figure 5: All sky Aitoff projection of the arrival probability density (Medina Tanco 1997b) under the assumption that the sources of UHECR are spatially distributed in the same way as the luminous matter in the nearby universe.

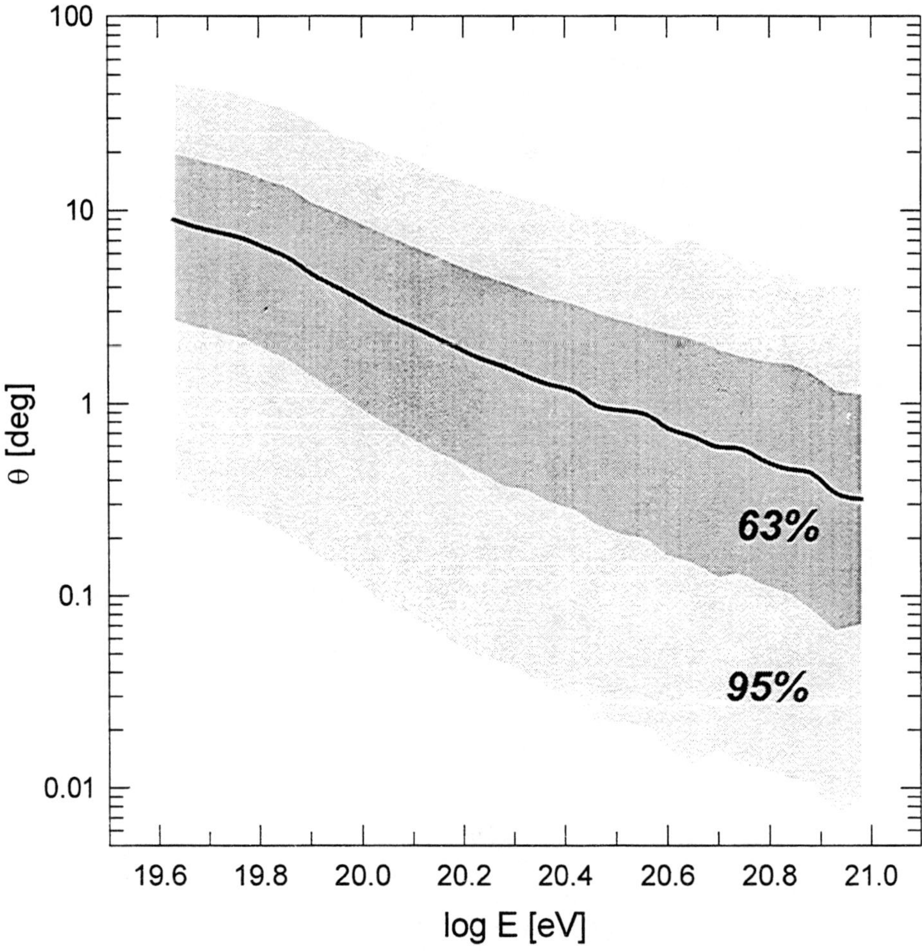

Figure 6: All sky median of the deflection of particles arriving at the detector, as a function of the arrival energy, for the same simulation in figure 5. the bands correspond to 63 and 95% confidence levels.

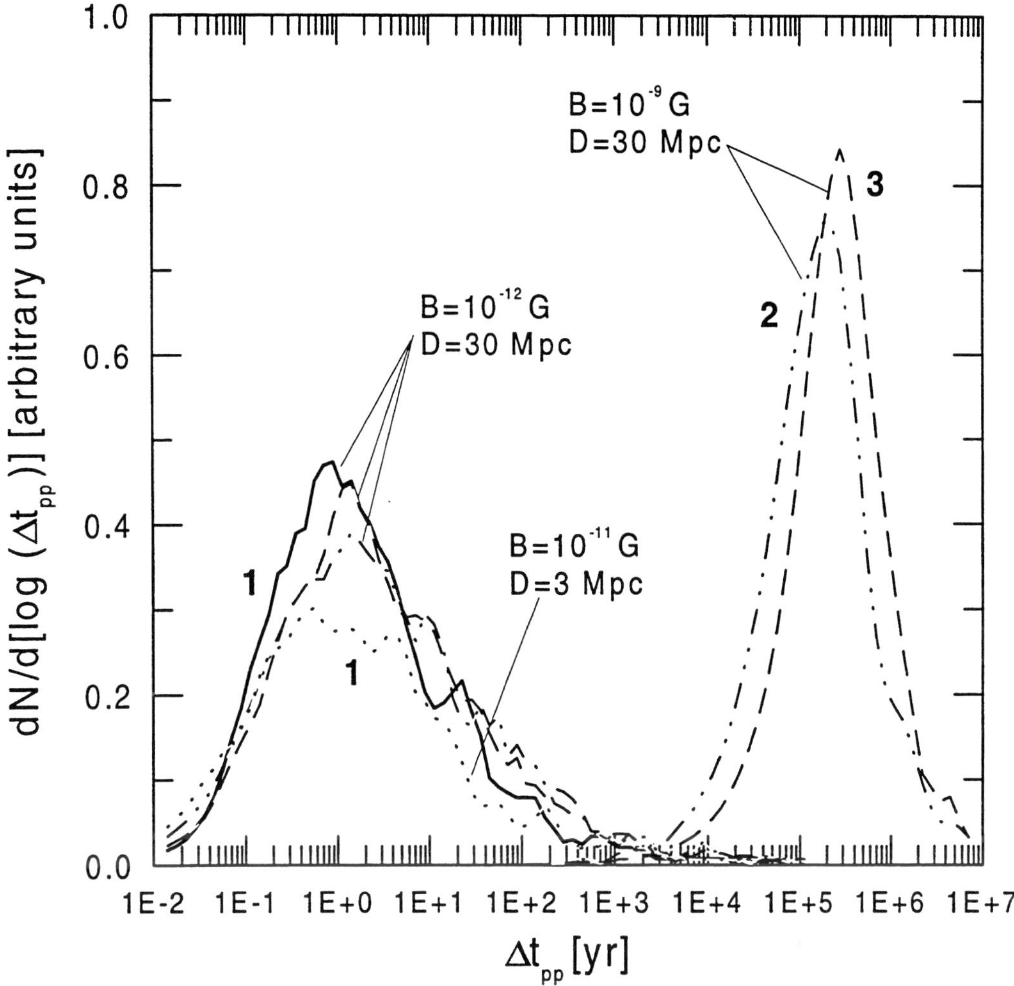

Figure 7.a: Distribution function of arrival time delays between the observed pair of protons in clusters 1, 2 and 3 due to propagation in the IGMF alone (i.e., at the external border of the galactic). halo). The simulations for pair 1 correspond to the two fiducial scenarios of Table II, and match the constrain in time delay imposed by the galactic portion of the tracks: BIGM=10-11G and D=3Mpc (dotted line) and BIGM = 10-12 G and D = 30 Mpc (continuous line). For pairs 2 (broken-dotted line) and 3 (broken line) two possible scenarios are explored: D = 30 Mpc and BIGM=10-12 G and BIGM=10-9G.

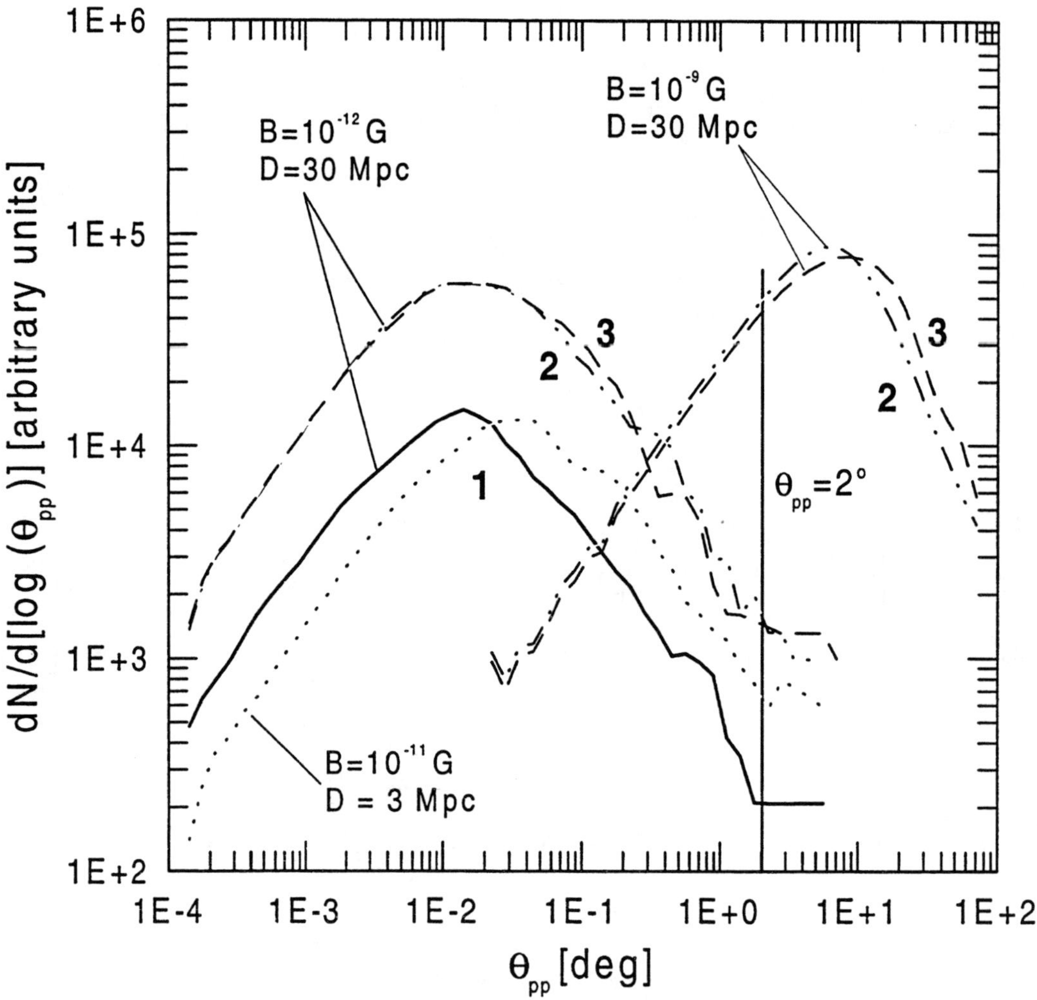

Figure 7.b: Distribution function of the angle between the momenta of the observed particles in each pair, at their arrival at the external border of the galactic halo after propagation through the IGMF. The conditions are the same as in figure 1. Also indicated is a separation angle of 2° typically obtained from the calculations of galactic propagation for all the three pairs.

Cosmic Ray Mass Composition at the Highest Energies

B.R. DAWSON

Department of Physics and Mathematical Physics
The University of Adelaide, Adelaide 5005, AUSTRALIA

Abstract

The mass composition of cosmic rays above 10^{17}eV is an important and challenging measurement. We discuss its importance in helping to discriminate between models of cosmic ray origin, and describe measurements taken to date. Despite some uncertainties related to models of hadronic interactions, there is beginning to be some consensus in the likely mass in this energy range. We look forward to future experiments (eg HiRes and the Pierre Auger Project) which will test these conclusions and extend measurements up to and beyond 10^{20}eV.

1. Introduction

The origin of the highest energy cosmic rays remains a mystery after more than thirty years of serious observations. Experimenters have encountered difficulties with the extremely low flux, and the necessity to view the cosmic rays indirectly through the extensive air showers (EAS) they produce in the atmosphere. Observations have yielded information on the cosmic ray energy spectrum, their arrival directions and clues to their atomic masses. Mass composition studies are some of the most difficult. Apart from our indirect view of the primary cosmic ray, our sampling of the EAS is incomplete even in the best experiments. We must cope with intrinsic fluctuations in the development of EAS which can be large even for showers initiated by cosmic rays of the same mass and energy. And, extraction of a likely mass from a measurement relies on models of hadronic interactions at energies well beyond accelerator experiments. In some ways, extrapolations from lower energy accelerator data can be constrained by the cosmic ray data, but there are still significant uncertainties.

These factors mean that mass resolution at the highest energies is extremely poor compared with direct measurements made at lower energies by balloon and satellite experiments. Workers at the highest energies typically talk in terms of a heavy composition (a flux dominated by nuclei like iron) or a light composition (one dominated by protons).

But even this information can be as important as the energy spectrum or anisotropy in determining cosmic ray origin. Quite clearly, one must know the likely mass range (and therefore charge range) of a particular data set before one can confidently interpret anisotropy information, given the influence of galactic and intergalactic magnetic fields. It is likely that 10^{20}eV cosmic ray protons will be deflected by only a few degrees in their passage through extragalactic fields from plausible sources. Real astronomy will be possible with the large apertures provided by current and future observatories, but we would like to know if the particles *are* protons at these energies.

Several origin models predict a protonic composition at the highest energies, either because of the production mechanism or because of propagation over large distances. Protons with energies above about 5×10^{19} eV will themselves suffer energy degradation through interactions with cosmic microwave background radiation. This limits source distances for 10^{20} eV protons to less than around 50 Mpc. But nuclei will also undergo interactions with photon fields. These interactions can occur in the source, and given the intensity of photon fields in some candidate sources, nuclei may be destroyed immediately. If not affected this way, nuclei will certainly interact with photons of the cosmic microwave background and infra-red radiation during propagation. Such a nucleus will lose 3-4 nucleons per Mpc when the nucleus energy exceeds about 2×10^{19} eV. A nucleus with energy above 10^{20} eV will not remain that energetic for a pathlength greater than 20 Mpc (Cronin 1992, Elbert and Sommers 1995).

In models where the sources are much closer, propagation effects are minimal and the composition expectation depends more on the physics within the source. Source models that include acceleration at strong shocks can favor heavy cosmic rays since the maximum attainable energy scales with the nuclear charge, Z (Cesarsky 1992). Still other models might predict a flux dominated by gamma-rays at the highest energies eg production of cosmic rays from the decay of topological defects (eg Sigl 1996).

Each class of model has implications in mass composition, the energy spectrum (eg is there a cut-off in the spectrum above 5×10^{19} eV?) and in anisotropy. All three types of measurements must be done in order to weed out specific models.

2. SOME EXPERIMENTAL RESULTS

EAS initiated by light nuclei like protons are quite different to those initiated by heavy nuclei. First, a heavy nucleus will have a shorter interaction length upon entering the atmosphere. Secondly, the heavy nucleus will immediately produce a larger hadronic cascade as its nucleus is broken up within the first few interactions. The larger hadronic cascade will produce a larger number of muons, and it will seed more electromagnetic cascades more quickly than a proton initiated shower. The EAS cascade will therefore develop more rapidly for iron showers, with the depth of shower maximum, $X_{\max}$ being approximately 100 g cm^{-2} shallower than in proton showers of the same energy.

The mass composition of cosmic rays between 10^{17} eV and 10^{19} eV has been tackled by several groups, with techniques including using atmospheric Cerenkov light from EAS, measuring the muon content of the cascades at ground level, and using fluorescence light to directly view shower development. These results have not all been consistent in their conclusions about mass composition, often because of incomplete access to EAS simulation code, or because of an inability to understand and overcome experimental biases. A review of measurements before the 1990's is given by Sokolsky, Sommers and Dawson (1992).

We can learn lessons from these earlier attempts. The best method is to choose a measurable parameter and relate it, through simulations, to the mass of the primary cosmic ray, carefully taking into account any experimental biases.

A more recent measurement of the composition in the energy range 10^{17}eV to 10^{19}eV has come from the Fly's Eye group, with direct measurements of air shower development profiles (Gaisser et al. 1993). The "stereo" Fly's Eye technique provided direct measurements of $X_{\max}$, and it was possible to study the behaviour of the mean value with energy, and use the fluctuations as a check on the conclusions. A study of three alternative hadronic models was undertaken. One model was ruled out because it could not explain early developing showers seen in the data. The other two models were more successful, and gave similar conclusions about the mass composition.

One of these models is the KNP model (Kopeliovich, Nikolaev and Potashnikova 1989), one of a class of QCD Pomeron models. The Fly's Eye data and the KNP model expectations for proton and iron EAS are shown in Figure 1. The so-called elongation rate for the data (the change in $X_{\max}$ per decade of energy) between 3×10^{17} and 10^{19}eV is $79 \pm 3\,\mathrm{g\,cm}^{-2}$ per decade. This is in stark contrast with the expectation of $50 \pm 3\,\mathrm{g\,cm}^{-2}$ per decade for a pure iron or pure proton composition. So even by itself, the elongation rate measurement supports a composition that is becoming lighter with energy. This view is strengthened by the absolute $X_{\max}$ measurements that show that the data are consistent with iron primaries at the lower energies, moving toward a protonic composition at higher energies.

The power of measuring the elongation rate as well as the absolute $X_{\max}$ is illustrated by these Fly's Eye measurements. The mean value of $X_{\max}$ at 3×10^{17}eV is clearly too shallow for models (such as simple scaling) that predict a rapid elongation rate of order $70\,\mathrm{g\,cm}^{-2}$ per decade. Such models predict deeper development at this energy, even for iron initiated EAS, and these models can be ruled out. What remains are models which predict slower elongation rates, accomplished with interaction cross-sections or inelasticities (or both) that increase with energy. The high elongation rate measured by the Fly's Eye then suggests a changing composition, one becoming lighter with energy. It is interesting that this feature in the mass spectrum is apparently correlated with the flattening of the energy spectrum (Bird et al 1993).

The Japanese AGASA array has published mass composition results based on measurements of the muon content of EAS in the same energy range as the Fly's Eye results (Hayashida et al. 1995). They compared the measured elongation rate with that expected from a scaling hadronic interaction model and concluded that the mass composition was not changing over the energy range, in contradiction with the Fly's Eye interpretation. This apparent contradiction can be resolved if the Fly's Eye and AGASA data are interpreted using the same hadronic interaction model. A study (Dawson, Meyhandan and Simpson 1998) using SIBYLL, a more modern hadronic model than used by AGASA and one that is more consistent with the KNP model used by the Fly's Eye group, shows that the experimental results are consistent. The AGASA results are illustrated in Figure 2 where the mass composition is measured using the muon density 600 m from the EAS core, and the lines are the predictions of the SIBYLL model. The composition at 10^{17}eV appears to be dominated by heavy nuclei, with a lighter composition becoming evident at the higher energies, in agreement with the trend seen by the Fly's Eye group.

These general conclusions are also supported by work of the Yakutsk group. Measurements of the atmospheric Cerenkov light from giant EAS (Dyakonov et al. 1993) and the muon content of these showers (Kalmykov et al. 1995) and application of the QGS-jet hadronic model, suggest a composition become lighter above 10^{16}eV.

3. NEW EXPERIMENTS

The High Resoultion Fly's Eye (HiRes) experiment (Loh 1993, Al-Seady et al. 1996, Abu-Zayyad et al. 1997) is applying the succesful Fly's Eye technique of viewing nitrogen fluorescence light from EAS, but with a detector with significantly better resolution and sensitivity. HiRes uses larger mirrors (2 m diameter as opposed to 1.5 m diameter) and smaller pixels (1° diameter as opposed to 5°) to view EAS over a very large area. The stereo technique has been maintained and strengthened with HiRes. All showers will be viewed with two eyes, with X_{max} resolution better than $20 \, \mathrm{g \, cm^{-2}}$ above 10^{18}eV. The larger mirrors and smaller fields of view of the phototubes improve the detector signal-to-noise by a factor of 7 over the Fly's Eye, and this allows the eyes to look over a much larger ground area. HiRes will have a stereo aperture exceeding $4000 \, \mathrm{km^2 sr}$ at 10^{19}eV extending to $10000 \, \mathrm{km^2 sr}$ at 10^{20}eV. The first site of HiRes, with 22 mirror units each containing cameras of 256 photo-multipliers is now operating at the Dugway Proving Grounds in Utah. The second site 13 km away is scheduled for operation from mid-1999. That site will contain 43 such mirror units.

HiRes will have the capacity to rapidly explore the decade in energy above 10^{19}eV, with an expectation of roughly 300 high quality events per year above that energy. Apart from studies of the structure of the energy spectrum in this decade, HiRes will be able to confirm and strengthen the mass composition results from the Fly's Eye detector. Already, some mass composition studies have been performed with HiRes. The prototype detector was set up to view the CASA and MIA ground arrays 3 km away. A set of events was collected over a two year period consisting of showers viewed by the HiRes prototype which also triggered the MIA muon array. These data span the important energy range from 10^{17} to 10^{18}eV where the mass composition is expected to begin to lighten. At the time of writing, this analysis is still underway, but it is exciting because it provides an opportunity to combine measurements from two experiments, each with mass composition sensitivity. An opportunity exists to check consistency of mass interpretation on a shower by shower basis, and to use these different features of the showers to (possibly) further constrain hadronic interaction models.

The Pierre Auger Project takes this idea of viewing EAS using two different techniques and expands it to studies of cosmic rays above 10^{19}eV. The Auger concept will be realised in Argentina starting in 1999 and later in the US. A giant array will be built at each of the two sites, consisting of 1600 water Cerenkov detectors spread over an area $3000 \, \mathrm{km^2}$. The arrays will operate 24 hrs a day, and will be augmented on clear moonless nights by atmospheric fluorescence detectors similar in concept to the HiRes experiment. The combination of fluorescence and ground array measurements of the same EAS is known as the Hybrid concept. Hybrid-viewed EAS will be important for many reasons (Dawson 1997). First, the techniques used by the ground array for assigning primary energy to an EAS will be checked and refined by the fluorescence technique, which is essentially a calorimetric measurement of the energy deposited by the shower in the atmosphere. Secondly, there will be on opportunity to extract interesting astrophysics from the Hybrid data set, a collection of EAS with information on longitudinal development (from the fluorescence detectors) and lateral structure (from the ground array). Mass composition studies will be especially interesting. The depth of maximum measurements can be combined with a number of measurements from the ground array. These are likely to include an estimate of the muon content of the EAS at ground level, and measurements of the time structure of the shower front at the ground (eg its curvature and its thickness) (Pryke 1997). Measurements with quite different measurement systematics can be compared to get a consistent mass composition picture. Simulations studies point to the expected power of the technique (Dawson and Pryke 1997).

4. CONCLUSION

There is a growing interest in the origin of the highest energy cosmic rays, as evidenced by the very large experiments being constructed and planned (including the Japanese Telescope Array (Teshima et al. 1996), an evolution of the Fly's Eye technique with plans to have a time averaged aperture rivalling that of the Auger Project). There is currently some consensus on the likely behaviour of mass composition at energies above 10^{17}eV, but we await new results to explore carefully the decade above 10^{19}eV, and beyond. The mass composition information will allow us to sensibly interpret the energy spectrum and anisotropy measurements in the process of discriminating between the many theories of cosmic ray origin. The pioneering experiments have answered many important questions, but in doing so have in some ways deepened the mystery of the origin of these particles. We all hope and trust that the new detectors will be able to unravel the mystery!

REFERENCES

Abu-Zayyad, T. et al., 1997, Proc. 25th Int. Cosmic Ray Conf., **5**, 325,321,329 and others.

Al-Seady, M. et al., 1996, in "Proceedings of International Symposium on Extremely High Energy Cosmic Rays: Astrophysics and Future Observatories", ed. M. Nagano, (ICRR, University of Tokyo, Tokyo), 191.

Bird, D.J. et al., 1993, Phys. Rev. Lett. **71**, 3401.

Cesarsky, C.J., 1992, Nucl. Phys. **28B** (Proc. Supp.), 51.

Cronin, J.W., 1992, Nucl. Phys. **28B** (Proc. Supp.), 213.

Dawson, B.R., 1997, in "Very High Energy Phenomena in the Universe", Proc. 32 nd Recontres de Moriond, Les Arcs France, eds. Y. Giraud-Heraud & J. Tran Thanh

Van (Editions Frontiers, Paris) 235.

Dawson, B.R. and Pryke, C.L., 1997, Proc. 25th Int. Cosmic Ray Conf., **5** 213.

Dawson, B.R., Meyhandan, R. and Simpson, K.M., 1998, "A Comparison of Cosmic Ray Composition Measurements at the Highest Energies", Astropart. Phys. (in press), astro-ph/9801260.

Dyakonov, M.N. et al., 1993, Proc. 23rd Int. Cosmic Ray Conf., **4**, 303.

Elbert, J.W. and Sommers, P., 1995, Ap. J. **441**, 151.

Gaisser, T.K. et al., 1993, Phys. Rev. D **47**, 1919.

Hayashida, N. et al., 1995, J. Phys. G. **21**, 1101.

Kalmykov N.N. et al., 1995, Proc. 24th Int. Cosmic Ray Conf., **1**, 123.

Kopeliovich, N.N., Nikolaev, N.N. and Potashnikova, I.K., 1989, Phys. Rev. D **39**, 769.

Loh, E.C., 1993, in "Proceedings of the Tokyo Workshop on Techniques for the Study of Extremely High Energy Cosmic Rays", ed. M. Nagano, (ICRR, University of Tokyo, Tokyo), 105.

Pryke, C.L., 1997, in "Very High Energy Phenomena in the Universe", Proc. 32 nd Recontres de Moriond, Les Arcs France, eds. Y. Giraud-Heraud & J. Tran Thanh Van (Editions Frontiers, Paris) 239.

Sigl, G., 1996, Space Sc. Rev. **75**, 375.

Sokolsky, P., Sommers, P. and Dawson, B.R., 1992, Physics Reports, **217**, 225-277.

Teshima, M. et al., 1996, in "Proceedings of International Symposium on Extremely High Energy Cosmic Rays: Astrophysics and Future Observatories", ed. M. Nagano, (ICRR, University of Tokyo, Tokyo).

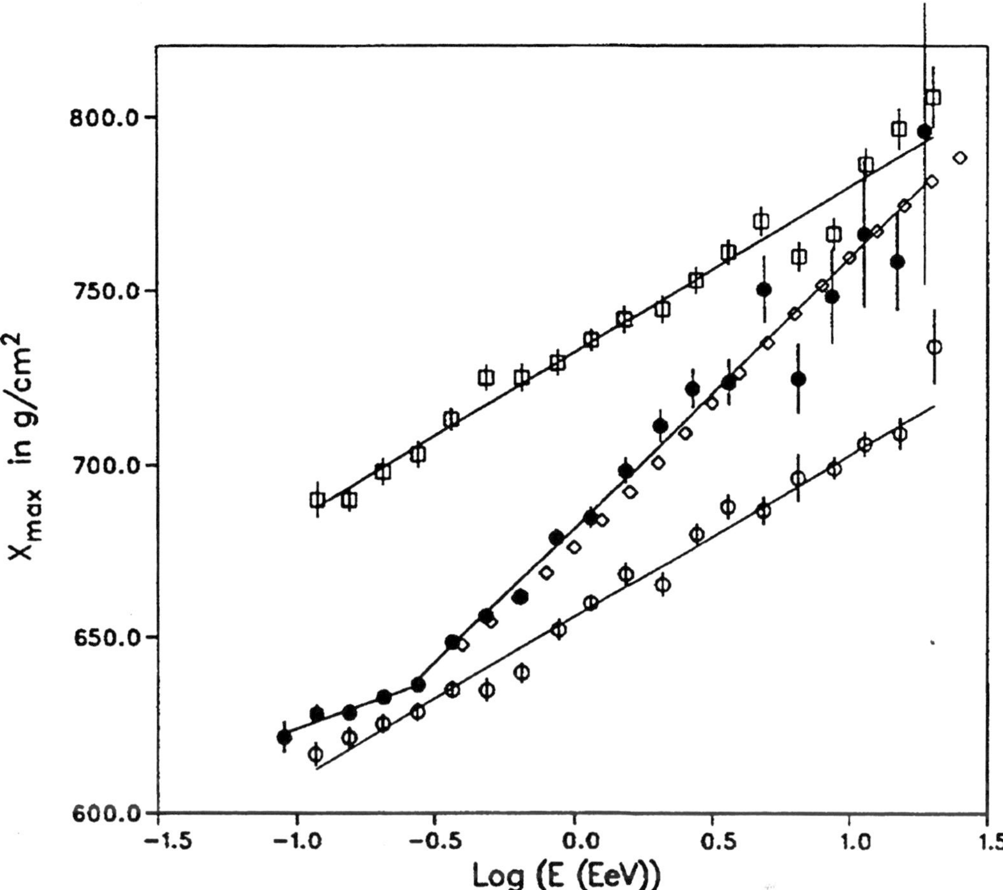

Figure 1: Fly's Eye mass composition results from Gaisser et al.(1993). The data are shown by the solid points. The open squares and open circles represent the KNP model predictions for protons and iron nuclei, respectively. The energy scale is in EeV, $1\,\text{EeV} = 10^{18}\text{eV}$.

 B.R. Dawson

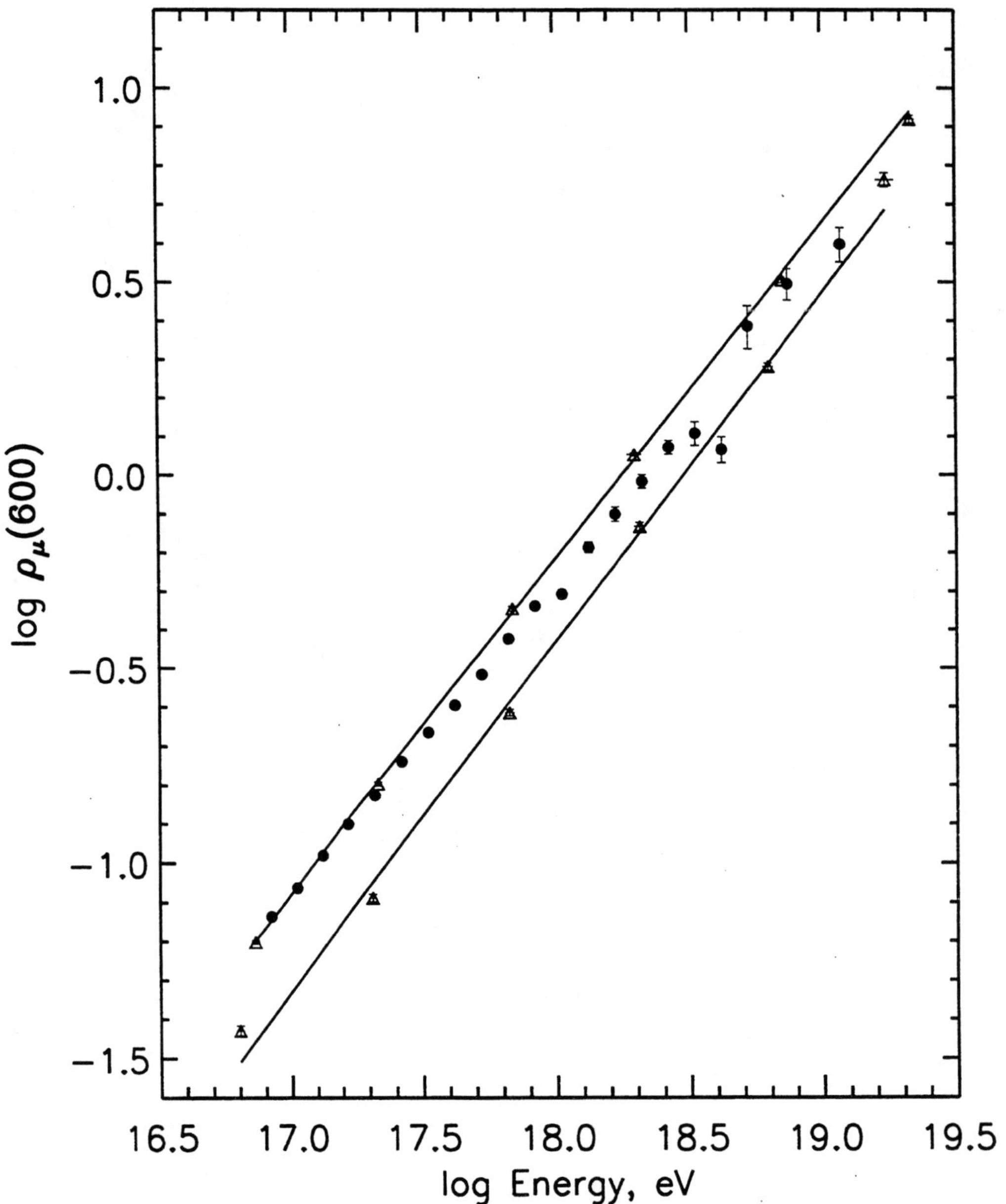

Figure 2: Data from the AGASA A1 array up to April 1993 (solid points) compared with MOCCA+SIBYLL simulations (triangles and solid lines). The top line represents iron primary particles, the lower line protons (from Dawson, Meyhandan and Simpson 1998).

VI. Annotated Bibliography

A Selected and Annotated Introductory Cosmic-Ray Bibliography

Older books
(Some can be hard to find)

H. Alfven and C. G. Falthammer, *Cosmical Electrodynamics* (Clarendon Press, Oxford, 1963)
> A masterpiece volume on the EM fields underlying cosmic-ray propagation.

Bell Science, *The Strange Case of the Cosmic Rays* (Bell Telephone, 1957)
> Produced, written, and directed by Frank Capra with animation by Shamus Culhane (of Betty Boop, Snow White, and Pinnochio fame). This movie is available in a VHS videotape release from Rhino Home Video (1991). Did I mention the puppetry?

T. E. Cranshaw, *Cosmic Rays* (Oxford University Press, London, 1963)

V. L. Ginzburg and J. L. Syrovatskii, *The Origin of Cosmic Rays* (Pergamon Press, Oxford, 1964)
> This is a classic work, still referenced and useful today. The leaky box model of Galactic cosmic ray propagation (new at the time) is given an excellent discussion.

L. T. Guild, *The Cosmic Ray in Literature* (Cokesbury Press, Tennessee, 1929)

S. Hayakawa, *Cosmic Ray Physics* (Wiley-Interscience, New York, 1969)

W. Heisenberg, Editor, *Cosmic Radiation* (Dover Publications, New York, 1946)
> A work that fully shows the interest in the field of cosmic rays during the 1940s.

A. M. Hillas, Editor, *Cosmic Rays* (Pergamon Press, Oxford, 1972)
> An excellent set of reprints with introductory material.

V. D. Hopper, *Cosmic Radiation and High Energy Interactions* (Prentice-Hall, Inc., New Jersey, 1964)

B. Lehnert, Editor, *Electromagnetic Phenomena in Cosmical Plasma*, IAU Symposium # 6 (North Holland, Amsterdam, 1958)

F. B. McDonald and C. E. Fichtel, Editors, *High Energy Particles and Quanta in Astrophysics* (MIT Press, Massachusetts, 1974)
> Good sections on cosmic-ray electrons, solar modulation, and solar energetic particles.

D. J. X. Montgomery, *Cosmic Ray Physics* (Princeton University Press, New Jersey, 1949)

H. Ogelman and J. R. Wayland, Editors, *Lectures in High-Energy Astrophysics*, NASA Special Publication SP-199 (1969)
> Covers photons, electrons, nuclei, and propagation theory as known at the time.

H. Ogelman and J. R. Wayland, Editors, *Introduction to Experimental Techniques of High-Energy Astrophysics*, NASA Special Publication SP-243 (1970)

E. N. Parker, *Interplanetary Dynamical Processes* (John Wiley and Sons, Inc., New York, 1963)

    Solar modulation, the solar wind, and the termination shock.

M. A. Pomerantz, *Cosmic Rays* (Van Nostrand, New York, 1971)

    Put out under the aegis of the Commission on College Physics, this is an undergraduate level survey (circa 1971) of the cosmic ray field.

S. Rosen, Editor, *Selected Papers on Cosmic Ray Origin Theories* (Dover Publications, New York, 1969)

    Includes great, historical papers from Fermi, Compton & Getting, and many others. Still useful as a collection of early papers.

B. Rossi, *High Energy Particles* (Prentice-Hall, Inc., New York, 1952)

    Covers many of the techniques and valuable information of the era's cosmic ray research.

B. Rossi, *Cosmic Rays* (McGraw-Hill, Inc., New York, 1964)

    A popular work filled with personality and a great deal of historical detail.

W. Sullivan, *Assault on the Unknown: The International Geophysical Year* (McGraw-Hill, New York, 1961)

J. T. Wilson, *I.G.Y.: The Year of the New Moons* (Alfred Knopf, New York, 1961)

    The above two are popular books on the International Geophysical Year (1957-8) which saw the first satellites and the first satellite observations of the cosmic radiation, the radiation belts, and the interplanetary magnetic fields.

A. W. Wolfendale, *Cosmic Rays* (Philosophical Library, Inc., New York, 1963)

Newer books

J. Arons, C. Max, and C. McKee, Editors, *Particle Acceleration Mechanisms in Astrophysics* (AIP, New York, 1979)

R. Clay and B. Dawson, *Cosmic Bullets: High Energy Particles in Astrophysics* (Addison-Wesley, Massachusetts, 1997)

    A very recent and up-to-date popular treatment of the cosmic rays, especially the highest energy cosmic rays. It's a well-written addition to the Frontiers of Science series of books edited by Paul Davies.

Committee on Cosmic-Ray Physics, *Opportunities in Cosmic-Ray Physics and Astrophysics* (National Academy Press, Washington, D.C., 1995)

    Presents an official statement of the status of the field and a roadmap for future work.

M. W. Friedlander, *Cosmic Rays* (Harvard University Press, Massachusetts, 1989)

    A popular historical treatment of the field at the level of Scientific American.

T. K. Gaisser, *Cosmic Rays and Particle Physics* (Cambridge University Press, Cambridge, 1990)

A fairly recent and authoritative treatment of the field, especially at higher energies.

S. S. Holt and G. Sonneborn, *Cosmic Abundances*: Astronomical Society of the Pacific Conference Series, Volume 99 (Astronomical Society of the Pacific, California, 1996)

Nucleosynthesis, abundance anomalies, cosmic-ray abundances, and the history of the light elements in the Galaxy.

W. V. Jones, F. J. Kerr, and J. F. Ormes, Editors, *Particle Astrophysics: The NASA Cosmic Ray Program for the 1990s and Beyond*, AIP Conference Proceedings # 203 (AIP, New York, 1990)

M. S. Longair, *High Energy Astrophysics*, Volume 1 (Cambridge University Press, Cambridge, 1994)

Features an overview (with references) of the early history of cosmic-ray research.

M. Nagano and F. Takahara, Editors, *ICRR International Symposium on Astrophysical Aspects of the Most Energetic Cosmic Rays* (World Scientific, Singapore, 1991)

J. F. Ormes, Editor, *Current Perspectives in High Energy Astrophysics*, NASA Reference Publication 1391, (1996)

Available from the NASA Center for AeroSpace Information, 800 Elkridge Landing Road, Linthicum Heights, MD 21090-2934, (301) 621-0390. Has a good series of articles covering a wide range of topics.

G. Setti, G. Spada, and A. W. Wolfendale, Editors, *Origin of Cosmic Rays*: IAU Symposium # 94 (D. Reidel Publishing, Dordrecht, 1981)

A wide selection of talks on the astrophysics of the cosmic-ray origin.

M. M. Shapiro and J. P. Wefel, Editors, *Genesis and Propagation of Cosmic Rays* (D. Reidel Publishing, Dordrecht, 1987)

M. M. Shapiro, R. Silberberg, and J. P. Wefel, Editors, *Cosmic Rays, Supernovae, and the Interstellar Medium* (Kluwer Academic Publishers, Dordrecht, 1991)

M. M. Shapiro, R. Silberberg, and J. P. Wefel, Editors, *Particle Astrophysics and Cosmology*: NATO ASI Series C, Volume 394 (Kluwer Academic Publishers, Dordrecht, 1993)

The previous three books are based on talks given at the Cosmic-ray Summer Schools at Erice, Italy. Cosmic rays, gamma rays, neutrinos, and cosmology. Many excellent review articles. The summer schools take place every other year.

P. Sokolsky, *Introduction to Ultrahigh Energy Cosmic Ray Physics* (Addison-Wesley, New York, 1988)

Excellent introduction to the highest energy cosmic rays, which has become quite an active field since this book was written.

C. J. Waddington, Editor, *Cosmic Abundances of Matter*: AIP Conference Proceedings # 183 (American Institute of Physics, New York, 1989)

The papers stick close to the focus on abundances: standard solar, anomalous cosmic-ray, Galactic cosmic-ray source, the solar wind, nucleosynthethic sources, the ISM, and ultraheavy cosmic rays.

G. P. Zank and T. K. Gaisser, Editors, *Particle Acceleration in Cosmic Plasmas*: AIP Conference Proceedings # 264 (American Institute of Physics, New York, 1992) A considerable amount of work and review on the theory of shock acceleration, both for heliospheric and cosmic-ray particles.

Review papers

Rapporteur talks and papers (reviewing presented papers at the conference along with the author's personal take on the field), invited papers, and the Victor Hess Memorial Lectures from each of the most recent International Cosmic Ray Conferences are a good starting point for review articles:

> M. S. Potgieter, B. C. Raubenheimer, and D. J. van der Walt, Editors, *Proceedings of the 25th ICRC (Durban, South Africa)*, **8** (World Scientific, Singapore, 1998)
>
> N. Iucci and E. Lamanna, Editors, *Proceedings of the 24th ICRC (Rome, Italy)*, **5**
>
> D. A. Leahy, R. B. Hicks, and D. Venkatesan, Editors, *Proceedings of the 23rd ICRC (Calgary, Canada)*, **5** (World Scientific, Singapore, 1994)
>
> *Proceedings of the 22nd ICRC (Dublin, Ireland)*, **5** (Dublin Institute for Advanced Study, Dublin, 1991)

P. Auger, *Rev. Mod. Phys.* **11**, 288 (1939)
> By the discoverer of extensive air showers, on the subject of the extensive air showers.

S. A. Colgate, *Adv. Space Res.* **4**, 367 (1984)
> Review of supernovae shock acceleration of the cosmic rays.

V. A. Dogel, *Radiofizika* **30**, 187 (1987)

T. K. Gaisser, F. Halzen, and T. Stanev, *Phys. Rep.* **238**, 173 (1995)
> Astrophysical neutrinos as implied by GeV-TeV gamma rays.

V. L. Ginzburg and V. S. Ptuskin, *Rev. Mod. Phys.* **48**, 161 (1976)

V. L. Ginzburg and V. S. Ptuskin, *Sov. Sci. Rev. E Astrophys. Space Phys.* **4**, 161 (1985)
> In many ways, these are condensed updates of the classic book by Ginzburg and Syrovatskii. Provides an excellent theoretical overview of models and theories of cosmic-ray origins and a very useful set of references.

J. P. Ostriker and C. F. McKee, *Rev. Mod. Phys.* **60**, 1 (1988)
> "Astrophysical blastwaves." Covered in full.

H. Reeves, *Rev. Mod. Phys.* **66**, 193 (1994)

An overview and very extensive reference list on the light element origins and abundances. Connects to cosmic-ray physics via the "X-Process" for the formation of LiBeB.

J. A. Simpson, *Ann. Rev. Nucl. & Part. Sci.* 33, 323 (1983)

An overview of the low energy cosmic rays, and elemental and isotopic analyses of the cosmic radiation in the energy regime in which detailed measurements are possible. Much of the data has been improved, but the theoretical framework for the observations remains similar.

J. Szabelski, *Astro-ph/9710191*, 17 October 1997

The focus is on the astrophysics of the highest energy cosmic rays.

G. Tarle and S. P. Swordy, *Scientific American* April 1998

A popular primer on the cosmic-ray antimatter.

Important books and papers for specific topics

Nucleosynthesis

W. D. Arnett, *Supernovae and Nucleosynthesis* (Princeton University Press, New Jersey, 1996)

W. D. Arnett and J. W. Truran, Editors, *Nucleosynthesis: Challenges and New Developments* (University of Chicago Press, Illinois, 1985)

D. N. Schramm and W. D. Arnett, *Explosive Nucleosynthesis* (University of Texas Press, Texas, 1973)

Heliospherics

T. Chang, Editor, *Ion Acceleration in the Magnetosphere and Ionosphere* (American Geophysical Union, Washington, DC, 1986)

L. A. Fisk et al., *Ap. J.* **190**, 417 (1974)

J. R. Jokipii, *Ap. J.* **146**, 480 (1966)

E. N. Parker, *Planet. Space Sci.* **13**, 9 (1965)

Three of the most cited and influential papers in the area of cosmic rays in the heliosphere. An excellent starting point.

B. T. Tsurutani and R. G. Stone, Editors, *Collisionless Shocks in the Heliosphere: Review of Current Research* (American Geophysical Union, Washington, DC, 1985)

Particle detectors

G. Bertolini and A. Coche, Editors, *Semiconductor Detectors* (North Holland, Amsterdam, 1968)

J. B. Birks, *Theory and Practice of Scintillation Counting* (Pergamon, Oxford, 1964)

G. Dearnaley and D. C. Northrop, *Semiconductor Detectors for Nuclear Radiations* (John Wiley & Sons, New York, 1966)

K. Kleinknecht, *Detectors for Particle Radiation* (Cambridge University Press, Cambridge, 1986)

> A new edition is scheduled for release in late 1998.

G. F. Knoll, *Radiation Detection and Measurement*, 2nd Edition (John Wiley & Sons, New York, 1989)

L. B. Loeb, *Basic Processes of Gaseous Electronics* (University of California Press, California, 1961)

P. Rice-Evans, *Spark, Streamer, Proportional, and Drift Chambers* (London, 1974)

F. Sauli, *Principles of Operation of Multiwire, Proportional, and Drift Chambers*, CERN Report 77-09 (1977)

J. F. Ziegler, J. P. Biersack, and U. Littmark, *The Stopping and Range of Ions in Solids*, Volume 1 (Pergamon Press, Oxford, 1985)

Detector geometry factors

J. D. Sullivan, *Nucl. Inst. Meth.* **95**, 5 (1971)

There are a couple of equation errors in the paper and even in various errata for the paper. Most significantly, there is an error in the form of the equation for the geometrical factor in the case of rectangular symmetry (as in many detector systems). The corrected equation 11 follows:

$$
\begin{aligned}
G = l^2 \ln & \left(\frac{l^2 + \alpha^2 + \delta^2}{l^2 + \alpha^2 + \beta^2} \frac{l^2 + \gamma^2 + \beta^2}{l^2 + \gamma^2 + \delta^2} \right) \\
& + 2\alpha \left(l^2 + \beta^2\right)^{1/2} \tan^{-1} \frac{\alpha}{\left(l^2 + \beta^2\right)^{1/2}} + 2\beta \left(l^2 + \alpha^2\right)^{1/2} \tan^{-1} \frac{\beta}{\left(l^2 + \alpha^2\right)^{1/2}} \\
& - 2\alpha \left(l^2 + \delta^2\right)^{1/2} \tan^{-1} \frac{\alpha}{\left(l^2 + \delta^2\right)^{1/2}} - 2\beta \left(l^2 + \gamma^2\right)^{1/2} \tan^{-1} \frac{\beta}{\left(l^2 + \gamma^2\right)^{1/2}} \\
& - 2\gamma \left(l^2 + \beta^2\right)^{1/2} \tan^{-1} \frac{\gamma}{\left(l^2 + \beta^2\right)^{1/2}} - 2\delta \left(l^2 + \alpha^2\right)^{1/2} \tan^{-1} \frac{\delta}{\left(l^2 + \alpha^2\right)^{1/2}} \\
& + 2\gamma \left(l^2 + \delta^2\right)^{1/2} \tan^{-1} \frac{\gamma}{\left(l^2 + \delta^2\right)^{1/2}} + 2\delta \left(l^2 + \gamma^2\right)^{1/2} \tan^{-1} \frac{\delta}{\left(l^2 + \gamma^2\right)^{1/2}}
\end{aligned}
\tag{11}
$$

A handful of web resources

Online edition of M. V. Zombeck's Handbook of Space Astronomy and Astrophysics: http://adsbit.harvard.edu/books/hsaa/

Nuclear Astrophysics Data online: http://isotopes.lbl.gov/isotopes/astro.html
Cosmic-ray / Gamma-ray / Neutrino experiment listings:
    http : // eu6.mpi-hd.mpg.de/CosmicRaySites.html
Directory of LeCroy modules: http://www-esd.fnal.gov/esd/catalog/main/lrsdir.htm
National Space Science Data Center (NSSDC): http://nssdc.gsfc.nasa.gov/
Cosmic-ray Deflection Society of North America:
    http://www.geocities.com/SunsetStrip/1483/

Original GZK cut-off papers

K. Greisen, *Phys. Rev. Letters* **16**, 748 (1966)
G. T. Zatsepin and V. A. Kuz'min, *JETP Letters* **4**, 78 (1966)

Fermi Acceleration

E. Fermi, *Ap. J.* **119**, 1 (1954)

National Geographic

A. Piccard, *Nat. Geo.*, March 1933, 353
    Heroic balloon flights.
F. B. Colton, *Nat. Geo.*, September 1946, 379
M. A. Pomerantz, *Nat. Geo.*, January 1953, 99
    On cosmic-ray experiments flown from the far north of Canada.
K. F. Weaver, *Nat. Geo.*, January 1969, 46

International Cosmic Ray Conferences

1.	Cracow, Poland, 1947	10.	Calgary, Canada, 1967
2.	Como, Italy, 1949	11.	Budapest, Hungary, 1969
3.	Bagneres di Bigorre, France, 1953	12.	Hobart, Australia, 1971
		13.	Denver, USA, 1973
4.	Guanjuato, Mexico, 1955	14.	Munich, Germany, 1975
5.	Varenna, Italy, 1957	15.	Plovdiv, Bulgaria, 1977
6.	Moscow, USSR, 1959	16.	Kyoto, Japan, 1979
7.	Kyoto, Japan, 1961	17.	Paris, France, 1981
8.	Jaipur, India, 1963	18.	Bangalore, India, 1983
9.	London, UK, 1965	19.	La Jolla, USA, 1985

20.   Moscow, USSR, 1987
21.   Adelaide, Australia, 1990
22.   Dublin, Ireland, 1991
23.   Calgary, Canada, 1993
24.   Rome, Italy, 1995
25.   Durban, South Africa, 1997
26.   Salt Lake City, USA, 1999

Index